Power Systems

For further volumes:
http://www.springer.com/series/4622

Hossein Seifi · Mohammad Sadegh Sepasian

Electric Power System Planning

Issues, Algorithms and Solutions

Springer

Prof. Hossein Seifi
Faculty of Electrical and Computer Engineering
Tarbiat Modares University
PO Box 14115-194
Tehran
Iran
e-mail: seifi_ho@modares.ac.ir

Dr. Mohammad Sadegh Sepasian
Power and Water University of Technology
PO Box 16765-1719
Tehran
Iran
e-mail: sepasian@pwut.ac.ir

ISSN 1612-1287

e-ISSN 1860-4676

ISBN 978-3-642-17988-4

e-ISBN 978-3-642-17989-1

DOI 10.1007/978-3-642-17989-1

Springer Heidelberg Dordrecht London New York

Cover design: eStudio Calamar, Berlin/Figueres

Printed on acid-free paper

Springer is part of Springer Science+Business Media (www.springer.com)

Preface

One of the largest, or perhaps, the largest scale system ever made, is the electric grid with its numerous components, called a power system. Over decades, power systems have evolved to the systems which may cover countries or even continents.

From one side, the behaviors, modeling and operation of the basic components of a power system should be understood and recognized. That is why so many books are published to address such issues.

On the other hand, once the system as a whole is observed, its analysis, operation and planning deserve special considerations. While analysis and to some extent, operation of power systems have received attention in literature and in terms of text books, power system planning is not rich from this viewpoint. This book is intended to cover this issue.

While the importance of power system planning can not be overstated, writing a text book on this issue is not an easy task due to some, but not limited to, reasons as follows

- Planning horizon is from short to long periods. The issues of concern are not the same; although some may be similar.
- Utilities and experts may think of a specific planning term quite differently. For instance, one may think of long-term power system planning to cover 20 years onward, while the other may consider it as 5–15 years.
- While the basics of say, load flow in a book on power system analysis, or Automatic Generation Control (AGC) in a book on power system operation, are essentially the same on similar books, the algorithms and the methodologies used in power system planning may be utility or even case dependent.

The book is intended to cover long-term issues of power system planning, mainly on transmission and sub-transmission levels. However, the reader would readily recognize that some of the chapters may also be used for mid-term or even short-term planning, perhaps with some modifications. In terms of the long-term planning itself, the algorithms presented are mainly so designed that they may be used for various time frames. However, enough input data should be available;

which may be unavailable for very long-term periods. Regarding the methodologies and the algorithms, the chapters are arranged in a case independent manner and the algorithms are formulated in the ways that the readers can readily modify them according to their wishes.

We envision two groups of audiences for this book. The first consists of final year BSc or graduate students with a major in power systems. The second group consists of professionals working in and around the power industry especially in planning departments.

To bridge the gap between formal learning of the algorithms and deep understanding of the materials, some Matlab M-file codes are generated and attached in Appendix L. They are based on the materials developed within the chapters and easy to follow. Once referred to any of the above codes within the chapters, it is shown as [#X.m; Appendix L: (L.Y)], where X stands for M-file name and Y stands for the relevant section number. These codes may be accessed through the publisher website, too. They are used to solve some of the examples within and some of the problems at the end of the chapters. However, we should emphasize that they are not designed as commercial software and the instructors may ask the students to modify them and the professionals may improve them to meet their special requirements.

Some numerical examples are solved within the chapters. Although we have tried to use realistic input parameters, especially economic parameters are quite case dependent. That is why, an artificial monetary unit abbreviated as R is used to refer to economic values.

We were fortune to make the most benefits of our both academic and professional positions in preparing the book. The first author is a professor of the Faculty of Electrical and Computer Engineering at Tarbiat Modares University (TMU) (Tehran/Iran). TMU is only involved in graduate studies. He has supervised or has under supervision more than 80 MSc and PhD students. At the same time, he has founded a National Research Center (Iran Power System Engineering Research Center, IPSERC) as an affiliated center to TMU, for which he is acting as the head. Over the last few years, IPSERC has been actively involved in more than 60 strategic planning studies for major Iranian electric utilities. His vast experiences within IPSERC are properly reflected in various chapters. Some commercial software is also developed, now used by some of Iranian utilities. The Iranian electric power industry ranks nearly 8th in the world, in terms of the generation capacity (roughly 57 GW, 2010) and his experiences are based on this rather large scale system.

The second author is a faculty member at Power and Water University of Technology (PWUT) and a senior expert in IPSERC since its foundation. PWUT is affiliated to the Ministry of Energy of the country with vast experiences in terms of practical issues.

Many individuals and organizations have made the writing of this book possible. We are deeply grateful to the experts in Iranian electric power industry who graciously discussed and helped our understanding of practical issues and their requirements. We enjoyed marvelous learning opportunity through carrying out the strategic planning studies for this industry. Mr Rae, Mr Akhavan (both from Tavanir), Dr Zangene, Mrs Zarduzi (both from Tehran Regional Electric Utility), Mr Zeraat-Pishe, Mr Asiae (both from Fars Regional Electric Utility), Mr Arjomand, Mr Torabi, Mr Ghasemi (all from Hormozgan Regional Electric Utility), Mr Mehrabi (from Yazd Regional Electric Utility), Mrs Ghare-Toghe (from Mazandaran Regional Electric Utility) are only a few among many others. Mr Saburi (from Tavanir) provided us some useful data for a part of Chap. 4.

However, we should especially thank Dr Ahmadian for his support in founding IPSERC from the Ministry of Energy viewpoint. Special thanks are due to Mr Mohseni Kabir, who was and is still acting as the deputy in planning affairs of Tavanir (Tavanir is the holding company of Iranian power industry). Besides very useful technical discussions with him, he also greatly helped bridge Tavanir with IPSERC.

Within IPSERC, many individuals have contributed developing the software; employed in the studies, discussing with the industry experts, etc. To name a few, Dr Akbari, Dr Yousefi, Dr Haghighat, Mr Khorram, Mr Elyasi, Mr Roustaei, Mrs Hajati, Mr Sharifzadeh, Mr Shaffee-Khah deserve special thanks.

Our gratitude also extends to all others who, somehow, participated in the development of the book-particularly our students who never cease to ask challenging questions-and to our friends who offered encouragement and support. Mr Daraeepour developed the Matlab M-files codes. Dr Sheikh-al-Eslam, Dr Akbari, Dr Dehghani, Mr Elyasi, Mrs Hajati, Mr Roustaei, Mr Khorram, Mr Velayati, Mr Sharif-Zadeh, Mr Karimi reviewed the chapters, solved some examples, devised some problems and provided us useful suggestions and comments. Mrs Najafi and Mrs Tehrani did an excellent job in typing the whole manuscript.

One name deserves special gratitude. We deeply owe Mr Elyasi for an excellent task of reviewing, typesetting, organizing the manuscript and careful editing of the book. He did a really marvelous task in a very nice and efficient manner.

Sincere thanks are due to Prof. Christoph Baumann and his colleagues, from Springer, for their support in the preparation of the book. Finally, we should thank our families who graciously accepted us as part-time family members during the course of this book.

We should mention that a review of the chapters is provided in Chap. 1. Although the book is intended to be a text book, power system planning is a research-oriented topic, too. That is why; we have also added a chapter, to cover research issues.

Finally, we should mention that although we have attempted to review the materials so that they are, hopefully, error free, some may still exist. Please feel free to email us feedback including errors, comments, opinions, or any other useful information. These suggestions from the readers for improving the book clarity and accuracy will be greatly welcomed.

Tehran, May 2011

Hossein Seifi
Mohammad Sadegh Sepasian

Contents

Chapter 1
Power System Planning, Basic Principles

1.1 Introduction

The electric power industry has evolved over many decades, from a low power generator, serving a limited area, to highly interconnected networks, serving a large number of countries, or even continents. Nowadays, an electric power system is one of the man-made largest scale systems; ever made, comprising of huge number of components; starting from low power electric appliances to very high power giant turbo-generators. Running this very large system is a real difficult task. It has caused numerous problems to be solved by both the educational and the industrial bodies. Lessons have to be learnt from the past. At the same time that the current situation should be run in an efficient manner, proper insights should be given to the future. As we will discuss it shortly, the word *operation* is the normal electric power term used for running the current situation. Referring to the future, the power system experts use the term *planning* to denote the actions required for the future. The past experiences are always used for efficient *operation* and *planning* of the system.

The word *planning* stems of the transitive verb *to plan,* meant as *to arrange a method or scheme beforehand for any work, enterprise, or proceeding.*[1] The aim here is to discuss the meanings of *method or scheme*, *beforehand* and *work, enterprise or proceeding* for a physical power system. In other words, we are going to discuss the *power system planning* problem in terms of the issues involved from various viewpoints; the methods to be used; the elements to be affected; the time horizon to be observed, etc.

We will shortly define and describe, in more details, these issues. Before that, however, a short review is provided for power system elements and structure (Sects. 1.2 and 1.3). To clarify the boundaries between various power system studies, a time-horizon perspective of such studies is given in Sect. 1.4. Power system planning issues may be looked at from various viewpoints. These are

[1] dictionary.reference.com.

H. Seifi and M. S. Sepasian, *Electric Power System Planning*, Power Systems,
DOI: 10.1007/978-3-642-17989-1_1, © Springer-Verlag Berlin Heidelberg 2011

discussed in more details in Sect. 1.5. Moreover, the emphasis is given to the long-term power system planning problem, dealt with in subsequent subsections. A review of chapters is provided in Sect. 1.6.

1.2 Power System Elements

As already noted, a typical power system is comprised of enormous number of elements. The elements may vary from a small lamp switch to a giant generator. However, the main elements of interest in this book are

- Generation facilities
- Transmission facilities
  - Substations
  - Network (lines, cables)
- Loads

As a matter of fact, in power system planning, the details of each element design are not of main interest. For instance, for a generation facility, the type (steam turbine, gas turbine, etc.), the capacity and its location are only determined.[2] In Sect. 1.3, we will see how these elements may be grouped in a typical power system structure.

1.3 Power System Structure

It is assumed that the reader is already familiar with the basic concepts of an electric power system. To highlight the elements affected in power system planning problems, Fig. 1.1 depicts a typical power system, comprising of the *generation*, the *interface* and the *load*. The generations and the loads are distributed throughout the system. As a result, some interfaces should be provided to transfer the generated powers to the loads. The generations may be in the form of a small solar cell or a diesel generator to a very giant nuclear power plant. The loads start, also, from a small shop/home to a large industrial complex. Due to both the technical and the economical viewpoints, the generation voltages may be as high as 33 kV or so, while the load voltages may be much lower. Moreover, the generation resources may be far away from load centers. To reduce the losses and to make the transmission possible, we have to convert the generation voltages to

[2] It is worth mentioning that following a basic generation planning such as the one noted above, a detailed power plant design is required in which the technical specifications of all elements are determined. This is in fact the *power plant design problem*, not to be dealt with in this book.

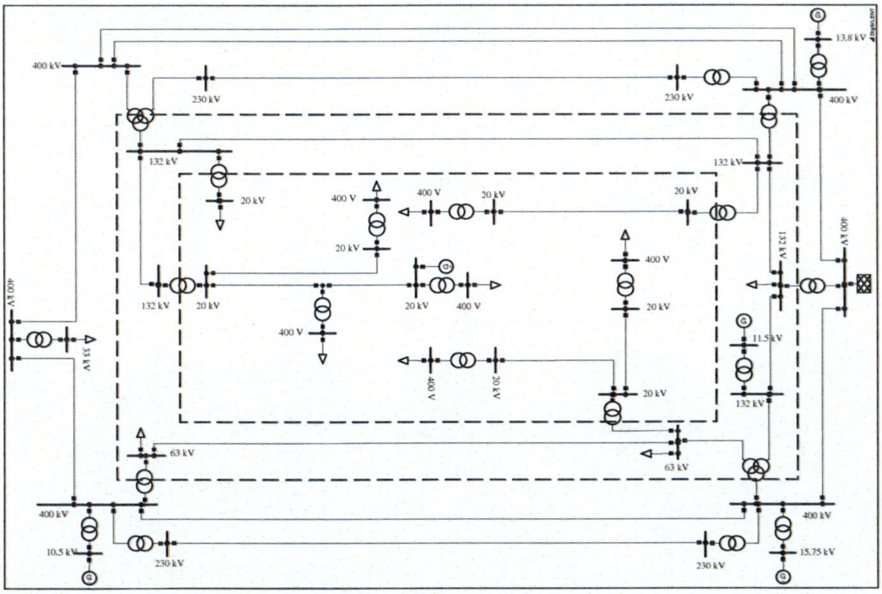

Fig. 1.1 A typical power system

much higher values and to reconvert them to lower ones at the receiving ends (load centers). As a result, the interfaces between the generations and the loads may comprise of several voltages, such as 20, 63, 132, 230, 400, 500 kV or even higher.[3] The available voltages depend much on each utility experiences within each country. However, regardless of what the available voltages are, it is of normal industrial practice to classify these voltages to

- Transmission (for example, 230 kV and higher)
- Sub-transmission (for example, 63, 132 kV, and similar)
- Distribution[4] (for example, 20 kV and 400 V).

Due to these various voltages, transformers are allocated throughout the network in the so called *substations*. For instance, a 400 kV substation[5] may comprise of four 400 kV:230 kV transformers. Each substation is also equipped with circuit breakers, current and potential transformers,[6] protection equipment, etc. The layout representation of a typical substation is shown in Fig. 1.2.

---

[3] The term Extra High Voltage (EHV) is normally used for voltages around 400–500 kV. UHV (Ultra High Voltage) is the term used for 735, 765 kV and higher voltages.

[4] For distribution systems, 400 V or so is defined as low voltage distribution, while 20 kV and similar are classified as medium voltage distribution.

[5] A substation is normally named based on the higher voltage level of its transformers.

[6] For measuring purposes.

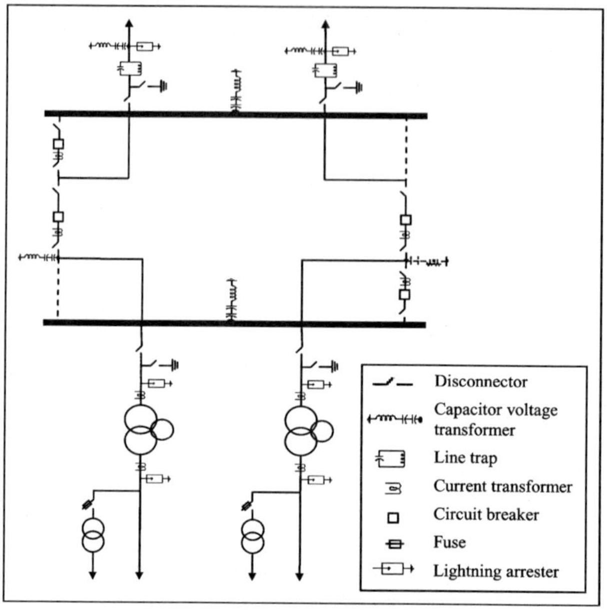

Fig. 1.2 The layout representation of a typical substation

1.4 Power System Studies, a Time-horizon Perspective

We briefly noted earlier that thinking of the current and the future states of a power system are called *operation* and *planning*, respectively. Let us now define these terms more precisely. Before that, however, we mention two typical studies that power system experts perform in real life.

First, suppose it is foreseen that the predicted load in 10 years from now, may be served provided that a new power plant is built. The expert has to decide on its required capacity, type and where the plant has to be connected to the network. Once decided properly, its constructing has to be started ahead of time, so that the plant is available in 10 years time. This is a typical long-term study of power systems (Fig. 1.3).

Second, suppose we are going to build a transmission line, passing through a mountainous area. Once built, the line may be subject to severe lightning. Lightning is such a very fast phenomena that it affects the system within nano-seconds. The designer should think of appropriate provisions on the line, by proper modeling the system in these very fast situations and performing enough studies, to make sure that the line does not fail, if such lightning happens in practice. This is a typical very short-term study of power systems.

Provided sufficient generation and transmission facilities are available for serving the loads, a power system decision maker[7] should perform a 1 week to

[7] The decision maker may be a utility, control center, system operator or similar.

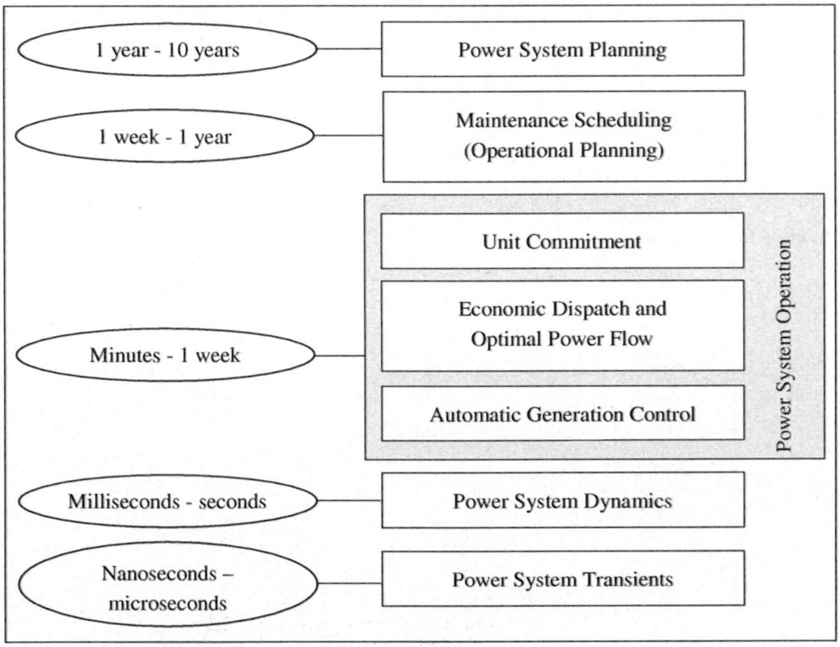

Fig. 1.3 A time-horizon perspective of power system studies

1 year[8] study to decide, in advance, on maintaining power system elements (power plants, transmission lines, etc.). This type of study is strictly required since if the plants are not maintained properly, they may fail in severe loading conditions. Moreover, the decision maker should know which elements are not available within the current year, so he or she can base his or her next decisions only on available elements. This type of study is called *maintenance scheduling*. Another term normally used is *operational planning*.

The *operational* phase starts from 1 week to minutes. These types of studies may be generally classified as[9]

- Hours to 1 week (for example, unit commitment),
- Several minutes to 1 h (for example, economic dispatch, Optimal Power Flow (OPF)),
- Minutes (for example, Automatic Generation Control (AGC)).

To discuss, briefly, the points mentioned above, suppose from ten power plants of a system, in the coming week, three are not available due to scheduled

[8] The time boundaries defined here are not crisp. They may change according to utilities experiences.

[9] Only some typical studies are mentioned in the operational phase. The actual studies may be more, but, they generally fall in the mentioned time periods.

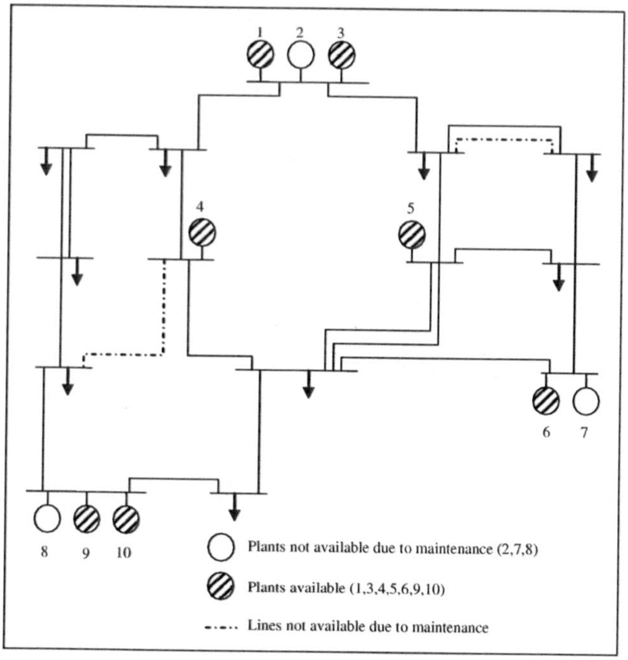

Fig. 1.4 The system available for system operation

maintenances (Fig. 1.4). The decision maker should decide on using the available plants for serving the *predicted* load for each hour of the coming week. Moreover, he or she should decide on the *generation level* of each plant, as the generation capacities of all plants may be noticeably higher than the predicted load. This type of study is commonly referred to as *unit commitment*. His or her decision may be based on some technical and/or economical considerations.[10] The final decision may be in the form of

- Commit unit 1 (generation level: 100 MW), unit 3 (generation level: 150 MW) and unit 6 (generation level: 125 MW), to serve the predicted load of 375 MW at hour 27 of the week (1 week = 168 h).
- Commit unit 1 (generation level: 75 MW) and unit 3 (generation level: 120 MW), to serve the predicted load of 195 MW at hour 35 of the week.

A complete list for all hours of the week should be generated. Once we come to the exact hour, the actual load may not be equal to the predicted load. Suppose, for instance, that the actual load at hour 27 to be 390 MW, instead of 375 MW. A further study has to be performed in that hour to allocate the actual load of 390 MW among the available plants at that hour (units 1, 3 and 6). This type of

[10] The reader may refer to any operational text books, either in regulated or market-based environments. See the list of the references at the end of the chapter.

study may be based on some technical and/or economical considerations and is commonly referred to as *economic dispatch* or *Optimal Power Flow* (OPF).[11]

Coming to the faster time periods, the next step is to automatically control the generation of the plants (for instance units 1, 3 and 6) via telemetry signals to required levels, to satisfy the load of 390 MW at hour 27. This task is normally referred to as *Automatic Generation Control* (AGC) and should be performed, periodically (say in minutes); as otherwise, the system frequency may undesirably change.

Further going towards the faster time periods, we come to *power system dynamics* studies, in milliseconds to seconds. In this time period, the effects of some components such as the power plants excitation systems and governors may be significant. Two typical examples are *stability* studies (for example, small signal, large signal, voltage stability, etc.) and *Sub-Synchronous Resonance (SSR)* phenomenon.[12]

The very far end of Fig. 1.3 consists of the very fast phenomenon of power system behaviors. It is the so called *power system transients* studies, involving studies on *lightning*, *switching transients* and similar. The time period of interest is from milliseconds to nanoseconds or even picoseconds.[13]

As *power system planning* is the topic of interest in this book, we will more discuss the subject in Sect. 1.5.

1.5 Power System Planning Issues

As described in Sect. 1.4, power system planning studies consist of studies for the next 1–10 years or higher. In this section, a more precise classification is given. Before that, it is worth mentioning that

> Power system planning is a process in which the aim is to decide on *new* as well as *upgrading existing* system elements, to adequately satisfy the loads for a foreseen future.

The elements may be

- Generation facilities
- Substations
- Transmission lines and/or cables
- Capacitors/Reactors
- Etc.

[11] OPF requires a more complex modeling of the problem. See the list of the references at the end of the chapter.

[12] The interested reader may refer to the text books available on the subject. See the list of the references at the end of the chapter.

[13] See Footnote no. 12.

The decision should be

- Where to allocate the element (for instance, the sending and receiving end of a line),
- When to install the element (for instance, 2015),
- What to select, in terms of the element specifications (for instance, number of bundles and conductor type).

Obviously, the loads should be adequately satisfied.[14] In the following sub-sections, some classifications of the subject are provided.

1.5.1 Static Versus Dynamic Planning

Let us assume that our task is to decide on the subjects given above for 2015–2020. If the peak loading conditions are to be investigated, the studies involve six loading conditions. One way is to, study each year separately irrespective of the other years. This type of study is referred to as *static planning* which focuses on planning for a single stage. The other is to focus on all six stages, simultaneously, so that the solution is found for all six stages at the same time. This type of study is named as *dynamic planning*.

Obviously, although the static planning for a specific year provides some useful information for that year, the process as given above leads to impractical results for the period as the solutions for a year cannot be independent from the solution from the preceding years. One way to solve the problem is to include the results of each year in the studies for the following year. This may be referred to as *semi-static*, *semi-dynamic*, *quasi-static* or *quasi-dynamic* planning. It is apparent that the dynamic planning solution can be more optimal in comparison with the semi-static planning solution.

We should mention that the word *dynamic* here should not be confused with *power system dynamics*, already noted in Sect. 1.4.

1.5.2 Transmission Versus Distribution Planning

We discussed earlier in Sect. 1.3 that we may distinguish three main levels for a power system structure, namely, transmission, sub-transmission and distribution. Distribution level is often planned; or at least operated, radially. Figure 1.5 depicts a typical distribution network, starting from a 63 kV:20 kV substation, ending to some types of loads, via both 20 kV and 400 V feeders. Note that switches A and B are normally open and may be closed if required. Switches C and D are normally

[14] In this book, we will see what *adequately* means in practice.

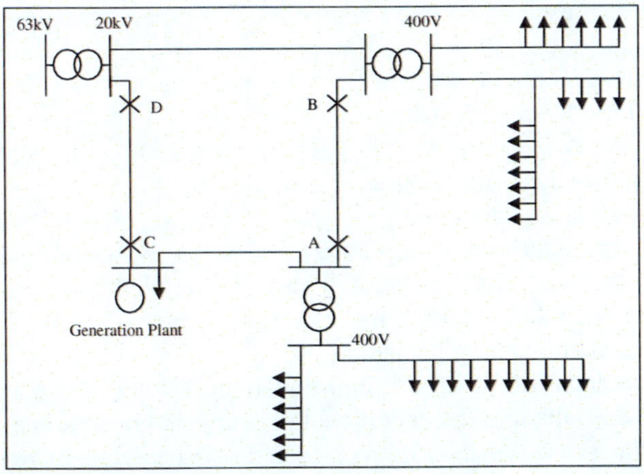

Fig. 1.5 A typical radial distribution network

closed and may be opened if required. A small generation is also connected to the network, as some types of local generations (named as Distributed Generations, or DGs) connected to the distribution systems, are of current industrial practices.

Looking at transmission and sub-transmission levels, these are generally interconnected, as already shown in Fig. 1.1. Normally both may be treated similarly, in terms of, the studies required and involved. From hereon, with *transmission*, we mean both transmission and/or sub-transmission levels, except otherwise specified.

As seen, both transmission and distribution networks comprise of lines/cables, substations and generations. However, due to specific characteristic of a distribution system (such as its radial characteristics), its planning is normally separated from a transmission system,[15] although much of the ideas may be similar.

In this book, we are mainly concerned with transmission planning.

1.5.3 Long-term Versus Short-term Planning

We mentioned in Sect. 1.4 that power system planning issues may cover a period of 1–10 years, or even more. Suppose that, for the peak loading condition of the coming year, a power system utility expert notices that from the two lines, feeding a substation, one would be overloaded by 10% of its rating, while, the other would be loaded by 60% of its rating. After careful studies, he or she finds out that if a

[15] To see some details on power system distribution, the reader may refer to available text books. See the list of the references at the end of the chapter.

control device[16] is installed on one line, the load distribution may be balanced on both lines. Once decided, the installation process of this device can be performed in such a way that no problem arises for the coming year. This is a typical short-term transmission planning decision.

Looking at the other extreme end, suppose that the load forecasting for the coming years shows that with all already available and planned generations, there would be a shortfall of generation in 9 years from now, onward. After a careful study, the planner decides on adding a new 2×500 MW steam power plant at a specific bus in that year. Its construction should start well in advance so that it would be available at the required time. His or her decision is a typical long-term (9-year) transmission planning decision.

There is no golden rule in specifying short-term or long-term planning issues. Normally, <1 year falls into the operational planning and operational issues (Sect. 1.4) in which the aim is typically to manage and operate available resources in an efficient manner. More than that falls into the planning stages. If installing new equipment and predicting system behavior are possible in a shorter time (for instance, for distribution systems, 1–3 years), the term of short-term planning may be used. More than that (3–10 years and even higher) is called long-term planning (typically transmission planning) in which predicting the system behavior is possible for these longer periods. Moreover, installing a new element (such as a 765 kV UHV line or a nuclear power plant) should be decided well in advance so that it would be available in due course.

Although the main focus of this book is on long-term power system planning, it is worth mentioning that the typical years mentioned above depend much on each utility experiences. The approaches presented are general enough to be applied to transmission planning issues (regardless of being short-term or long-term), but not necessarily to distribution planning issues (although the general ideas, may be used).

1.5.4 Basic Issues in Transmission Planning

With due attention to all points mentioned in previous sections, we come now to our main interest of *transmission planning*. The term commonly used in literature is *Transmission Expansion Planning* (TEP), to show that we focus on long-term issues.

Before going further, we should point out that, in this book, to avoid confusion between the distribution planning and the planning issues involving high voltages, we have used the terminology TEP to emphasize the fact that the transmission and the sub-transmission levels are considered. We may use the general term of power system planning, noting the fact that distribution planning is excluded from our

[16] Such as a phase shifting transformer.

discussions. Sometimes, the terminology of *Network Expansion Planning* (NEP) is also used to point out the same concepts. As we use NEP for the expansion studies of the network (lines, cables, etc.), we have not followed this idea. In Sect. 1.5.4.1 through 1.5.4.6, the topics of interest in TEP (or more properly, power system planning; excluding distribution planning) are introduced. We do not use the terminology TEP much often in this book. Instead, the issues are considered.

In Sect. 1.6, we will talk how the book chapters are organized to cover the points.

1.5.4.1 Load Forecasting

The first crucial step for any planning study is to predict the consumption for the study period (say 2015–2020), as all subsequent studies will be based on that. This is referred to as *load forecasting*. The same term is used for operational purposes, too. However, it is understood that a *short-term load forecasting*, used for operational studies, is significantly different from the *long-term* one used in planning studies. In a short-term load forecasting, for predicting the load for instance, of the next week, we come across predicting the load for each hour of the coming week. It is obvious that the determining factors may be weather conditions, special TV programs and similar.

In a *long-term load forecasting* which is of the main interest of this book, we normally wish to predict the peak loading conditions of the coming years. Obviously, the determining factors are different here. Population rate increase, GDP (Gross Domestic Product)[17] and similar terms have dominant effects.

1.5.4.2 Generation Expansion Planning

After predicting the load, the next step is to determine the generation requirements to satisfy the load. An obvious simple solution is to assume a generation increase equal to load increase. If, for instance, in year 2015, the peak load would be 40,000 MW and at that time, the available generation is 35,000 MW, an extra generation of 5,000 MW would be required. Unfortunately, the solution is not so simple at all. Some obvious reasons are

- What types of power plants do we have to install (thermal, gas turbine, nuclear, etc.)?
- Where do we have to install the power plants (distributed among 5 specific buses, 10 specific buses, etc.)?
- What capacities do we have to install (5×1000 MW, or 2×1000 MW and 6×500 MW, or …)?
- As there may be an outage on a power plant (either existing or new), should we install extra generations to account for these situations? If yes, what, where and how?

[17] See Chap. 3 for the description of the economical terms.

Still there are other points to be observed, to be discussed later in this book. This is a very complex problem, commonly referred to as *Generation Expansion Planning* (GEP) problem.

1.5.4.3 Substation Expansion Planning

Once the load is predicted and the generation requirements are known, the next step is to determine the substation requirements, both, in terms of

- Expanding the existing ones,
- Installing some new ones.

This is referred to as *Substation Expansion Planning* (SEP). SEP is a difficult task as many factors are involved such as

- Those constraints due to the upward grid, feeding the substations,
- Those constraints due to the downward grid, through which the substation supplies the loads,
- Those constraints due to the factors to be observed for the substation itself.

1.5.4.4 Network Expansion Planning

Network Expansion Planning (NEP) is a process in which the network (transmission lines, cables, etc.) specifications are determined. In fact, the network is a media for transmitting the power, efficiently and in a reliable manner from generation resources to the load centers. We will see in this book that what *efficiently* and *reliable manner* mean in practical terms. We will see how these factors influence our decision so that we have to decide from an enormous number of alternatives.

As inputs to the NEP problem, GEP and SEP results are assumed to be known.

1.5.4.5 Reactive Power Planning

In running NEP, the voltages are assumed to be flat (i.e. 1 p.u.) and reactive power flows are ignored. The main reason is the fact that constructing a line is not considered as a main tool for voltage improvement. Moreover, the running time of NEP can be exceptionally high or even the solution may not be possible if AC Load Flow (ACLF) is employed. That is why in practice, NEP is normally based on using Direct Current Load Flow (DCLF).[18] Upon running GEP, SEP and NEP,

[18] In Appendix A, we have briefly formulated DCLF.

the network topology is determined. However, it may perform unsatisfactorily,[19] if a detailed AC Load Flow (ACLF) is performed, based on existing algorithms.[20] To solve such a difficulty, static reactive power compensators, such as capacitors and reactors may be used. Moreover, some more flexible reactive power resources such as SVCs[21] may also be required. The problem is, however

- Where to install these devices?
- What capacities do we have to employ?
- What types do we have to use?

These types of studies are commonly referred to as *Reactive Power Planning* (RPP) and are clear required steps in a power system planning process.

1.5.4.6 Planning in Presence of Uncertainties

The electric power industry has drastically changed over the last two decades. It has moved towards a market oriented environment in which the electric power is transacted in the form of a commodity. Now the generation, transmission and distribution are unbundled and may belong to separate entities.[22] The planner can not, for instance, dictate where the generation resources have to be allocated. In this way, NEP problem is confronted by an uncertain GEP input. So, how NEP can be solved, once the input data is uncertain?

This was a simple example of the problems that current power system planners face. Obviously, some types of solutions have to be found.

1.6 A Review of Chapters

Nearly most decision makings noted above require some types of optimization problems to be solved. This topic is addressed in Chap. 2. Moreover, we face some economic decisions in this book. Some basic economic principles are dealt with in Chap. 3. Load forecasting is covered in Chap. 4. While GEP is treated in Chaps. 5 and 6, SEP is addressed in Chap. 7. Chapters 8 and 9 are devoted to NEP. RPP is discussed in Chap. 10. Planning in the presence of uncertainties is discussed in Chap. 11. The research trends are given in Chap. 12. A comprehensive example is demonstrated in Chap. 13.

---

[19] Unacceptable voltages of some buses, etc.

[20] Newton–Raphson, Fast Decoupled, etc.

[21] We will, later on, talk about it in this book.

[22] To become familiar with power system restructuring, see the list of the references at the end of the chapter.

References

For a detailed description on various issues of power system operation; including unit commitment, economic dispatch and optimal power flow; [1] may be consulted for a regulated (traditional) power system. [2] covers the same points in a deregulated (restructured) environment. While power system stability issues are discussed in [3], in [4] fast transient issues are covered. A good reference for electric distribution planning is [5]. Fundamental aspects of power system economics and deregulations are described in [6–9]. Basic power system issues are covered in many references. Some are introduced in [10–13]. A book devoted to some aspects of power system planning problem is [14] while some other issues, especially in a deregulated environment, are covered in [15].

1. Wood AJ, Wollenberg BF (1996) Power generation, operation, and control, 2nd edn. Wiley, New York
2. Shahidehpour M, Yamin H, Li Z (2002) Market operations in electric power systems: forecasting, scheduling, and risk management. Wiley-IEEE Press, New York
3. Grigsby LL (ed) (2007) Power system stability and control, 2nd edn. CRC Press, Taylor & Francis group, Florida
4. Greenwood A (1991) Electrical transients in power systems, 2nd edn. Wiley, New York
5. Willis HL (2004) Power distribution planning reference book, 2nd edn. Marcel Dekker, New York
6. Ilic M, Galiana F, Fink L (1998) Power system restructuring: engineering and economics. Kluwer Academic, Boston
7. Stoft S (2002) Power system economics, designing markets for electricity. Wiley-IEEE Press, New York
8. Armstrong M, Cowan S, Vickers J (1994) Regulatory reform. MIT Press, Cambridge
9. Kirschen D, Strbac G (2004) Fundamentals of power system economics. Wiley, Chichester
10. Saadat H (2002) Power system analysis. McGraw Hill, Boston
11. Weedy BM, Cory BJ (1998) Electric power systems, 4th edn. Wiley, Chichester
12. Grainger J, Stevenson JW (1994) Power system analysis. McGraw Hill, New York
13. Elgerd O, Van Der Puije PD (1997) Electric power engineering. Kluwer Academic Press, New York
14. Wang X, McDonald J (1994) Modern power system planning. McGraw Hill, New York
15. Mazer A (2007) Electric power planning for regulated and deregulated markets. Wiley-IEEE Press, New Jersey

Chapter 2
Optimization Techniques

2.1 Introduction

In everyday life, all of us are confronted with some decision makings. Normally, we try to decide for the *best*. If someone is to buy a commodity, he or she tries to buy the *best* quality, yet with the *least* cost. These types of decision makings are categorized as *optimization problems* in which the aim is to find the *optimum* solutions; where the *optimum* may be either the *least* or the *most*.

The aim of this chapter is to review briefly the basics of optimization problems. Obviously, the details are beyond the scope of this book and should be followed from available literature. However, a simple example is devised and solved using some of the approaches; as detailed in Appendix B.

2.2 Problem Description

Most of the operational and planning problems consist of the following three major steps

- Definition
- Modeling
- Solution algorithm

In the following subsections, we discuss them in some details.

2.2.1 Problem Definition

In any optimization problem, the decision maker should decide on the following items

H. Seifi and M. S. Sepasian, *Electric Power System Planning*, Power Systems,
DOI: 10.1007/978-3-642-17989-1_2, © Springer-Verlag Berlin Heidelberg 2011

- Decision (independent) and dependent variables
- Constraints functions
- Objective functions

2.2.1.1 Decision and Dependent Variables

Decision variables are the independent variables; the decision maker has to determine their optimum values and based on those, other variables (dependent) can be determined. For instance, in an *optimum generation scheduling problem*, the active power generations of power plants may be the decision variables. The dependent variables can be the total fuel consumption, system losses, etc. which can be calculated upon determining the decision variables. In a *capacitor allocation problem*, the locations and the sizing of the capacitor banks are the decision variables, whereas the dependent variables may be bus voltages, system losses, etc.

An n-decision variable problem results in an n-dimensional solution space in which any point within that space can be a solution. A two-dimensional case is shown in Fig. 2.1.

2.2.1.2 Constraints Functions

In a real-life optimization problem, some limitations may apply to the solution space. These are typically technical, economical, environmental and similar limitations; named as *constraints* which either directly or indirectly divide the solution space into acceptable (*feasible*) and unacceptable (*non-feasible*) regions. The decision maker should find a solution point within the feasible region. For

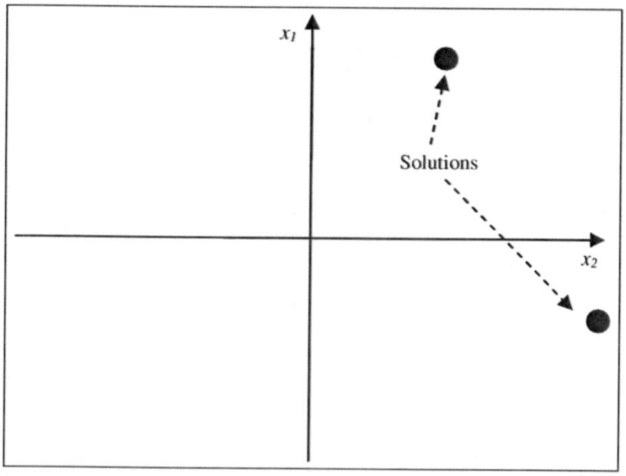

Fig. 2.1 The solution space for a two-dimensional case

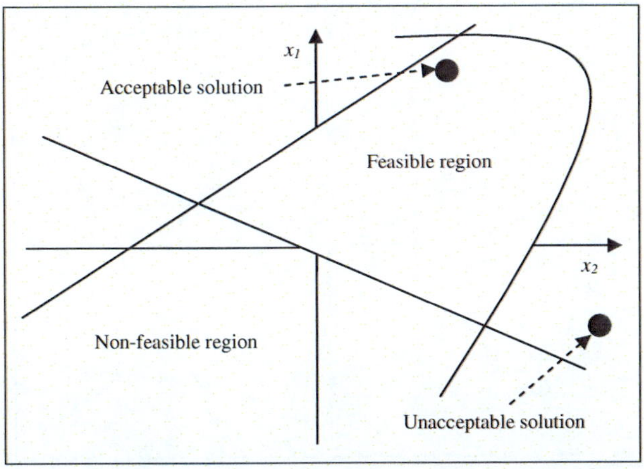

Fig. 2.2 Feasible and non-feasible regions due to constraints

instance, in an *optimum generation scheduling* problem, the active power generations of the power plants should be within their respective maximum and minimum values; or, the total generation of the plants should satisfy total load and a specified reserve. In a *capacitor allocation problem*, a technical constraint may be the maximum number of the capacitor banks which may be employed for a specific bus. An economical constraint may be a limit on the total practical investment cost which should not be violated. The way the constraints behave in a two-dimensional case is shown in Fig. 2.2.

2.2.1.3 Objective Functions

From the numerous points within the feasible region of a problem, the decision maker should select the most desirable. The desirable should, however, be somehow defined. For instance, in a classroom, a teacher may select a student as the best if morality is the main concern. He or she may select another if enthusiasm is observed. In fact, an *objective function* is a function in terms of the decision variables by which the decision maker shows his or her desirable solution. In Fig. 2.3, if the *objective function* is defined as maximizing x_1, the solution ends up in point A, whereas, if minimizing x_2 is the objective function, point B would be the final solution. In an *optimum generation scheduling* problem, the objective function may be chosen as the total fuel cost to be minimized. In a *capacitor allocation problem*, the objective function may be the investment cost or the system losses or both (to be minimized). The problem is considered to be *single-objective* if just one objective function is to be optimized. It is in contrast to *multi-objective* optimization problems in which several functions are to be simultaneously, optimized.

In a practical case, an optimization problem may have many maximum and minimum points. For instance, consider the case depicted in Fig. 2.4 in which the

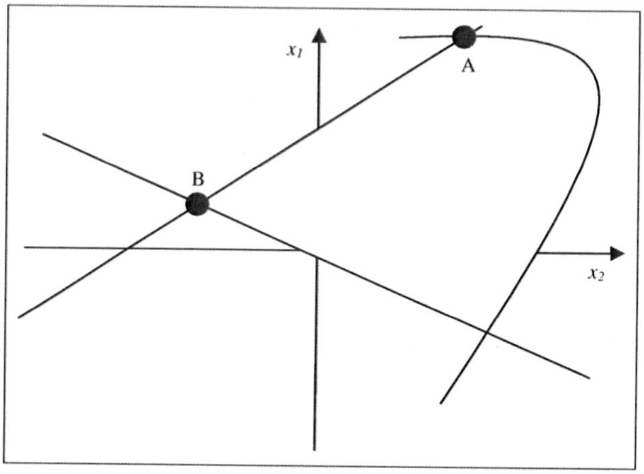

Fig. 2.3 Optimum points in a two-dimensional case

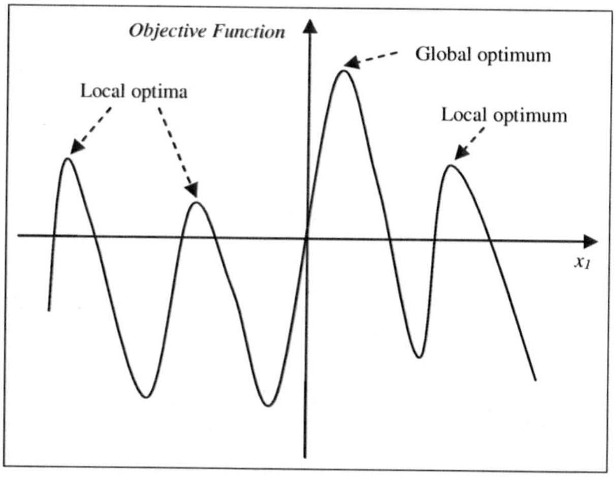

Fig. 2.4 Local and global optimum points

objective function is considered to be a function of only x_1 and is to be maximized. As shown, there are some local optima in the sense that they are optimum in the vicinity of nearby points. From those *local optimum points*, one is the *global optimum*.

2.2.2 Problem Modeling

Once the decision variables, the constraints and the objective function terms are decided, the decision maker should model the problem in a proper form to be solved. The modeling depends much on the available tools and the algorithms for the problem solving, the accuracy required, the simplifications possible, etc. A generic optimization problem model would be in the form given by

$$\begin{gathered}\text{Minimize or Maximize } C(x)\\ \text{Subject to } g(x) \leq b\end{gathered} \tag{2.1}$$

where x is the decision variable, $C(x)$ is the objective function and $g(x) \leq b$ is the *inequality* constraint.

The decision variables may be either real or integer. For instance, in an *optimum generation scheduling* problem, the active power generations are real while in a *capacitor allocation problem*, the number of capacitor banks to be installed in a specific bus is integer.

C and g may be either continuous or discrete functions of the decision variable in an explicit or implicit form; linear or nonlinear. Based on those, the optimization problem is appropriately named. For instance an *integer linear optimization* problem is a problem in which both C and g are linear functions of integer decision variables.

Generally speaking, as

- Maximizing C is equivalent to minimizing $(-C)$.
- We can name the *equality* constraint as $f(x)$, to separate it from $g(x)$.
- $g(x) > g^{lo}$ (or $(g(x) - g^{lo}) > 0$) is equivalent to $-(g(x) - g^{lo}) < 0$.
- There may be more than just one $f(x)$ or one $g(x)$.
- There may be more than just one independent variable x (instead, a vector of $\mathbf{x}$).

The general optimization problem may be stated as

$$\underset{\mathbf{x}}{\text{Min}}\ C(\mathbf{x}) \tag{2.2}$$

$$\text{s.t. } \mathbf{f}(\mathbf{x}) = 0 \tag{2.3}$$

and

$$\mathbf{g}(\mathbf{x}) \leq 0 \tag{2.4}$$

2.3 Solution Algorithms, Mathematical Versus Heuristic Techniques

The constrained optimization problem as stated by (2.2), (2.3) and (2.4) may be solved by some available optimization techniques. These techniques may be generally classified as *mathematical* and *heuristic*. Both have received attention in power system literature. These are reviewed in the following subsections.

2.3.1 Mathematical Algorithms

A mathematical optimization technique formulates the problem in a mathematical representation; as given by (2.2) through (2.4). Provided the objective function and/or the constraints are nonlinear, the resulting problem is designated as *Non Linear optimization Problem* (NLP). A special case of NLP is *quadratic programming* in which the objective function is a quadratic function of **x**. If both the objective functions and the constraints are linear functions of **x**, the problem is designated as a *Linear Programming* (LP) problem. Other categories may also be identified based on the nature of the variables. For instance, if **x** is of *integer* type, the problem is denoted by *Integer Programming* (IP). Mixed types such as MILP (*Mixed Integer Linear Programming*) may also exist in which while the variables may be both real and integer, the problem is also of LP type.

For mathematical based formulations, some algorithms have, so far, been developed; based on them some commercial software have also been generated. In the following subsections, we briefly review these algorithms. We should, however, note that generally speaking, a mathematical algorithm may suffer from numerical problems and may be quite complex in implementation. However, its convergence may be guaranteed but finding the global optimum solution may only be guaranteed for some types such as LP.

There is no definite and fixed classification of mathematical algorithms. Here, we are not going to discuss them in details. Instead, we are going to introduce some topics which are of more interest in this book and may be applicable to power system planning issues.[1] Some topics, such as *game theory*, which are of more interest for other power system issues (such as market analysis of power systems), are not addressed here.

2.3.1.1 Calculus Methods

These types of methods are the traditional way of seeking optimum points. These are applicable to continuous and differentiable functions of both objective and constraints terms. They make use of differential calculus in locating the optimum points.

Based on the basic differential calculus developed for finding the optimum points of *C*(**x**) (see (2.2)), the method of *Lagrange Multipliers* has been developed in finding the optimum points; where equality constraints (2.3) may also apply. If inequality constraints (2.4) are also applicable, still the basic method may be used; however, the so called *Kuhn-Tucker* conditions should be observed. The solution is not so straightforward in that case.

---

[1] The optimum seeking methods are generally known as *programming techniques* or *operations research*; a branch of mathematics. For more details, the interested reader may consult the list of the references at the end of the chapter.

2.3.1.2 Linear Programming (LP) Method

As already noted, LP is an optimization method in which both the objective function and the constraints are linear functions of the decision variables. This type of problem was first recognized in the 1930s by the economists in developing methods for the optimal allocation of resources.

Noting the fact that

- Any LP problem can be stated as a minimization problem; due to the fact that, as already described, maximizing $C(\mathbf{x})$ is equivalent to minimizing $(-C(\mathbf{x}))$.
- All constraints may be stated as equality type; due to the fact that any inequality constraint of the form given by

$$a'_1x_1 + a'_2x_2 + \cdots + a'_nx_n < b' \tag{2.5}$$

or

$$a''_1x_1 + a''_2x_2 + \cdots + a''_nx_n > b'' \tag{2.6}$$

can be transformed to equality constraints, given by

$$a'_1x_1 + a'_2x_2 + \cdots + a'_nx_n + x'_{n+1} = b' \tag{2.7}$$

$$a''_1x_1 + a''_2x_2 + \cdots + a''_nx_n - x''_{n+1} = b'' \tag{2.8}$$

respectively, where x'_{n+1} and x''_{n+1} are nonnegative variables, known as *surplus* variables.
- All decision variables can be considered nonnegative, as any x_j, unrestricted in sign, can be written as $x_j = x'_j - x''_j$ where

$$x'_j \geq 0 \quad \text{and} \quad x''_j \geq 0 \tag{2.9}$$

It can be seen that x_j will be negative, zero or positive depending on whether x''_j is greater than, equal to or less than x'_j.

The problem can be stated in a form known as *canonical*. Then, a solution known as the *simplex method,* first devised in 1940s, may be used to solve the problem.

Using the simplex method normally requires a large amount of computer storage and time. The so called *revised simplex method* is a revised method in which less computational time and storage space are required.

Still another topic of interest in LP problems is the *duality theory*. In fact, associated with every LP problem, a so called *dual* problem may be formulated. In many cases, the solution of an LP problem may be more easily obtained from the dual problem.

If the LP problem has a special structure, a so called *decomposition principle* may be employed to solve the problem in which less computer storage is required. In this way, the problem can be solved more efficiently.

Transportation problems are special LP problems, occurring often in practice. These problems can be solved by some algorithms which are more efficient than the simplex method.

2.3.1.3 Non Linear Programming (NLP) Method

We noted earlier that if the objective function and/or the constraints are nonlinear functions of the decision variables, the resulting optimization problem is called NLP.

Before proceeding further on NLP problems, we should note that most practical problems are of constrained type in which some constraint functions should be satisfied. As for constrained problems, however, some algorithms work on the principle of transforming the problem into a unconstrained case, we initially review some existing algorithms on solving unconstrained problems.

The solution methods for unconstrained problems may be generally classified as *direct* search (or *non-gradient*) methods and *descent* (or *gradient*) methods. The former methods do not use the partial derivatives of the objective function and are suitable for simple problems involving a relatively small number of variables. The latter methods require the evaluations of the first and possibly, the higher order derivatives of the objective function. As a result, these methods are generally more efficient than the direct methods.

All the unconstrained optimization methods are iterative in nature and start from an initial trial solution; moving stepwise in a sequential manner towards the optimum solution. The *gradient* methods have received more attention in power system literature. For instance, in the so called *steepest descent* method; widely used in power system literature, the gradient vector is used to calculate the optimum step length along the search direction so that the algorithm efficiency is maximized.

Let us come back to the constrained case. Two types of methods, namely, *direct* and *indirect* methods apply. In the former methods, the constraints are handled in an explicit manner, while in most of the latter methods; the constrained problem is converted into a sequence of unconstrained problems and solved through available algorithms.

As an example of the *direct* methods, in the so called *constraint approximation* method, the objective function and the constraints are linearized about some point. The resulting approximated LP problem is solved using LP techniques. The resulting solution is then used to construct a new LP problem. The process is continued until a convergence criterion is satisfied.

As an example of the *indirect* methods, the so called *penalty function* method, works on the principle of converting the problem into an unconstrained type. It is, in turn, classified as *interior* and *exterior* penalty function methods. In the former, the sequence of unconstrained minima lie in the feasible region while in the latter, they lie in the infeasible region. In both, they move towards the desired solution.

1	2	3	4		15		23	24
1011	1011	0011	1011	....	1101	....	1111	1110

Fig. 2.5 Units combinations over the 24-h period

2.3.1.4 Dynamic Programming (DP) Method

Dynamic Programming is a widely used technique in power system studies. It is, in fact, a mathematical technique used for multistage decision problems; originally developed in 1950s.

A multistage decision problem is a problem in which optimal decisions have to be made over some stages. The stages may be different times, different spaces, different levels, etc. The important point is that the output of each stage is the input to the next serial stage.

The overall objective function is to be optimized over all stages. It is normally a function of the decision variables (x_i) of all stages. The important fact is that one can not start from optimizing the first stage; moving forward toward the final stage; as there may be some correlations between the stages, too.

To make the problem clear, let us express a power system example. Suppose we are going to minimize the generation cost of a power system over a 24-h period. Some information is as follows

- There are four generation units available; each of which may be either *off* or *on* (so that various combinations are possible, such as, 1111, 1101, 1001, 0011,...).
- The unit efficiencies are different; so that if the system load is low and say, two units can meet the load, we should use the higher efficient units to supply the load.
- The load varies throughout the 24-h period; changing at each hour (stage).

The multistage decision problem is, in fact, deciding on the units to be *on* at each stage so that the overall generation cost over the 24-h period is minimized. We note that if no other constraint was imposed, we should optimize our problem at each stage and sum it over all stages. In other words, 24 single stage optimization problems[2] have to be solved to find the final solution.

Suppose that the final solution looks like Fig. 2.5 in which the unit combinations are shown at each stage.

As shown, unit 1 is *on* at hours 1 and 2, *off* at hour 3, and *on* again at hour 4. Now what happens if a constraint is imposed expressing the fact that if unit 1 is turned *off*, it can not be turned on unless a 5-h period is elapsed.[3] So, our above solution is not practical. Now, how can we find the solution?

---

[2] It is important to note that the problem to be used at each stage is irrelevant to our discussion here. In fact, it may be LP, NLP or any other problem.

[3] This type of constraint is called *minimum down time* of a unit.

One can check that at each stage, for the above four unit case, the number of combinations is $2^4 - 1 = 15$.[4] For the 24-h period, the number of combinations would be $(15)^{24}$. What happens if the number of the units is, say, 100 and the number of stages is, say, 168 (a week). The number of the overall combinations would be $(2^{100} - 1)^{168}$ [5]!

In DP technique, a multistage decision problem is decomposed into a sequence of single stage problems; solved successively. The decomposition should be done in such a way that the optimal solution of the original problem can be obtained from the optimal solution of single stage problems.

2.3.1.5 Integer Programming Method

In the algorithms discussed so far, each of the decision variables may take any real value. What happens if a decision variable is limited to take only an *integer* value? For instance, if the decision variable is the number of generation units, taking a real value is meaningless. The optimization algorithms developed for this class of problems are classified as IP methods. If all decision variables are of integer type, the problem is addressed as IP problem. If some decision variables are of integer type while some others are of non-integer type, the problem is known as *mixed integer programming* problem.

Moreover, based on the nature of the original problem, both *integer linear programming* and *integer nonlinear programming* methods have been developed. As a result, in power system literature, some terms such as MILP have appeared.

2.3.2 Heuristic Algorithms

Most mathematical based algorithms can guarantee reaching an optimal solution; while do not necessarily guarantee reaching a global optimum. Global optimality may be only reached, checked or guaranteed for simple cases.

On the other hand, many practical optimization problems do not fall in strict forms and assumptions of mathematical based algorithms. Moreover, if the problem is highly complex, we may not readily be able to solve them, at all, through mathematical algorithms. Besides, finding global optimum is of interest, as finding a local one would be a major drawback.

Heuristic algorithms are devised to tackle the above mentioned points. They, normally, can solve the combinatorial problems, sometimes very complex, yet in a reasonable time. However, they seek good solutions, without being able to guarantee the optimality, or even how close the solutions are to the optimal point. Moreover,

---

[4] The combination 0000 is considered infeasible.

[5] The so called, *curse of dimensionality*, in DP problems.

some modified heuristic algorithms have been developed in literature by which improved behaviors are attained, claiming that the optimal solutions are guaranteed.

A simple heuristic algorithm may be devised based on some types of sensitivity analysis. For instance, in a *capacitor allocation problem*, the sensitivities of the objective function may be determined by the application of a capacitor bank in a bus. Once done, the capacitor is added to the most sensitive bus and the procedure is repeated until no further improvement is achieved in terms of the objective function.

However, most heuristic algorithms are based on some biological behaviors. Basically, all start from either a point or a set of points, moving towards a better solution; through a guided search. Few have been developed so far, some are worth mentioning here

- Genetic Algorithm (GA), based on genetics and evolution,
- Simulated Annealing (SA), based on some thermodynamics principles,
- Particle Swarm (PS), based on bird and fish movements,
- Tabu Search (TS), based on memory response,
- Ant Colony (AC), based on how ants behave.

Still, other techniques may be cited. However, we limit our discussions here to the above algorithms. The interested reader should consult the references at the end of this chapter.

2.3.2.1 Genetic Algorithm

In nature, each species is confronted by a challenging environment and should adapt itself for the maximum likelihood of survival. As time proceeds, the species with improved characteristics survives. In fact, the so called *fittest* type is survived. This type of phenomenon which happens in nature is the basis of the evolutionary based GA.

Genetic Algorithm was mainly developed by Holland. The decision variables to be found are *binary-coded*, *real value-coded* or *integer-coded*, in the form of a string of *genes*. This string is called the problem *chromosome,* selected from the so called set of *populations*. The objective function is calculated for this chromosome as the problem *fitness function*. After setting an initial population, selecting a chromosome and calculating its fitness, a next population is generated; based on the procedure outlined afterwards. Initial chromosomes are called as *parents* and the regenerated chromosomes are called *offspring*. As we will see, the regeneration results in chromosomes with better fitness values. The algorithm proceeds until no further improvement is achieved in fitness function.

We note that GA uses only the objective function information and not the derivatives. As it randomly, but in a guided way, searches the feasible space, the likelihood of reaching at the vicinity of the global optimum is high; although converging onto the global optimum itself is not very likely. *Selection*, *crossover* and *mutation* as the three main GA operators are described next.

- *Selection*. Based on the chromosome structure defined, a population of chromosomes is initially generated, either, randomly or intelligently. 30–100 chromosomes may be considered. Then, we may *select* two chromosomes as *parents* for further process. The fitness value is used as the criterion for *parents* selection.
- *Crossover*. Once parents are selected, we should generate new strings; *offsprings*, through two types of operators. The so called *crossover* works on the principle of interchanging the values after a specific position. For instance if A and B are the initial two selected chromosomes

$$\begin{array}{l} \qquad\qquad\qquad\qquad\quad\downarrow \\ A: \; 0 \; 1 \; 1 \; 0 \; 1 \; 0 \; 1 \; 1 \; 1 \; 0 \\ B: \; 1 \; 0 \; 1 \; 0 \; 1 \; 1 \; 1 \; 0 \; 0 \; 1 \end{array}$$

and crossover operator is applied at position 6, the resulting offsprings look like

$$\begin{array}{l} A': \; 0 \; 1 \; 1 \; 0 \; 1 \; 0 \; 1 \; 0 \; 0 \; 1 \\ B': \; 1 \; 0 \; 1 \; 0 \; 1 \; 1 \; 1 \; 1 \; 1 \; 0 \end{array}$$

This type of regeneration is done randomly at various positions. As a result, a new population of chromosomes is generated in which, again, the *selection* process may be restarted.
- *Mutation*. An inherent drawback of the crossover operator is the fact that at some particular position, the value of the *gene* may not change at all. To avoid this problem, the *mutation* operator tries to alter the value of a gene, randomly from 1 to 0 and vice versa. We should mention, however, that this is done quite infrequently.

We should mention that the operators defined above are the simplest types. In practice, more sophisticated operators are developed to improve GA performance. Currently, GA has received extensive attention in power system literature.

2.3.2.2 Simulated Annealing

Simulated Annealing is a flexible algorithm in dealing with combinatorial optimization problems. It may be applied to complex problems, involving even non-differentiable, discontinuous and non-convex functions.

Annealing is the natural process of cooling a molten material; from a high temperature. If the cooling process is performed under thermal equilibrium conditions, annealing results in formation of crystals. The formation of a perfect crystal is equivalent to a state of minimum energy.

It was in the 1980s that the principles cited above were first appeared as an algorithm in solving optimization problems. It was noted that a correspondence may be defined between the physical states of a matter and the solution space of an optimization problem. The free energy of the matter may correspond to the objective function of the optimization problem.

Before proceeding further, we should first discuss the *Metropolis* algorithm as the basis of SA algorithm.

- *Metropolis algorithm*. The particles forming a material have different levels of energy, according to a probability distribution and based on their temperature (T). The Metropolis algorithm works on the principle of generating a new state S_j; from a given initial state S_i; with energy E_i. This new state is generated by a mechanism, consisting of a small perturbation in the original state. The perturbation is, in fact, obtained by moving one of the particles chosen by the Monte Carlo method.[6]
For the energy of the new state, E_j (found probabilistically), the difference $E_j - E_i$ is checked to be less than or equal to zero in order to accept the new state S_j. If this difference is positive, still S_j is accepted; but with a probability given by

$$p = e^{(E_i - E_j)/k_B T} \tag{2.10}$$

where T is the temperature of the material and k_B is the Boltzmann constant. The process given above normally requires a large number of state transitions in reaching the state with the lowest energy level.

The above principles are followed in solving an optimization problem. SA consists basically of two main mechanisms. One is *the generation of alternatives (states)* and the other is *an acceptance rule*. Initially, for a given temperature T_0, a sequence of configurations is generated (N_0). The initial configuration S_i is then chosen. T_k is the control parameter. Initially T is large; then is reduced based on a *cooling schedule*. The acceptance criterion is as discussed in *Metropolis* algorithm.

Initial temperature T_0, the number of transitions performed at each temperature level (N_k), final temperature, T_f (as the stopping criterion) and the cooling sequence (given by $T_{k+1} = g(T_k) \cdot T_k$; where $g(T_k)$ is a function which controls the temperature), are four main SA parameters. Appropriate determinations of the above parameters have received attention in literature.

2.3.2.3 Particle Swarm

Some natural creatures such as fishes and birds behave as a swarm. Each individual coordinates its movement with the others in such a way that it does not collide with the others, moves towards the destination and moves to the center of the group (swarm).

It was mid 1990s that the basic idea of PS was formulated as an optimization algorithm.

The characteristics of each individual (the so called *agent*) are shown in a two-dimensional space by its position (x and y) and its velocity vector (v_x and v_y). Each agent optimizes its movement towards the destination. In doing so, it tracks

[6] For details, see the list of the references at the end of the chapter.

- The best value of the objective function which it has achieved so far (the so called *pbest*),
- The best value of the objective function which the other agents have achieved so far (the so called *gbest*).

So, the agent modifies its position, noting

- Its current position,
- Its current velocity,
- The distances between the current position with *pbest* and *gbest*.

Mathematically speaking, new position of an agent i in iteration $k+1$ (s_i^{k+1}) can be determined from its current (iteration k) position (s_i^k); knowing its velocity at iteration $k+1$ (v_i^{k+1}).[7] (v_i^{k+1}) can be determined as

$$v_i^{k+1} = wv_i^k + C_1 rand_1 (pbest_i - s_i^k) + C_2 rand_2 (gbest - s_i^k) \tag{2.11}$$

where w is a weighting factor, C_1 and C_2 are weighting coefficients and $rand_1$ and $rand_2$ are two random numbers between 0 and 1.

The first term results in agent movement in the same direction as before; as a result exploring new search space. That is why, w, is called the *diversification* coefficient. Usually it is defined as[8]

$$w = \overline{w} - \left(\frac{\overline{w} - \underline{w}}{\overline{iter}} \right) iter \tag{2.12}$$

$\bar{w}$ and $\underline{w}$ are typically selected to be 0.9 and 0.4, respectively. With (2.12), initially diversification is heavily weighted and is reduced towards the end of the search procedure. On the other hand, the second and the third terms of (2.11) result in the so called *intensification*. C_1 and C_2 may be typically selected to be 2.0.

The steps involved in a PS optimization algorithm can be generally described as

(a) Generate the initial condition for each agent
(b) Evaluate the searching point of each agent
(c) Modify each searching point

The procedure is repeated for a maximum number of iterations.

It should be mentioned that some variations of PS optimization method have been developed, so far, to account for some practical combinatorial optimization problems.

---

[7] $s_i^{k+1} = s_i^k + v_i^{k+1}$.

[8] $\overline{w}$, $\underline{w}$ and $\overline{iter}$ are the maximum w, the minimum w and the maximum number of iterations, respectively.

2.3.2.4 Tabu Search

Tabu means *forbidden* to search or to consider. Unlike other combinatorial approaches, TS is not related to physical phenomena. It was initially proposed in the early 1980s. It is an iterative procedure which starts from an initial solution and tends to move to new solution space in a more aggressive or greedier way than GA or SA. The neighborhood, from which the next solution/move is to be selected, is modified by classifying some moves as *tabu*,[9] others as desirable.

At each iteration of the algorithm, a neighborhood structure is defined; a move is then made to the best configuration. To escape from local optimum points, some transitions to the configurations with higher costs are also allowed. Similar to the PS algorithm, using *intensification* and *diversification* result in a more comprehensive exploration of attractive regions and, at the same time, moving to previously unvisited regions. These help avoiding trapping in local optimum points.

The steps involved in a TS optimization algorithm may be summarized as

(a) Generate an initial solution,
(b) Select move,
(c) Update the solution. The next solution is chosen from the list of neighbors which is either considered as *desired* (*aspirant*) or not *tabu* and for which the objective function is optimum.

The process is repeated based on any stopping rule proposed. Unlike other heuristic algorithms, there is not enough theoretical background for tailoring TS to a practical problem at hand and the users have to resort to their practical experiences.

2.3.2.5 Ant Colony

The AC optimization technique is a combinatorial optimization technique, initially developed in the early 1990s. It is based on the behaviors of insects, especially the ants.

The ants have wonderful ability in finding the shortest distance from a food to their nest. Even if an obstacle is put in between, they again find the shortest distance.

The scientists have discovered that the main tool of this phenomenon is the so called *pheromone* used as the basic communication media among the individuals.

Upon walking, each ant deposits a chemical substance, called pheromone, as a trail on the ground. Initially, all ants move around in a random manner to search for food. If they are considered to have the same speed, the one finding the food more quickly (i.e. with the shortest distance) returns to the nest sooner and deposits pheromone on coming back. The path will be richer in pheromone. Other ants will

---

[9] Those with undesirable (higher for a minimization problem) objective functions.

soon recognize it as a promising path and all follow it. Based on the above, some AC algorithms have been developed. Basically, the steps are as follows:

- *Initialization* in which the problem variables, are encoded and initial population is generated; randomly within the feasible region. They will crawl to different directions at a radius not greater than R.
- *Evaluation* in which the objective function is calculated for all ants.
- *Trail adding* in which a trail quantity is added for each ant; in proportion to its calculated objective function (the so called *fitness*).
- *Ants sending* in which the ants are sent to their next nodes, according to the *trail density* and *visibility*.
- We have already described *trail density* as the pheromone is deposited. The ants are not completely blind and will move to some extent based on node *visibilities*. These two actions resemble the steps involved in PS and TS algorithms (*intensification* and *diversification*) to avoid trapping in local optimum points.
- *Evaporation* in which the trail deposited by an ant is eventually evaporated and the starting point is updated with the best combination found.

The steps are repeated until a stopping rule criterion is achieved.

References

Extensive books are published on optimizations techniques. Some typical ones are introduced here. The reader may consult specific books on detailed applications and algorithms of each subject.

1. Hillier FS, Lieberman GJ (2010) Introduction to operations research, 9th edn. McGraw Hill, New York
2. Rao SS (2009) Engineering optimization, 4th edn. Wiley, New Jersey
3. Chong EKP, Zak SH (2001) An introduction to optimization, 2nd edn. Wiley, New York
4. Branke J, Deb K, Miettinen K, Slowinski R (eds) (2008) Multiobjective optimization. Springer, Berlin
5. Bartholomew-Biggs M (2008) Nonlinear optimization with engineering applications. Springer, Boston
6. Burke EK, Kendall G (2005) Search methodologies: introductory tutorials in optimization and decision support techniques. Springer, New York

Chapter 3
Some Economic Principles

3.1 Introduction

All of us are familiar with *economics*; although we are not, necessarily, able to define it in scientific terms. It affects our daily lives as we earn money and expend it afterwards. Economics is, in fact, the study of *how a society decides on what, how and for whom to produce*. While the so called *microeconomic analysis* focuses on a detailed treatment of individual *decisions about some particular commodities*, the so called *macroeconomic analysis* emphasizes the interactions in the *economy as a whole*.

Similar to any other social science, economics has appeared in power system field, too. Like any other man-made industry, electric power industry is confronted with revenues and costs; resulting in economic principles to be continuously observed. The emerged electric power markets have resulted in full involvements of this industry in economic based theories, applications and principles.

The subject of economics is quite vast. We are not, here, to investigate its principles. We do not want to be involved in those aspects of economics which, somehow, interact with electric markets, too. Instead, we want to, shortly, review the definitions of some basic terms used in power system planning field and especially in this book. The terms defined are not, necessarily, related to each other. Later on, they will be used throughout the book, once needed. The cash-flow concept is reviewed in Sect. 3.3. The methods for economic analysis are covered in Sect. 3.4.

3.2 Definitions of Terms

- **Revenue**
  Revenue is the money that a company earns by providing services in a given period such as a year.
- **Cost**
  Cost is the expense incurred in providing the services during a period.

H. Seifi and M. S. Sepasian, *Electric Power System Planning*, Power Systems,
DOI: 10.1007/978-3-642-17989-1_3, © Springer-Verlag Berlin Heidelberg 2011

- **Profit**
  Profit is the excess of revenue over the cost.
- **Investment cost**[1]
  Investment cost is the cost incurred in investing on machinery equipment and buildings used in providing the services.
- **Operational cost**
  Operational cost is the cost incurred on running a system to provide the services. Wages, resources (fuel, water, etc.), taxes are such typical costs.
- **Depreciation**
  Depreciation is the loss in value resulting from the use of machinery and equipment during the period. During a specific period, the cost of using a capital good is the depreciation or loss of the value of that good, not its purchase price. *Depreciation rate* is the rate of such a loss in value.
- **Nominal interest rate**
  Nominal interest rate is the annual percentage increase in the nominal value of a financial asset. If a lender makes a loan to a borrower, at the outset, the borrower agrees to pay the initial sum (the *principal*) with *interest* (at the rate determined by *interest rate*) at some future date.
- **Inflation rate**
  Inflation rate is the percentage increase per a specific period (typically a year) in the average price of goods and services.
- **Real interest rate**
  Real interest rate is the *nominal interest rate* minus the inflation rate.
- **Present value**
  Present value of some money at some future date is the sum that if lent out today, would accumulate to x by that future date. If this present value is represented by P and the annual interest rate is termed i, after N years we would have (F)

$$F = P(1 + i)^N \tag{3.1}$$

or

$$P = \frac{1}{(1 + i)^N} F \tag{3.2}$$

- **Discount factor**
  Discount factor is the factor used in calculating present values. It is equal to $1/(1 + i)^N$ (see (3.2)).

[1] Sometimes called *capital* or *capital investment* cost. In Chap. 5, we differentiate a little bit more, between these two terms. However, we mainly use *investment cost* as the most common term.

- **Salvation value**[2]
  Salvation value is the real value of an asset/equipment, remaining, at a specific time and after considering the depreciation rate.
- **Gross Domestic Product (GDP)**
  GDP measures the output produced by factors of production located in a domestic economy regardless of who owns these factors. GDP measures the value of output produced within the economy. While most of this output would be produced by domestic factors of production, there may be some exceptions.
- **Gross National Product (GNP) or Gross National Income (GNI)**
  GNP (or *GNI*) measures the total income earned by domestic citizens regardless of the country in which their factor services are supplied. GNP (or GNI) equals GDP plus net property income from abroad.
- **Nominal GNP**
  Nominal GNP measures GNP at the prices prevailing when income is earned.
- **Real GNP**
  Real GNP or *GNP* at constant prices adjusts for inflation by measuring GNP in different years at the prices prevailing at some particular calendar data known as the *base year*.
- **Per capita income (or per capita real GNP)**
  Per capita real GNP is real GNP divided by the total population. It is real GNP per head.

3.3 Cash-flow Concept

The flow of money, both the *inputs* and the *outputs*, resulting from a *project* is called cash-flow. In order to understand this concept, we should first define the *time value ofmoney*.

3.3.1 Time Value of Money

Any one easily understands that *money makes money*. In other words, if we invest an amount of *X*, we expect some percent to be added at the end of the year. In other words ₹ *X* at *present* worths more in the *future*. This concept is used if some one *invests* or *borrows* money.

---

[2] This term is not a very common economic term. However, as it is used in WASP package (see Chap. 5) for GEP problem, it is introduced here.

Example 3.1 If someone invests R 100 on a project with a 5% predicted return, he or she would gain R 105 at the end of the year. In other words, R 100 at present would worth R 105 in one-year time.

Example 3.2 Assume someone borrows R 100 to pay it back within a year with an annual interest rate of 10%. He or she would have to return R 110 at the end of the year.

In practice, the cases are more complex than the cases cited above. For economic analysis of a decision or a project, we should, first, define some of the economic terms as follows.

3.3.2 Economic Terms

For a project, the cash flows are of the following two types

- Inflows (such as an income)
- Outflows (such as a cost)

Both types may occur at *present* or in a specific time in the *future*. We should, then, define the *present value of money* (P) and the *future value of money* (F). The number of periods is assumed to be n while the interest rate is assumed to be i (%).

A value of P at present in n-year time worths as follows

$$
\begin{array}{ll}
F_1 = P + P \times i = P(1+i) & \text{at the end of the first year} \\
F_2 = F_1 + F_1 \times i = F_1(1+i) \ = P(1+i)^2 & \text{at the end of the second year} \\
\vdots & \vdots \\
F = F_{n-1} + F_{n-1} \times i = P(1+i)^n & \text{at the end of the } n\text{th year}
\end{array}
$$

In other words if we have R F in n-year time, it would worth $F/(1+i)^n$ at present. $(1+i)^n$ is named as *compound amountfactor* and is denoted by $(F/P, i\%, n)$. $1/(1+i)^n$ is named as *present worth factor* and is denoted by $(P/F, i\%, n)$.

Example 3.3 If we repeat example 3.1 for a 5-year period, the investor gains R $(1+0.05)^5 \times 100 =$ R 127.6 at the end of the fifth year.

Example 3.4 What happens if we repeat example 3.2 for a 10-year period. In other words, the borrower has to pay back the money in 10-year time. We can readily check that the borrower should return a total amount of R $(1+0.1)^{10} \times 100 =$ R 259.4 at the end of the 10-year time.

Regarding example 3.3, the investor may get equal annual payments and not the total amount at the end of the fifth year. In example 3.4, the borrower may have to pay the money back in equal annual amounts. As cash flows occur in different times, how should we calculate them?

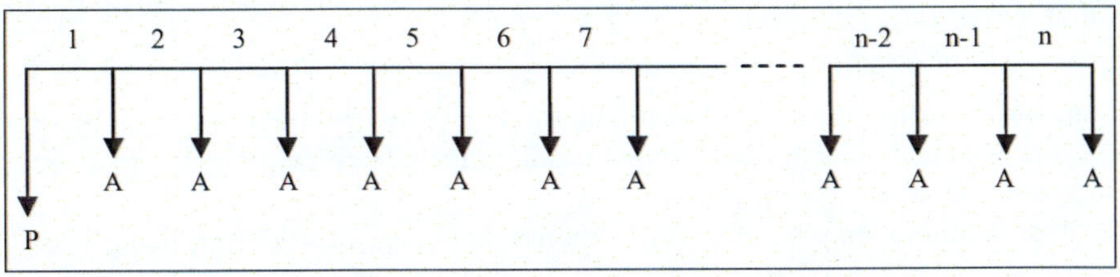

Fig. 3.1 Uniform payments in *n*-year time

As shown in Fig. 3.1, a present ₹ P is paid back in a regular amount of ₹ A at the end of each year. As a payment of ₹ A in n-year time worths $(1/(1 + i)^n)\, A$ at present, we would have

$$\begin{aligned} P &= \left[\left(\frac{1}{(1+)}\right)A + \left(\frac{1}{(1+i)^2}\right)A + \cdots + \left(\frac{1}{(1+i)^n}\right)A\right] \\ &= \left[\left(\frac{1}{(1+i)}\right) + \left(\frac{1}{(1+i)^2}\right) + \cdots + \left(\frac{1}{(1+i)^n}\right)\right]A \end{aligned} \tag{3.3}$$

As from elementary calculus

$$x + x^2 + x^3 + \cdots + x^n = \frac{x(1-x^n)}{1-x} \tag{3.4}$$

then

$$P = \left[\frac{(1+i)^n - 1}{i(1+i)^n}\right]A \tag{3.5}$$

or

$$A = \left[\frac{i(1+i)^n}{(1+i)^n - 1}\right]P \tag{3.6}$$

$[((1+i)^n - 1)/(i(1+i)^n)]$ is named as *uniform series present worthfactor* and is denoted by (P/A, i%, n). $[(i(1+i)^n)/((1+i)^n - 1)]$ is named as *capital recovery factor* and is denoted by (A/P, i%, n). It is easy to verify that

$$A = \left[\frac{i}{(1+i)^n - 1}\right]F \tag{3.7}$$

$$F = \left[\frac{(1+i)^n - 1}{i}\right]A \tag{3.8}$$

where the brackets in (3.7) and (3.8) are named as *sinking fund factor* and *series compound amount factor*, respectively; denoted by (A/F, i%, n) and (F/A, i%, n), respectively.

3.4 Economic Analysis

From various solutions available for a problem, a planner should select the best, in terms of both technical and economic considerations. Here we are going to discuss the economic aspect of a problem.

Three methods may be used for economic appraisal of a project, namely as

- Present worth method
- Annual cost method
- Rate of return method

In evaluating a project, we should note that various plans may be different in terms of effective economic life. Sometimes, it is assumed that the economic life of a plan is infinite ($n \to \infty$).

3.4.1 Present Worth Method

In this method, all input and output cash flows of a project are converted to the *present* values. The one with a net negative flow (*Net Present Worth*, NPW) is considered to be viable. From those viable, the one with the lowest net flow is the best plan.

In this method, if the economic lives of the plans are different, the study period may be chosen to cover both plans in a fair basis. For instance, if the economic lives of two plans are 3 and 4 years, respectively, the study period may be chosen to be 12 years.

Example 3.5 Consider two plans A and B with the details shown in Table 3.1.

With an interest rate of 5%, NPW_A and NPW_B are calculated as

$$NPW_A = 1000 + 50 \times (P/A, 5\%, 25) - 100 \times (P/A, 5\%, 25) \\ - 300 \times (P/F, 5\%, 25) = \text{R}\,206.71$$

$$NPW_B = 1300 + 70 \times (P/A, 5\%, 25) - 150 \times (P/A, 5\%, 25) \\ - 500 \times (P/F, 5\%, 25) = \text{R}\,24.83$$

Table 3.1 Details of plans A and B

Items	A	B
Investment cost (R)	1000	1300
Operational cost (R/year)	50	70
Profit (R/year)	100	150
Salvation value[a] (R)	300	500
Economic life (year)	25	25

[a] The value left at the end of the 25th year

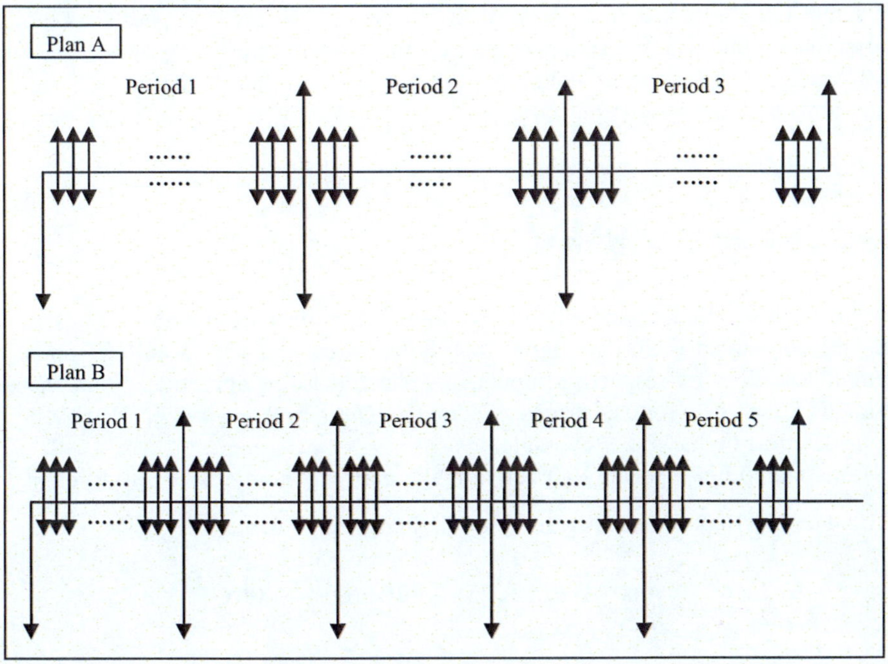

Fig. 3.2 Unequal economic lives for plans A and B

As both NPWs are positive, we can conclude that for both plans, the costs are more than the profits and none is a good choice. However, plan B is more attractive if we have to choose a plan.

Example 3.6 Repeal example 3.5, if the economic life of plan B is 15 years.

As already noted, we need to evaluate the plans for a 75-year period; to cover both plans in a rational basis. The case is depicted in Fig. 3.2. NPW_A and NPW_B are calculated as

$$\begin{aligned} NPW_A = {} & 1000 + 1000 \times (P/F, 5\%, 25) + 1000 \times (P/F, 5\%, 50) \\ & + 50 \times (P/A, 5\%, 75) - 100 \times (P/A, 5\%, 75) \\ & - 300 \times (P/F, 5\%, 25) - 300 \times (P/F, 5\%, 50) \\ & - 300 \times (P/F, 5\%, 75) = \text{R}\,285.78 \\ NPW_B = {} & 1300 + 1300 \times (P/F, 5\%, 15) + 1300 \times (P/F, 5\%, 30) \\ & + 1300 \times (P/F, 5\%, 45) + 1300 \times (P/F, 5\%, 60) \\ & + 70 \times (P/A, 5\%, 75) - 150 \times (P/A, 5\%, 75) \\ & - 500 \times (P/F, 5\%, 15) - 500 \times (P/F, 5\%, 30) \\ & - 500 \times (P/F, 5\%, 45) - 500 \times (P/F, 5\%, 60) \\ & - 500 \times (P/F, 5\%, 75) = \text{R}\,430.11 \end{aligned}$$

We find out the fact that if we have to choose a plan anyway, plan A is more attractive in this case. We should emphasize that considering a 75 year period does not mean that the actual economic lives of the plans are longer in this case and is used only for comparison purposes.

3.4.2 Annual Cost Method

In this method, all input and output cash flows of a project are converted to a series of uniform annual input and output cash flows. A project with a uniform annual output less than its respective input is considered to be attractive. From those attractive, the one with the least *Net Equivalent Uniform Annual Cost* (NEUAC) is considered to be the most favorable.

This method is especially attractive if the plans economic lives are different.

Example 3.7 Repeat example 3.5 with the annual cost method.

$$\begin{aligned} NEUAC_A &= 1000\,(A/P, 5\%, 25) + 50 - 100 - 300(A/F, 5\%, 25) \\ &= \text{R}\ 14.66/\text{year} \\ NEUAC_B &= 1300(A/P, 5\%, 25) + 70 - 150 - 500(A/F, 5\%, 25) \\ &= \text{R}\ 1.76/\text{year} \end{aligned}$$

Therefore, plan B is more attractive.

Example 3.8 Repeat example 3.6 with the annual cost method.

$$\begin{aligned} NEUAC_B &= 1300(A/P, 5\%, 15) + 70 - 150 - 500(A/F, 5\%, 15) \\ &= \text{R}\ 22.07/\text{year} \end{aligned}$$

Comparing $NEUAC_A = 14.66$ and $NEUAC_B = 22.07$ results in choosing plan A.

3.4.3 Rate of Return Method

There are some input and output cash flows during the economic life of a project. If we consider an interest rate at which these cash flows are equal (i.e., the net is zero), the resulting rate is named as *Rate Of Return* (ROR). ROR should be compared with the *Minimum Attractive Rate Of Return* (MAROR). Provided ROR is greater than MAROR, the plan is attractive. From those attractive, the one with the highest ROR is the most favorable.

ROR can be calculated using one of the methods outlined in Sects. 3.4.1 or 3.4.2. A trial and error approach may be used to find out the solution.

Example 3.9 Calculate ROR of example 3.5 using the method outlined in Sect. 3.4.1.

$$\begin{aligned}
&PWC_A = PWB_A \\
&\quad 1000 + 50 \times (P/A, ROR\%, 25) = 100 \times (P/A, ROR\%, 25) \\
&\quad\ + 300 \times (P/F, ROR\%, 25) \Rightarrow ROR = 3.1\% \\
&PWC_B = PWB_B \\
&\quad 1300 + 70 \times (P/A, ROR\%, 25) = 150 \times (P/A, ROR\%, 25) \\
&\quad\ + 500 \times (P/F, ROR\%, 25) \Rightarrow ROR = 4.8\%
\end{aligned}$$

where *PWC* is Present Worth Cost and *PWB* is Present Worth Benefit.

If MAROR is considered to be 5%, none is attractive. If we have to choose a plan anyway, plan B is more attractive due to its higher ROR.

3.4.4 A Detailed Example

For supplying the loads in a utility, new generation facilities, namely, 400 and 600 MW, in 5-year and 10-year times, respectively, are required. Three scenarios are investigated as follows

- **Scenario 1**
  The utility may install a 400 MW natural gas fueled unit in the first period and a 600 MW hydro unit in the second period. However, transmission lines with 1500 MVA km equivalent capacity should be constructed (500 MVA km in the first period and 1000 MVA km in the second period), while no new natural gas piping is required in either of the periods.
- **Scenario 2**
  Installing two natural gas fueled units at the heavy load area (400 MW for the first period and 600 MW for the second period) is another choice by which no new transmission line is required. However, new natural gas piping is required, as the heavy load area is confronted by natural gas deficiency. The piping should provide full capacity requirement for each unit. Assume that 2×10^6 m^3 km is required for the 400 MW generation, while 3×10^6 m^3 km is needed for the 600 MW one.
- **Scenario 3**
  The utility has a third option in which the generation requirements may be fulfilled through neighboring systems. However, an equivalent of 500 MVA km transmission line should be constructed within the second period.

Table 3.2 Generation units characteristics

Type	Investment cost (R/kW)	Operational cost (R/kW year)	Fuel cost (R/MWh)	Life (year)
Hydro	1000	5	–	50
Gas fueled	250	20	30	25

Table 3.3 Piping and transmission line characteristics

Type	Investment cost	Operational cost	Life (year)
Natural gas piping	R 15/m^3 km	R 0.15/m^3 km year	50
Transmission line	R 5/kVA km	R 0.025/kVA km year	50

The studies have shown that for the above scenarios, the system losses would be increased by 40, 4 and 12 MW, respectively, in the first period and by 60, 6 and 18 MW, respectively, in the second period.[3]

Assuming the interest rate to be 15%, the cost of the losses to be R 800/kW,[4] the cost of meeting the loads through neighboring systems to be R 0.1/kWh and R 0.07/kWh for the first and the second periods, respectively, and the load factor to be 0.8 for both periods, find out the best scenario using the cost terms as outlined in Tables 3.2 and 3.3. In the evaluation process, assume the costs would be increased based on annual inflation rate. Moreover, assume the investment costs to be incurred at year 3 and year 7, in the first and the second periods, respectively. Consider the study period to be 15 years.

Defining the following variables

C_{IG}: The generation unit investment cost,
C_{IL}: The transmission line investment cost,
C_{IP}: The piping (natural gas) investment cost,
C_{OG}: The generation unit operational cost,
C_{OL}: The transmission line operational cost,
C_{OP}: The piping operational cost,
C_L: The cost of the losses,
C_F: The fuel cost.

and assuming

- The costs are incurred as shown in Fig. 3.3 (All costs assumed to be incurred at the end of each year).
- The concept of NPV to be used. As the elements life times are not identical and are larger than the study period, the investment costs are initially converted to an

[3] Resulting in new facilities to be installed for compensating such losses.
[4] For each year.

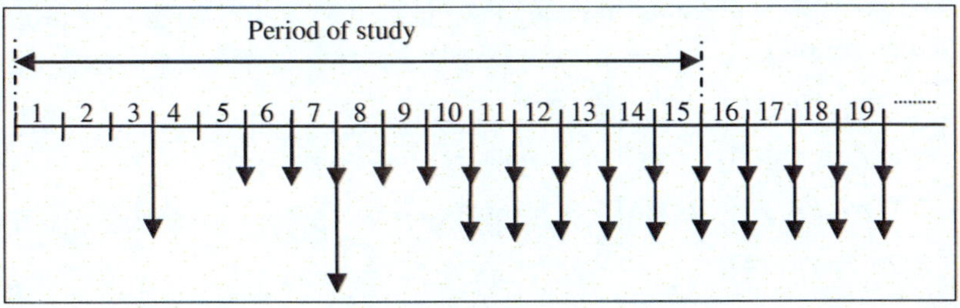

Fig. 3.3 The costs incurred during the study period

annual basis and added to all other annual terms. The costs after the study period (up to the life times) are converted to the base year and considered as negative costs (i.e., income or, in fact, asset).

The details for the scenarios are as follows[5]

- **Scenario 1**

$$
\begin{aligned}
C_{IG}^{1} &= \text{R}\,400 \times 250 \times 10^{3}\\
C_{IG}^{2} &= \text{R}\,600 \times 1000 \times 10^{3}\\
C_{IL}^{1} &= \text{R}\,500 \times 5 \times 10^{3}\\
C_{IL}^{2} &= \text{R}\,1000 \times 5 \times 10^{3}\\
C_{OG}^{1} &= \text{R}\,400 \times 20 \times 10^{3}/\text{year}\\
C_{OG}^{2} &= \text{R}\,600 \times 5 \times 10^{3}/\text{year} + \text{R}\,400 \times 20 \times 10^{3}/\text{year}\\
C_{OL}^{1} &= \text{R}\,500 \times 0.025 \times 10^{3}/\text{year}\\
C_{OL}^{2} &= \text{R}\,1000 \times 0.025 \times 10^{3}/\text{year} + \text{R}\,500 \times 0.025 \times 10^{3}/\text{year}\\
C_{F}^{1} &= \text{R}\,0.8 \times 400 \times 8760 \times 30/\text{year}\\
C_{F}^{2} &= \text{R}\,(0.8 \times 1000 \times 8760 - 600 \times 8760) \times 30\ /\text{year}\\
C_{L}^{1} &= \text{R}\,40 \times 800 \times 10^{3}/\text{year}\\
C_{L}^{2} &= \text{R}\,60 \times 800 \times 10^{3}/\text{year}
\end{aligned}
$$

In terms of C_F^1 and C_F^2, it is assumed that the energy requirement of the first period is produced by the gas fueled unit; while in the second period, some part is generated by the hydro unit (at its full capacity due to low operation cost) and the rest is generated by the gas fueled unit.

[5] Superscripts 1 and 2 denote periods 1 and 2, respectively.

Now based on the points already described, the values should be properly modified as follows

$$C_{IG} = C_{IG}^{1}(P/F,15\%,3) + C_{IG}^{2}(P/F,15\%,7) - C_{IG}^{1}(A/P,15\%,25)(P/A,15\%,15)(P/F,15\%,15) - C_{IG}^{2}(A/P,15\%,50)(P/A,15\%,45)(P/F,15\%,15) = \text{R}\,206529190.1$$

$$C_{IL} = C_{IL}^{1}(P/F,15\%,3) + C_{IL}^{2}(P/F,15\%,7) - C_{IL}^{1}(A/P,15\%,50)(P/A,15\%,40)(P/F,15\%,15) - C_{IL}^{2}(A/P,15\%,50)(P/A,15\%,45)(P/F,15\%,15) = \text{R}\,2603205.4$$

$$C_{OG} = C_{OG}^{1}(F/A,15\%,5)(P/F,15\%,10) + C_{OG}^{2}(F/A,15\%,5)(P/F,15\%,15) = \text{R}\,22447524.4$$

$$C_{OL} = C_{OL}^{1}(F/A,15\%,5)(P/F,15\%,10) + C_{OL}^{2}(F/A,15\%,5)(P/F,15\%,15) = \text{R}\,51905.2$$

$$C_{F} = C_{F}^{1}(F/A,15\%,5)(P/F,15\%,10) + C_{F}^{2}(F/A,15\%,5)(P/F,15\%,15) = \text{R}\,183706824.6$$

$$C_{L} = C_{L}^{1}(F/A,15\%,5)(P/F,15\%,10) + C_{L}^{2}(F/A,15\%,5)(P/F,15\%,15) = \text{R}\,93104503.6$$

Therefore, for scenario 1

$$C_{TOTAL} = C_{IG} + C_{IL} + C_{OG} + C_{OL} + C_{F} + C_{L} = \text{R}\,508443153.4 \tag{3.9}$$

- **Scenario 2**

  Similar to the above, for scenario 2

$$C_{TOTAL} = C_{IG} + C_{IP} + C_{OG} + C_{OP} + C_{F} + C_{L} = \text{R}\,475313882.3 \tag{3.10}$$

where

$$C_{IG} = C_{IG}^{1}(P/F,15\%,3) + C_{IG}^{2}(P/F,15\%,7) - C_{IG}^{1}(A/P,15\%,25)(P/A,15\%,15)(P/F,15\%,15) - C_{IG}^{2}(A/P,15\%,25)(P/A,15\%,20)(P/F,15\%,15) = \text{R}\,93175249.6$$

$$C_{IP} = C_{IP}^{1}(P/F,15\%,3) + C_{IP}^{2}(P/F,15\%,7) - C_{IP}^{1}(A/P,15\%,50)(P/A,15\%,40)(P/F,15\%,15) - C_{IP}^{2}(A/P,15\%,50)(P/A,15\%,45)(P/F,15\%,15) = \text{R}\,27441104.6$$

$$C_{OG} = C_{OG}^{1}(F/A,15\%,5)(P/F,15\%,10) + C_{OG}^{2}(F/A,15\%,5)(P/F,15\%,15) = \text{R}\,29904937.7$$

$$
\begin{aligned}
C_{OP} &= C_{OP}^1(F/A, 15\%, 5)(P/F, 15\%, 10) \\
&\quad + C_{OP}^2(F/A, 15\%, 5)(P/F, 15\%, 15) = \text{R}\,224287.0 \\
C_F &= C_F^1(F/A, 15\%, 5)(P/F, 15\%, 10) \\
&\quad + C_F^2(F/A, 15\%, 5)(P/F, 15\%, 15) = \text{R}\,314360704.9 \\
C_L &= C_L^1(F/A, 15\%, 5)(P/F, 15\%, 10) \\
&\quad + C_L^2(F/A, 15\%, 5)(P/F, 15\%, 15) = \text{R}\,9310450.4
\end{aligned}
$$

in which

$$
\begin{aligned}
C_{IG}^1 &= \text{R}\,400 \times 250 \times 10^3 \\
C_{IG}^2 &= \text{R}\,600 \times 250 \times 10^3 \\
C_{IP}^1 &= \text{R}\,2 \times 10^6 \times 15 \\
C_{IP}^2 &= \text{R}\,3 \times 10^6 \times 15 \\
C_{OG}^1 &= \text{R}\,400 \times 20 \times 10^3/\text{year} \\
C_{OG}^2 &= \text{R}\,600 \times 20 \times 10^3/\text{year} + \text{R}\,400 \times 20 \times 10^3/\text{year} \\
C_{OP}^1 &= \text{R}\,2 \times 0.15 \times 10^6/\text{year} \\
C_{OP}^2 &= \text{R}\,3 \times 0.15 \times 10^6/\text{year} + \text{R}\,2 \times 0.15 \times 10^6/\text{year} \\
C_F^1 &= \text{R}\,0.8 \times 400 \times 8760 \times 30/\text{year} \\
C_F^2 &= \text{R}\,(0.8 \times 1000 \times 8760) \times 30/\text{year} \\
C_L^1 &= \text{R}\,4 \times 800 \times 10^3/\text{year}
\end{aligned}
$$

- **Scenario 3**

  In this scenario, we would have

$$
\begin{aligned}
C_{IL}^1 &= \text{R}\,0 \\
C_{IL}^2 &= \text{R}\,500 \times 5 \times 10^3 \\
C_{OL}^1 &= \text{R}\,0/\text{year} \\
C_{OL}^2 &= \text{R}\,500 \times 0.025 \times 10^3/\text{year} \\
C_L^1 &= \text{R}\,12 \times 800 \times 10^3/\text{year} \\
C_L^2 &= \text{R}\,18 \times 800 \times 10^3/\text{year}
\end{aligned}
$$

If C_S^1 and C_S^2 denote the costs of providing the electricity through the neighboring systems in the first and the second periods, respectively, we would have

$$
\begin{aligned}
C_S^1 &= \text{R}\,0.8 \times 400 \times 8760 \times 10^3 \times 0.1/\text{year} \\
C_S^2 &= \text{R}\,0.8 \times 1000 \times 8760 \times 10^3 \times 0.07/\text{year}
\end{aligned}
$$

Table 3.4 Summary of the results

Scenario	C_{TOTAL} (R)
1	508,443,153.4
2	475,313,882.3
3	902,238,444.6

Therefore

$$\begin{aligned} C_{IL} &= C_{IL}^2(P/F, 15\%, 7) \\ &\quad - C_{IL}^2(A/P, 15\%, 50)(P/A, 15\%, 45)(P/F, 15\%, 15) = \text{R}\,632893.4 \\ C_{OL} &= C_{OL}^2(F/A, 15\%, 5)(P/F, 15\%, 15) = \text{R}\,10357.5 \\ C_S &= C_S^1(F/A, 15\%, 5)(P/F, 15\%, 10) \\ &\quad + C_S^2(F/A, 15\%, 5)(P/F, 15\%, 15) = \text{R}\,873663842.6 \\ C_L &= C_L^1(F/A, 15\%, 5)(P/F, 15\%, 10) \\ &\quad + C_L^2(F/A, 15\%, 5)(P/F, 15\%, 15) = \text{R}\,27931351.1 \end{aligned}$$

and

$$C_{TOTAL} = C_{IL} + C_{OL} + C_L + C_S = \text{R}\,902238444.6 \tag{3.11}$$

The results for scenarios are reported in Table 3.4. As seen, scenario 2 is the best choice in terms of economical considerations.

References

The books published on principles of economics are quite high. Three typical ones are introduced in [1–3]. Reference [4] is the typical book extensively used for power system economics.

1. Salvatore D, Diulio EA (1996) Schaum's outline of theory and problems of principles of economics. McGraw-Hill, New York
2. Bishop M (2004) Essential economics. Profile Books Ltd, London
3. McDowell M, Thom R, Frank R, Bernanke B (2006) Principles of economics. McGraw-Hill, Boston
4. Kirschen D, Strbac G (2004) Fundamentals of power system economics. Wiley, Chichester

Chapter 4
Load Forecasting

4.1 Introduction

In this chapter we are going to talk about *load forecasting*, as one of the basic and perhaps the most important module of power system planning issues. Although some other words, such as, demand and consumption are also used instead of load, we use load as the most common term. The actual term is electric load; however, electric is omitted here and assumed to be obvious. It is well understood that both the energy (MWh, kWh) and the power (MW, kW) are the two basic parameters of a load. By load, we mean the power. However, if energy is required in our analyses, we will use the energy demand or simply the energy, to refer to it. Obviously if the load shape is known, the energy can be calculated from its integral.

Forecasting refers to the prediction of the load behavior for the future. In this chapter, we discuss the load forecasting issue from various viewpoints.

The load characteristics are dealt with in Sect. 4.2. The load driving parameters are described in Sect. 4.3. There, we discuss load forecasting from various time-frames, including Short-Term Load Forecasting (STLF), Mid-Term Load Forecasting (MTLF) and Long-Term Load Forecasting (LTLF). As the main concern of this book, LTLF methods are discussed in Sect. 4.5. However, before that, Sect. 4.4 is devoted to an important topic of interest in LTLF, namely, spatial load forecasting. Some numerical examples are given in Sect. 4.6.

4.2 Load Characteristics

Let us start from a low power appliance such as a refrigerator, turning on and off, irregularly. At the same time, there are other appliances at a home, which, somehow, and to some extent, smooth out the load fluctuation of that home. Now, what happens to the load fluctuation of a distribution substation feeding several

H. Seifi and M. S. Sepasian, *Electric Power System Planning*, Power Systems,
DOI: 10.1007/978-3-642-17989-1_4,

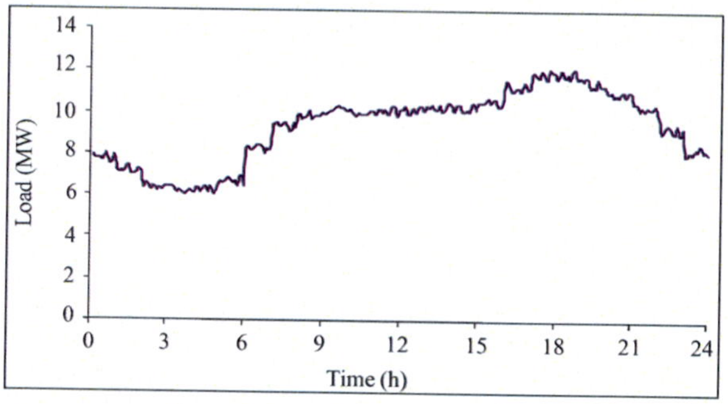

Fig. 4.1 The daily load of a distribution substation

homes. Still, the smoothing becomes more apparent. The daily load of a distribution substation may look like the one shown in Fig. 4.1.

On the other hand, the distribution substations are supplied, through subtransmission and transmission networks; from transmission substations. The daily load curve of a transmission substation has a general shape similar to Fig. 4.1. The same is true for the whole network consisting of several transmission substations.

Instead of focusing on the actual level,[1] let us now focus on the load shape of Fig. 4.1. Suppose we are going to know the load shape of the last week. Obviously we should gather the minute-by-minute data required. To simplify the task, let us assume that the load does not vary in each hour. In that case, the load shape may be drawn as shown in Fig. 4.2. It is evident that the load shape of a working day is significantly different from that of a weekend day. Moreover, even the load shape of working days may be different due to, say, the weather conditions. Let us now, go further towards the load shape of the past year. If the time step used is still 1-hour, $365 \times 24 = 8760$ data are required. It is evident that the task may be accomplished for a year or even for the last 10 years or more. These load shapes may be used for the detailed calculation of energy demands. However, they are of less use in planning studies, as we will see in this book.

Let us, now, focus on the future, without any available data. We are going to forecast the hour-by-hour daily (or weekly) load of our test case. This load curve is used by the system operator to decide the necessary actions. On the other hand, if for evaluating the generation deficiency, the load shape of the coming summer is to be forecasted, is it really possible to do so? In other words, it is possible to predict, the hour-by-hour load for several months from now? Basically if we have to forecast the hour-by-hour load,[2] we should accept the uncertainties involved.[3] However, we will see in this book that we often require less detailed, but as accurately as possible, the

[1] Here we are not involved with the level. In Sect. 4.4, we will come back to the point.

[2] For some types of studies such as fuel and water managements.

[3] See Chap. 11 for the uncertainties involved in power system planning problem.

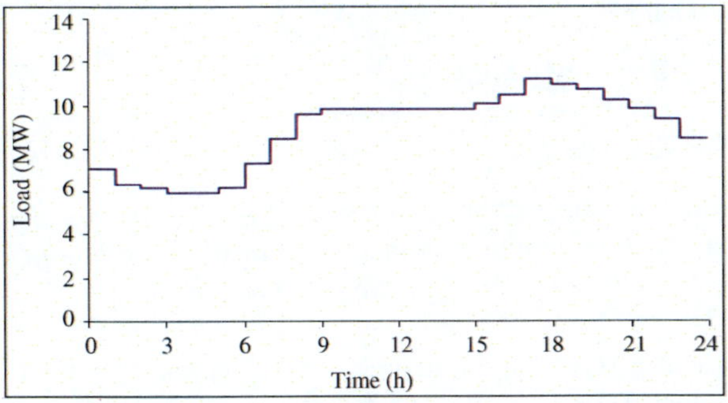

Fig. 4.2 The discretized load of the distribution substation

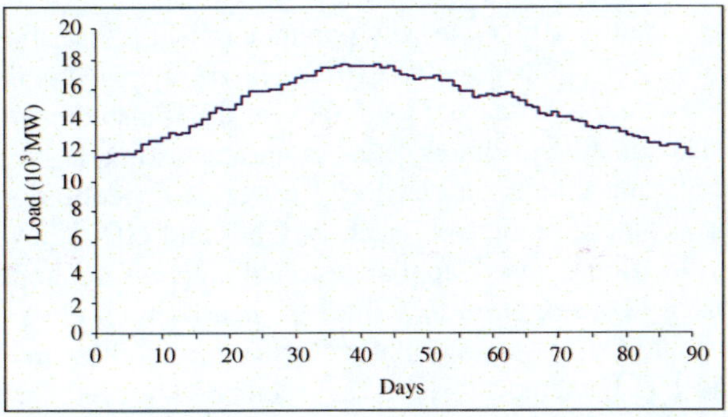

Fig. 4.3 A typical seasonal peak load curve

load shape. For instance, we may want to know the variation of daily peak loads of the coming summer. In other words, only 90 data are required. Such a seasonal load curve is shown in Fig. 4.3. We have not assumed that the load is flat in each day. Instead, we have only focused on the peak values.

If for planning purposes, we are going to predict the load variations for the next several years (say 10 years or more), we do not bother the daily variations.[4] Instead we may have to predict, say, the summer and the winter peaks, of the coming years. It means that for a 10-year prediction, only 20 data are required.

4.3 Load Driving Parameters

Once we have talked about the various load shapes in Sect. 4.2, in this section we focus on the parameters affecting the forecasted load of future. These driving parameters are quite a few. Some typical ones are as follows

[4] Even if we bother, who can predict the daily variations of say, 5 years from now?

- Time factors such as
  - Hours of the day (day or night)
  - Day of the week (week day or weekend)
  - Time of the year (season)
- Weather conditions (temperature and humidity)
- Class of customers (residential, commercial, industrial, agricultural, public, etc.)
- Special events (TV programmes, public holidays, etc.)
- Population
- Economic indicators (per capita income, Gross National Product (GNP), Gross Domestic Product (GDP), etc.)
- Trends in using new technologies
- Electricity price

The reader may readily add some new parameters to the list above.

For instance, it is well understood that if the *electricity price* is predicted to be high, it results in a reduced forecasted load. Obviously, it also depends on weather conditions; the class of the customers, etc. As another example, special TV programmes have dominant effects on electricity usage of residential sector. On the other hand, if the economic indicators such as GNP and GDP show a promising future and new electricity based appliances/technologies are appearing in the market, the electricity consumption may increase nearly in all class of customers.

For the reasons cited so for, we normally classify load forecasting methods into STLF, MTLF and LTLF methods. The STLF methods are used for hour-by-bour predictions while LTLF may be used for the peak seasonal predictions. STLF may be used for 1 day to 1 week, while LTLF may be used for several years. In this way, some driving parameters may be ineffective or ignored[5] for each of the above categories. For instance, GDP may have strong effects on LTLF; while ineffective in STLF. On the other hand, TV programmes are effective in STLF but ineffective in LTLF.

Figure 4.4 shows a schematic diagram in which the driving parameters are distributed among various load forecasting time frames. STLF normally results in hour-by-hour forecast (for 1 day to 1 week). MTLF normally results in daily forecast (for several weeks to several months). Normally the peak of the day is forecasted. LTLF focuses on monthly or seasonal forecasts (the peak of the month or the season) for several years from now.

It should be noticed as we move towards longer time frames, the accuracies of some driving parameters drop. For instance, the price forecast for STLF is more accurate than that of MTLF. The same is true for weather forecast. Due to inaccuracies involved in long-term driving parameters, it is of common practice to perform LTLF for several scenarios (such as various GDPs, weather forecasts, etc.). As the main concern of this book, we will focus on LTLF as detailed in Sect. 4.5.

Another point of interest is the geographical distribution of loads. This issue is commonly referred to *spatial load forecasting* and addressed in Sect. 4.4.

[5] Either assumed to be fixed or ineffective in our model.

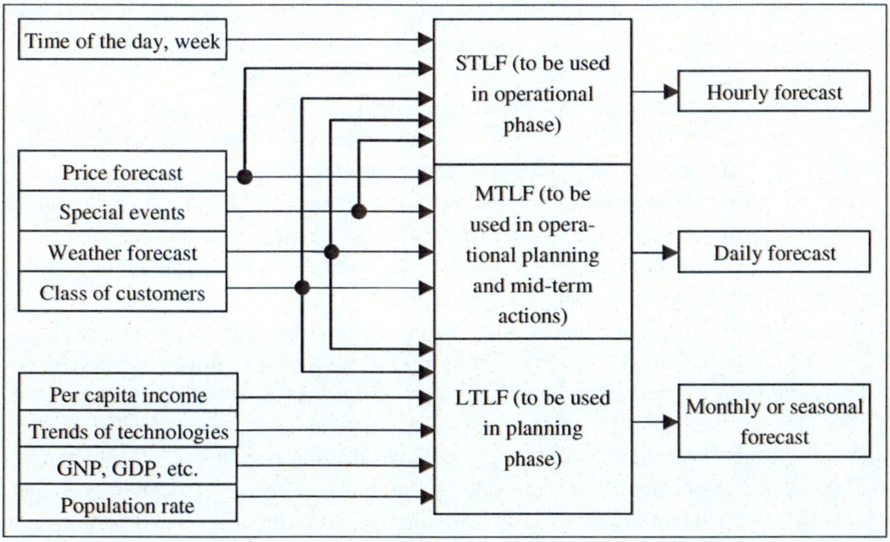

Fig. 4.4 The driving parameters

4.4 Spatial Load Forecasting

As earlier highlighted and discussed in Chap. 1 and in this chapter, planning for the future expansion of a power system involves determining both the *capacities* and the *locations* of future components; namely, generation facilities, transmission/sub-transmission/distribution lines and/or cables and various substations. As we will see, later on, in this book, this requires forecasting the future loads with geographic details (locations and magnitudes). In power system context, this topic is addressed as *spatial load forecasting*.

In Sect. 4.3, we classified load forecasting methods into STLF, MTLF and LTLF. There, we did not focus on the actual points for which their loads are to be predicted. Instead, we focused on the time frames and each category applications.

Suppose a power system operator is going to use STLF results for secure operation of the system. Obviously, he or she does not bother the exact details of small area loads, but is more interested in knowing the possible loads of substations. This type of forecasting is readily handled by existing STLF methods, beyond the scope of this book.[6]

Now let us move towards LTLF. We talked about its driving parameters in Sect. 4.3. For the future, we even may not know the details of the locations and the capacities of the future substations. Instead, we have to predict, initially, the small area loads (locations and magnitudes) in order to plan (location, capacity and

[6] See the references at the end of this chapter.

possible loading) for the future substations (see Chap. 7).[7] In fact, we have to use the methods discussed in Sect. 4.5 for small area loads. Once done, we may move upwards to predict the magnitudes and the locations of higher level loads. Spatial load forecasting is accomplished by dividing utility system into a number of small areas and forecasting the load of each. In some cases, the small areas used may be irregular in shape or size, corresponding to the service areas assigned to particular delivery system components such as substations or feeders. A simple choice is to use a gird of square cells that covers the region to be studied.

Once the load of each cell is predicted, the electric load of the system (or a larger geographical area) can be predicted.

An important aspect of electric load is that *cells* (small areas) do not simultaneously demand their peak powers. The *coincidence factor* defined as the ratio of *peak system load* to the *sum of small areas peak loads* is, normally in the range of 0.3–0.7.[8]

We earlier discussed about the long-term load driving parameters. For instance, GDP and population rate were mentioned there as two affecting parameters. Now, if we focus on a small area, is it really possible to predict the above two parameters for a small area?[9] Moreover what happens to load predication based on various classes of customers.[10] Later on, we will provide more details.

4.5 Long Term Load Forecasting Methods

The LTLF methods are basically *trend analysis*, *econometric modeling*, *end-use analysis* and *combined analysis*. These are briefly discussed in the following subsections.

4.5.1 Trend Analysis

The trend extrapolation method uses the information of the past to forecast the load of the future. A simple example is shown in Fig. 4.5, in which load is shown for the last 10 years and predicted to be 2906 MW in 2015. A curve fitting approach may be employed to find the load of the target year. This approach is simple to understand and inexpensive to implement. However, it implicitly assumes that the trends in various load driving parameters remain unchanged during the study period. For instance, if there is a substantial change in economic growth, the approach fails to forecast the future load, accurately. In a modified

---

[7] Once substations are decided, we move towards other steps of the planning procedure.

[8] This ratio depends on the system under study and may be estimated using historical data.

[9] They are normally predicted for larger geographical areas.

[10] A small area may be dominantly residential, while another may be industrial or combinatory.

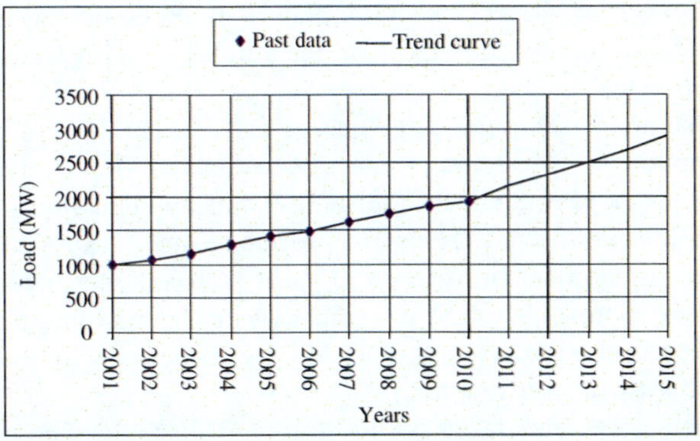

Fig. 4.5 Trend analysis

method, more weights may be attached to the loads towards the end of the past period. In this way, the prediction may be improved.

4.5.2 *Econometric Modeling*

In this approach, initially the relationship between the load and the driving parameters (Sect. 4.3) is estimated. The relationship may be nonlinear, linear; additive or in the form of multiplication. This relationship is established based on available historical data. Various driving parameters may be checked to find the ones that have the dominant effects.

A typical nonlinear estimation is

$$D_i = a(\text{per capita income})_i^b\,(\text{population})_i^c\,(\text{electricity price})_i^d \qquad (4.1)$$

where i denotes the year and a, b, c and d are the parameters to be determined from the historical data.

Once this relationship is established, the future values of the driving variables (i.e. per capita income, population, electricity price, etc.) should be projected. D_i for a future year can then be determined.

This approach is widely used and may be applied to various customer classes (residential, commercial, etc.) and to the system as a whole. It is relatively simple to apply. The drawback is the assumption of holding the relationship established for the past to be applicable for the future. In this way, the influence of any new driving parameter cannot be taken into account.

4.5.3 *End-use Analysis*

This type of analysis is mostly confined to residential loads but may be applied with some modifications to other load classes, too. As a simple example, if refrigerator is concerned, based on the number of households and estimating the percent of households having a refrigerator, the number of refrigerators for a future year may be estimated.

Following that and based on average energy use of such an appliance, the total energy consumption of refrigerators may be estimated. It is obvious that the average energy use is dependent on the intensity of appliance use, its efficiency and thermal efficiency of homes. The same procedure may be applied to other type of appliances and equipment in order to forecast the total energy requirement.

As evident, this approach explicitly predicts the energy consumption. If the load is to be estimated, some indirect approaches have to be used to convert the predicted energy to load (power demand).

This approach may lead to accurate results if its extensive accurate data requirements can be provided. Various driving parameters effects may be taken into account.

4.5.4 Combined Analysis

The end-use and econometric methods may be simultaneously used to forecast the load. It has the advantages and disadvantages of both approaches.

4.6 Numerical Examples

In this section, we try to demonstrate the steps involved in load forecasting through two case studies; namely, for a regional utility based on end-use analysis (Sect. 4.5.3) and for a large utility (or even a country) through econometric modeling (Sect. 4.5.2).

4.6.1 Load Forecasting for a Regional Utility

Figure 4.6 shows the region for which the load is to be forecasted. It consists of eight subregions (area). Each area consists of some subareas, supplied through some substations, either existing or new.[11] A summary of the data is shown in Table 4.1.

The substations are both at transmission and sub-transmission levels. The numbers shown are not of practical use here and represent typical values for an actual system. An area is normally designated by observing the fact that it is within

[11] New in the study year for which the load is to be predicted.

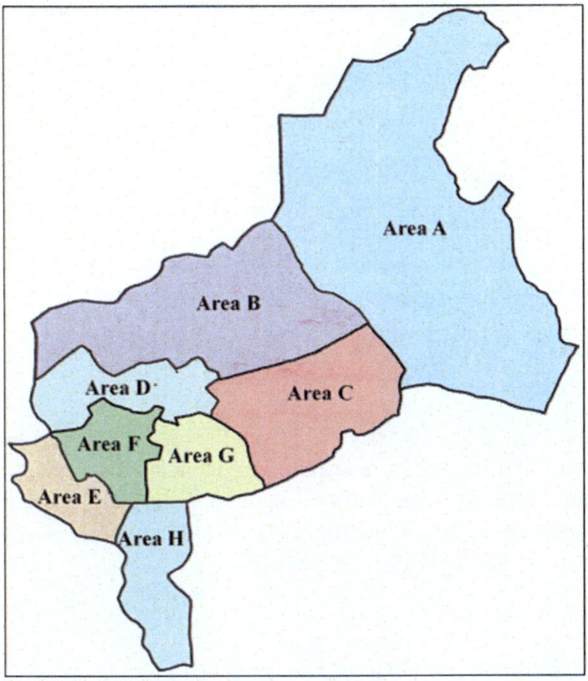

Fig. 4.6 Geographical distribution of the areas in the region

Table 4.1 Data summary

Area	Number of subareas	Number of existing substations
A	8	2
B	7	4
C	6	3
D	5	14
E	4	1
F	10	3
G	5	3
H	3	2

the service territory of some sub-transmission substations. Sometimes a metropolitan is considered as an area.

The aim is to predict the peak load, as well as, the energy demand of the regional utility for 10 years from the current year; with a time step of 1 year. The process starts from the subareas; moving upwards to reach the load for the region. It is assumed that the geographical characteristics of the subareas as well as their load data for the last 10 years are known. Before presenting numerical data, some basic definitions and concepts are described first.

4.6.1.1 Definitions and Concepts

It is assumed that each subarea consists of the following three types of loads

- Urban
- Rural
- Large customers

The urban loads, typically, consist of

- Residential
- Commercial
- Public
- Small industrial
- Distribution losses

Historical data as well as extensive data from the regional departments, in charge of the above mentioned sections, are required to reach at reasonable predictions. An urban load is not, actually, concentrated at a specific geographical point and is distributed throughout the urban territory. These points have to be observed.

The rural types of loads, mainly consist of

- Residential
- Agricultural
- Others (small industrial, public, etc.)

The residential part may be estimated based on the estimated number of homes and the estimated power consumption of each home. The latter is, itself, determined based on its existing figure and the possible increase in usage due to various reasons (say, new appliances and technologies appearing in rural areas).

The agricultural part is determined based on the estimated number of wells, their average depths and their average water flows. For instance, in a subarea, there may be a total number of 491 deep well, with 75 meter average depth and 25 l/s average flow. These figures may be 2735, 36 and 15, respectively, for semi-deep wells in the same subarea. Based on these figures, the agricultural load of the subareas and, as a result, the area may be determined.

The remaining part of the rural types of the loads should also be estimated. If difficult, sometimes, a fixed percentage (say 25%) may be considered.

The large customers are considered separately, as they do not obey any specific rule, in terms of, the forecasted loads. They may be either existing or new. The future loads of existing types may be estimated based on their previous and foreseen performances. The loads of new types are determined based on the demands of their respective contracts with the utility. Both require extensive data gathering and communications with the large customers and the departments in charge of large customers. They are typically the customers with more than 1 MW demand.

Based on the above, for each subarea, the peak forecasted load is determined for each class of the loads. The coincidence factors should then be used to find out the forecasted load of the area and then the region. These factors may be determined based on both historical data and some engineering judgments.

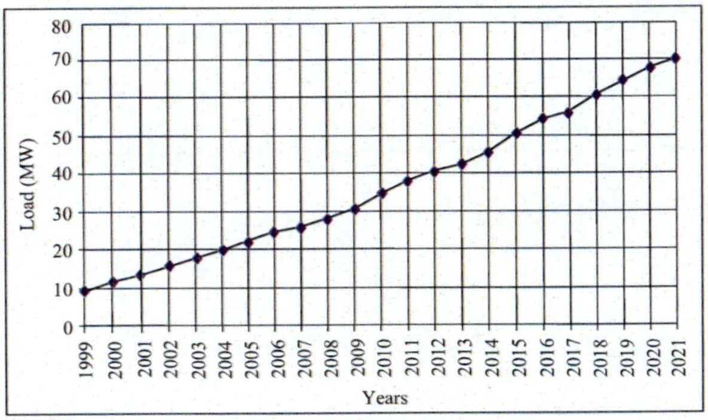

Fig. 4.7 The load of area E

4.6.1.2 Numerical Data

To save space, for the region under consideration, we show the results for area E (Table 4.1). This area consists of four subareas. There are two main urban loads in subareas 1 and 2 (UE1 and UE2). The load of the area is estimated to be varying as shown in Fig. 4.7. The details are as follows.

The results for the urban parts are shown in Table 4.2 for some selected years. Moreover, the results for the agricultural part of the area are shown in Table 4.3. The rural load data are provided in Table 4.4. Finally, the results for the area are tabulated in Table 4.5. Table 4.6 is the subarea-based results for the same area.

This procedure is repeated for all areas. Once done, the results should be combined to reach at the results for the region as follows

- The results for the rural load of the region are shown in Table 4.7.
- The results for the agricultural load of the region are shown in Table 4.8.

These results as well as the results for the urbans and large customers, are summarized in Table 4.9.

Moreover, if the annual *Load Factor* (LF) is defined as

$$LF = \frac{\text{Total energy (in MWh)}}{\text{Peak load (in MW)} \times 8760} \tag{4.2}$$

based on the historical load factors of the region and an estimation for these values for the coming years, total energy may also be forecasted. The results are detailed in Table 4.10.[12]

[12] Note that, some specific loads are also added in this table. These may be of the same nature of large customers without having a contract. For instance, a large residential complex may be considered as a specific load.

Table 4.2 Results for urban loads of area E

Urban name	Year						
	2011 (MW)	2012 (MW)	2013 (MW)	2014 (MW)	2016 (MW)	2019 (MW)	2021 (MW)
UE1	4.7	5	5.4	5.8	6.6	8	9
UE2	13.5	14.4	15.4	16.5	18.9	22.6	25.6
Total	18.2	19.4	20.8	22.3	25.5	30.6	34.6

4.6.2 Load Forecasting of a Large Scale Utility

Suppose the load of a large scale utility, composed of some regional utilities, is to be forecasted. Obviously, one way is to combine the results obtained from the regional utilities; observing coincidence factors; to generate the results for the main utility. Sometimes, we may look at the problem as a whole and want to predict the overall consumption of the utility, without having to be involved much in details of the regions. A typical case is described in this section. Obviously, the procedure outlined is not unique and may be adjusted based on available data. Moreover, the application is not merely for large scale utilities and may be applied to any scale system, provided the required data are available.

4.6.2.1 Definitions and Concepts

First, let us review a basic definition as follows

- Total Demand (TD)[13] is the sum of the Supplied Demand (SD), Load Curtailment (LC), Import/Export Transactions (IET), Frequency Drop term (FD), Interrupted Loads (IL),[14] System Losses (SL) and Auxiliary Demand (AD) of the power plants, i.e.

$$TD = SD + LC + IET + FD + IL + SL + AD \tag{4.3}$$

Using a standard software and based on historical data, we should, initially, find out the driving parameters for the load. For instance, GDP,[15] population, per capita demand and average electricity price may be four main driving parameters. However, other parameters may also be tried and checked. If not considered, we have, implicitly, assumed that they are either non-driving parameters or there are some types of correlations between them and those already observed.

[13] We assume that for the historical data, the demand supplied is not the actual demand required (TD). In fact if, for instance, we have some load curtailments (LC) or the system operator has intentionally dropped the frequency to compensate, somewhat, the generation deficiency (FD), we have to add these and similar terms to find out the actual demand (TD). All terms are in MW.

[14] The loads interrupted based on some types of contracts.

[15] For definition of GDP, see Chap. 3.

Table 4.3 Results for the agricultural load of area E

Subarea	Year	Deep wells					Semi-deep wells					Total		
		Total number	Electrified	Average depth (m)	Average flow (l/s)	Load (kW)	Total number	Electrified	Average depth (m)	Average flow (l/s)	Load (kW)	Total	Electrified	Load (kW)
1	2010	21	20	85	17	1084	64	60	40	11	1320	85	80	2404
	2011	21	20	85	17	1129	64	60	40	11	1550	85	80	2679
	2016	21	20	85	17	1632	64	60	40	11	1627	85	80	3259
	2021	21	20	85	17	1943	64	60	40	11	2071	85	80	4014
2	2010	16	14	95	14	698	49	45	45	10	1215	65	59	1913
	2011	16	14	95	14	904	49	45	45	10	1304	65	59	2208
	2016	16	14	95	14	1165	49	45	45	10	1596	65	59	2761
	2021	16	14	95	14	1290	49	45	45	10	1852	65	59	3142
3	2010	126	114	90	15	4232	207	188	50	12	4834	333	302	9066
	2011	126	120	90	15	4874	207	190	50	12	4897	333	310	9771
	2016	126	122	90	15	7204	207	198	50	12	6785	333	320	13,989
	2021	126	124	90	15	7749	207	204	50	12	6991	333	328	14,740
4	2010	17	15	85	9	861	144	126	48	12	5184	161	141	6045
	2011	17	15	85	9	930	144	130	48	12	5332	161	145	6262
	2016	17	15	85	9	972	144	132	48	12	5451	161	147	6423
	2021	17	15	85	9	988	144	141	48	12	5806	161	156	6794
Total for the area	2010	180	163	89	15	6875	464	419	47	12	12,553	644	582	19,428
	2011	180	169	89	15	7837	464	425	47	12	13,083	644	594	20,920
	2016	180	171	89	15	10,973	464	435	47	12	15,459	644	606	26,432
	2021	180	173	89	15	11,970	464	450	47	12	16,720	644	623	28,690

Table 4.4 Results for rural load of area E

Year	No. of homes	Percent electrified	No. of homes, electrified	Per home consumption (W)	Residential load (kW)	Residential load and others (kW)
1999	3389	99	3355	353	1184	1628
2000	3431	99	3397	369	1253	1723
2001	3475	99	3440	384	1321	1820
2002	3522	99	3487	401	1398	1926
2003	3575	99	3539	418	1479	2037
2004	3623	100	3623	430	1558	2143
2005	3670	100	3670	435	1596	2195
2006	3721	100	3721	440	1637	2253
2007	3740	100	3740	448	1676	2304
2008	3762	100	3762	456	1715	2360
2009	3972	100	3972	464	1843	2534
2010	4159	100	4159	472	1963	2700
2011	4181	100	4181	480	2007	2762
2012	4225	100	4225	488	2062	2837
2013	4268	100	4268	497	2121	2917
2014	4312	100	4312	505	2178	2997
2015	4356	100	4356	514	2239	3082
2016	4401	100	4401	523	2302	3166
2017	4447	100	4447	532	2366	3255
2018	4492	100	4492	541	2430	3344
2019	4540	100	4540	551	2502	3442
2020	4586	100	4586	560	2568	3536
2021	4634	100	4634	570	2641	3635

Whatever the approach is used, we should use a procedure for checking the method accuracy. If, say, the historical data is available for the last 15 years, we may use the results of the first 10 years for producing the model. Thereafter, its prediction behavior may be checked for the next 5 years, using actual data. Once done and approved, the best model may be used to forecast the loads of the coming years.

Various scenarios may be checked. For instance, one scenario may be considered as the load being dependent on GDP and population, only. Other combinations may be tried as new scenarios. Various fitting procedures and models may also be checked.[16] These are, typically, available in commercial software.[17]

Even new scenarios may be generated with weighted driving parameters. For instance, a driving parameter may also be given a higher weighting in comparison with another. A scenario may also be generated by a combination of already generated scenarios, weighted based on their respective accuracies which are already checked.

---

[16] For more informations on available models (AR, ARMA, etc.), see, the list of the references at the end of the chapter.

[17] Eviews and SPSS are two typical available software.

Table 4.5 Results for area E (details)

Year	Urban (MW)	Rural (MW)		Large customers (MW)			Total load (MW)			Coincidence factor (%)
		Agricultural	Residential and others	Existing	Future	Total	Non coincident	Coincident[a]	Increase rate (%)	
1999	8	0.8	1.6	0	0	0	10.4	9.5	–	91
2000	9.6	0.9	1.7	0	0	0	12.2	11.1	16.8	91
2001	10.4	1	1.8	0	0	0	13.2	12	8.1	91
2002	11.2	4	1.9	0	0	0	17.1	15.6	30	91
2003	11.6	5.8	2	0	0	0	19.4	17.7	13.5	91
2004	11.7	9.1	2.1	0	0	0	22.9	19.9	12.4	87
2005	12.2	10	2.2	0	0	0	24.4	22	10.6	90
2006	13.1	13	2.3	0	0	0	28.4	25.6	16.4	90
2007	13.4	13.2	2.3	0	0	0	28.9	26	1.6	90
2008	14.6	14.2	2.4	0	0	0	31.2	28.1	8.1	90
2009	16.6	15.4	2.5	0	0	0	34.5	31.1	10.7	90
2010	17	19.4	2.7	0	0	0	39.1	35.2	13.2	90
2011	18.2	20.9	2.8	0	0	0	41.9	38.1	8.2	91
2012	19.4	22	2.8	0	0	0	44.2	40.2	5.5	91
2013	20.8	23.1	2.9	0	0	0	46.8	42.6	6	91
2014	22.3	24.2	3	0	0	0	49.5	45	5.6	91
2015	23.8	25.3	3.1	0	3	3	55.2	50.2	11.6	91
2016	25.5	26.4	3.2	0	5	5	60.1	54.7	9	91
2017	27.1	26.9	3.3	0	5	5	62.3	56.7	3.7	91
2018	28.8	27.3	3.3	0	7	7	66.4	60.4	6.5	91
2019	30.6	27.8	3.4	0	9	9	70.8	64.4	6.6	91
2020	32.5	28.2	3.5	0	10	10	74.2	67.5	4.8	91
2021	34.6	28.7	3.6	0	11	11	77.9	70.1	3.9	90

[a] Figure 4.7 is drawn based on the values of this column

Table 4.6 Results for area E (Subarea-based)

Subarea	Year	Urban load (MW)	Rural (MW)		Large customers (MW)		Non coincident (MW)	Coincident (MW)	Coincidence factor (%)
			Agricultural	Residential and others	Existing	Future			
1	2011	4.7	2.68	0.5	0	0	7.88	7.1	90
	2016	6.6	3.26	0.5	0	0	10.36	9.3	90
	2021	9	4.01	0.5	0	0	13.51	12.2	90
2	2011	13.5	2.21	0.6	0	0	16.31	14.7	90
	2016	18.9	2.76	0.7	0	2	24.36	21.9	90
	2021	25.6	3.14	0.8	0	6	35.54	32	90
3	2011	0	9.77	1	0	0	10.77	9.7	90
	2016	0	13.99	1.2	0	3	18.19	16.4	90
	2021	0	14.74	1.3	0	5	21.04	18.9	90
4	2011	0	6.26	0.7	0	0	6.96	6.6	95
	2016	0	6.42	0.8	0	0	7.22	6.9	95
	2021	0	6.79	1	0	0	7.79	7.4	95
Subarea total	2011	18.2	20.92	2.8	0	0	41.92	38.1	91
	2016	25.5	26.43	3.2	0	5	60.13	54.5	91
	2021	34.6	28.68	3.6	0	11	77.88	70.5	91

Table 4.7 Results for rural load of the region

Year	No. of homes	% of homes, electrified	No. of homes, electrified	Per home consumption (W)	Residential load (kW)	Residential and others (kW)
1999	39,538	95	37,561	403	15,137	20,763
2000	40,405	95	38,385	413	15,853	21,782
2001	41,284	95	39,220	434	17,021	23,359
2002	42,160	96	40,474	442	17,890	24,528
2003	43,160	96	41,434	458	18,977	26,007
2004	44,671	97	43,331	468	20,279	27,854
2005	45,648	97	44,279	477	21,121	28,989
2006	45,785	98	44,869	483	21,672	29,722
2007	47,740	98	46,785	492	23,018	31,601
2008	48,711	98	47,737	502	23,964	32,875
2009	50,560	99	50,054	508	25,427	34,931
2010	52,008	99	51,488	519	26,722	36,673
2011	52,283	99	51,760	529	27,381	37,624
2012	52,823	99	52,295	538	28,135	38,651
2013	53,367	100	53,367	541	28,872	39,669
2014	53,916	100	53,916	550	29,654	40,711
2015	54,472	100	54,472	559	30,450	41,817
2016	55,030	100	55,030	568	31,257	42,909
2017	55,601	100	55,601	577	32,082	44,022
2018	56,172	100	56,172	586	32,917	45,168
2019	56,750	100	56,750	595	33,766	46,359
2020	57,307	100	57,307	605	34,671	47,560
2021	57,900	100	57,900	614	35,551	48,816

4.6.2.2 Numerical Data

For a typical system, assume TD (see (4.3)), GDP (see Chap. 3) and population for the last 31 years are as shown in Table 4.11. The aim is to predict the load for 2011–2017. What we do is to use various approaches in which the historical data of the years 1980–2006 are employed to predict the loads of the year 2007–2017. Based on the prediction behavior observed for 2007–2010 (comparing the predicted load with the actual load), we may then select the best approach.

The following approaches are tested

- Linear Curve Fitting (LCF) as follows

$$TD = a + bx \tag{4.4}$$

where x is the *year number* (1 through 27 in Table 4.11).

- Second order Curve Fitting (SCF) as follows

$$TD = a + bx + cx^2 \tag{4.5}$$

Table 4.8 Results for agricultural load of the region

Year	Deep wells						Semi-deep wells						Total		
	Total No.	Electrified	Average depth	Average flow (l/s)	Load (kW)	Increase rate (%)	Total No.	Electrified	Average depth	Average flow (l/s)	Load (kW)	Increase rate (%)	Total No.	Electrified	Load (kW)
2010	1307	1181	107	22	65,257	–	1110	986	39	9	26,584	–	2417	2167	91,841
2011	1307	1246	107	22	71,000	8.8	1110	1019	39	9	28,912	8.8	2417	2265	99,912
2016	1307	1270	107	22	81,709	15.1	1110	1060	39	9	35,858	24	2417	2330	117,567
2021	1307	1291	107	22	86,990	6.5	1110	1091	39	9	41,170	14.8	2417	2382	128,160

Table 4.9 Overall results for the region

Year	Urban load		Rural load				Large customers			Total load				Coincidence factor (%)
	Load (MW)	Increase rate (%)	Agricultural (MW)	Increase rate (%)	Residential and others (MW)	Increase rate (%)	Existing (MW)	Existing and new (MW)	Increase rate (%)	Non coincident (MW)	Increase rate (%)	Coincident (MW)	Increase rate (%)	
1999	184.9	0	18.5	0	20.8	0	40.1	40.1	0	264	0	235	–	88.9
2000	206.8	11.9	20.1	8.6	21.8	4.9	44.6	44.6	11.2	293	11	271	15.3	92.4
2001	222.2	7.5	25.1	24.9	23.4	7.2	52.7	52.7	18.2	323	10.3	311	14.8	96.2
2002	229.1	3.1	33.1	31.9	24.5	5	96	96	82.2	383	18.3	366	17.7	95.6
2003	256.3	11.9	40.9	23.6	26	6	133.1	133.1	38.6	456	19.2	385	5.2	84.4
2004	272.4	6.3	52.2	27.6	27.9	7.1	190	190	42.7	542	18.9	464	20.5	85.7
2005	294.2	8	57.1	9.4	29	4.1	203.7	203.7	7.2	584	7.7	504	8.4	86.3
2006	316	7.4	64	12.1	29.7	2.5	220.9	220.9	8.5	631	8	526	4.4	83.4
2007	325.4	5.3	69.8	9.1	31.6	6.3	264.9	264.9	19.9	692	9.7	568	8	82.1
2008	342.6	4.3	74.9	7.3	32.9	4	281.1	281.1	6.1	731	5.7	610	7.4	83.5
2009	357.3	2.3	82.7	10.4	34.9	6.3	346.9	347	23.4	822	12.4	666	9	81
2010	365.4	11	91.8	11.1	36.7	5	384.8	384.8	10.9	879	6.9	735	10.4	83.6
2011	405.7	9	99.9	8.8	37.6	2.6	402.9	403.7	4.9	947	7.7	786	6.9	83
2012	442.4	9	103.4	3.5	38.7	2.7	427	433	7.3	1017	7.4	854.7	8.7	84
2013	482.4	9	107	3.4	39.7	2.6	457.5	484.2	11.8	1113	9.4	935.1	9.4	84
2014	526	9.1	110.5	3.3	40.7	2.6	492.2	589.7	21.8	1267	13.8	1064.2	13.8	84
2015	573.6	9.1	114	3.2	41.8	2.7	529.5	781.1	32.5	1511	19.2	1268.9	19.2	84
2016	625.7	7.8	117.6	3.1	42.9	2.6	561.4	910.9	16.6	1697	12.3	1425.5	12.3	84
2017	674.4	7.8	119.7	1.8	44	2.6	592.2	1057.4	16.1	1896	11.7	1592.3	11.7	84
2018	727	7.8	121.8	1.8	45.2	2.6	602.4	1161.5	9.8	2056	8.4	1726.7	8.4	84
2019	783.8	7.8	123.9	1.7	46.4	2.6	612.4	1269.2	9.3	2223	8.2	1867.6	8.2	84
2020	844.9	7.8	126	1.7	47.6	2.6	612.4	1376.4	8.4	2395	7.7	2011.7	7.7	84
2021	911	7.8	128.2	1.7	48.8	2.6	612.4	1465.4	6.5	2553	6.6	2144.9	6.6	84

Table 4.10 Loads and energy demands for the region

Area	Year	2004	2009	2010	2011	2012	2013	2014	2015	2016	2017	2019	2021
A	MW	18	26	29	31	33	35	39	46	49	53	58	66
	GWh	66	105	116	129	145	158	174	210	238	260	301	359
B	MW	58	96	117	127	134	161	190	232	271	312	386	454
	GWh	262	517	631	671	714	824	983	1212	1401	1588	1931	2275
C	MW	18	40	42	47	51	55	107	182	212	239	291	358
	GWh	87	153	174	192	206	235	394	649	738	901	1115	1409
D	MW	228	293	302	357	394	429	470	518	575	623	704	793
	GWh	1068	1580	1736	1820	2031	2212	2506	2874	3259	3577	4185	4715
E	MW	20	31	35	38	40	43	45	50	54	56	64	71
	GWh	95	158	168	184	198	213	225	253	279	290	330	357
F	MW	15	25	28	31	33	35	37	43	47	50	57	68
	GWh	74	125	139	150	167	186	201	238	260	279	322	376
G, H	MW	35	61	68	74	79	85	91	97	104	129	156	175
	GWh	158	313	309	368	392	437	471	512	554	638	754	862
P1[a]	MW	53	56	68	68	69	71	73	74	77	78	82	85
	GWh	264	362	399	413	420	435	448	469	490	499	524	544
P2	MW	50	70	70	70	70	70	70	90	100	120	140	170
	GWh	93	241	265	282	282	282	282	365	403	471	540	651
P3	MW	3	28	28	32	36	40	43	47	51	55	55	55
	GWh	0	49	96	112	125	139	152	174	192	207	207	207
P4	MW	0	0	0	0	0	0	0	10	20	30	50	50
	GWh	0	0	0	0	0	0	0	24	51	79	131	140
Region	MW[b]	465	666	735	786	855	935	1064	1269	1426	1592	1868	2145
	GWh	2167	3603	4033	4321	4680	5121	5836	6980	7865	8789	10340	11895

[a] P1-P4 are specific loads
[b] Coincidence factors are observed among the areas

- Third order Curve Fitting (TCF) as follows

$$TD = a + bx + cx^2 + dx^3 \tag{4.6}$$

- Exponential Curve Fitting (ECF) as follows

$$TD = a(1 - e^{-bx}) \tag{4.7}$$

- Univariate ARMA[18] (UARMA)
- Multivariate ARMA (MARMA)

[18] For some details on ARMA, see Appendix C.

Table 4.11 Historical data for the last 31 years

No.	Year	Actual load (MW)	GDP (10^6 R)	Population/1000
1	1980	2934	219,191	36,393
2	1981	3242	209,919	37,814
3	1982	3773	178,149	39,291
4	1983	3741	170,281	40,826
5	1984	4171	191,667	42,420
6	1985	4884	212,877	44,077
7	1986	5625	208,516	45,798
8	1987	6672	212,686	47,587
9	1988	7487	193,235	49,445
10	1989	7999	191,312	50,662
11	1990	8738	180,823	51,909
12	1991	9184	191,503	53,187
13	1992	10,276	218,539	54,496
14	1993	11,205	245,036	55,837
15	1994	12,064	254,822	56,656
16	1995	13,383	258,601	57,478
17	1996	14,369	259,876	58,331
18	1997	15,251	267,534	59,187
19	1998	16,109	283,807	60,055
20	1999	17,465	291,769	61,070
21	2000	18,821	300,140	62,103
22	2001	19,805	304,941	63,152
23	2002	21,347	320,069	64,219
24	2003	23,062	330,565	65,301
25	2004	24,750	355,554	66,300
26	2005	27,107	379,838	67,315
27	2006	29,267	398,234	68,345
28	2007	32,217	413,765	69,254
29	2008	34,107	437,344	70,313
30	2009	34,894	464,308	71,410
31	2010	37,639	496,313	72,483

Using a standard software such as Eviews (For some details, see Appendix D), based on the historical data, various parameters are calculated as follows

- LCF; $a = -549.50$, $b = 950.83$
- SCF; $a = 1336.42$, $b = 425.60$, $c = 20.50$
- TCF; $a = 614.35$, $b = 917.92$, $c = -29.70$, $d = 1.28$
- ECF; $a = 659{,}561$, $b = 0.001416$

Results for various approaches are shown in Table 4.12 for 2007–2017. The actual load for 2007–2010 are also shown. Note that GDP and population for year 2010, onwards, are considered to be increased at the rates of 3.9 and 1.33%, respectively.

Table 4.12 Results for various approaches

No.	Year	Actual load (MW)	GDP (10^6R)	Pop./1000	Forecast (MW)					
					LCF	SCF	TCF	ECF.	UARMA	MARMA
28	2007	32,217	413,765	69,254	26,074	29,325	31,130	25,639	31,088	31,102
29	2008	34,107	437,344	70,313	27,025	30,919	33,474	26,536	33,086	33,251
30	2009	34,894	464,308	71,410	27,975	32,554	35,982	27,431	35,163	35,718
31	2010	37,639	496,313	72,483	28,926	34,231	38,661	28,326	37,319	38,502
32	2011	–	507,728	73,840	29,877	35,948	41,518	29,219	39,553	41,375
33	2012	–	527,529	74,822	30,828	37,706	44,562	30,111	41,865	44,228
34	2013	–	548,103	75,717	31,779	39,505	47,800	31,002	44,255	47,135
35	2014	–	569,479	76,826	32,730	41,345	51,239	31,891	46,723	50,149
36	2015	–	591,689	77,848	33,680	43,226	54,888	32,779	49,267	53,305
37	2016	–	614,764	78,883	34,631	45,148	58,754	33,666	51,887	56,628
38	2017	–	638,740	79,932	35,582	47,111	62,845	34,552	54,581	60,144

Table 4.13 Prediction behavior

No.	Year	LCF	SCF	TCF	ECF	UARMA	MARMA
28	2007	19.07	8.98	3.37	20.42	3.50	3.46
29	2008	20.77	9.35	1.86	22.2	2.99	2.51
30	2009	19.83	6.70	3.12	21.39	0.77	2.36
31	2010	23.15	9.06	2.71	24.74	0.85	2.29
Average error (%)		20.7	8.52	2.77	22.19	2.03	2.66

Expressing the prediction behavior in terms of the error as follows

$$\text{Error} = \left|\frac{\text{Forecasted} - \text{Actual}}{\text{Actual}}\right| \times 100\% \tag{4.8}$$

The errors observed for various approaches are shown in Table 4.13. As shown, UARMA, MARMA and TCF are ranked as the best choices, in terms of, the prediction behavior.

References

References [1] and [2] are two books published on some aspects of load forecasting in an electric power system.

Short term load forecasting has received much attention in literature. Some of them are covered in [3–7]. Some details on the models discussed in Sect. 4.6.2, are provided in [8]. References [9] and [10] emphasize spatial load forecasting. References [11] and [12] are devoted to load forecasting bibliography at the time of publication. The publications on long term load forecasting are also quite a few. Some of them are given in [13–19].

1. Willis HL (2002) Spatial electric load forecasting. Marcel dekker, New York
2. Weron R (2006) Modeling and forecasting electricity loads and prices: a statistical approach (the Wiley finance series). Wiley, Chichester

3. Gross G, Galiana FD (1987) Short-term load forecasting. Proc IEEE 75(12):1558–1573
4. El-Magd A, Sinha MA, Naresh K (1982) Short-term load demand modeling and forecasting: A review. IEEE Trans Syst Man Cybernet 12(3):370–382
5. Moghram I, Rahman S (1989) Analysis and evaluation of five short-term load forecasting techniques. IEEE Trans Power Syst 4(4):1484–1491
6. Hippert HS, Pedreira CE, Souza RC (2001) Neural networks for short-term load forecasting: a review and evaluation. IEEE Trans Power Syst 16(1):44–55
7. Daneshdoost M, Lotfalian M, Bumroonggit G, Ngoy JP (1998) Neural network with fuzzy set-based classification for short-term load forecasting. IEEE Trans Power Syst 13(4):1386–1391
8. Soliman SA, Al-Kandari AM (2010) Electrical load forecasting: modeling and model construction. Elsevier Inc, Burlington
9. Willis HL, Engel MV, Buri MJ (1995) Spatial load forecasting. IEEE Comput Appl Power 8(2):40–43
10. Willis HL, Northcote-Green JED (1983) Spatial electric load forecasting: a tutorial review. Proc IEEE 71(2):232–253
11. IEEE Committee Report. (1980) Load forecast bibliography phase I. IEEE Trans Power Apparatus and Syst PAS-99(1):53–58
12. IEEE Committee Report (1981) Load forecast bibliography, phase II. IEEE Trans Power Apparatus Syst PAS-100(7):3217–3220
13. Carpinteiro OAS, Leme RC, de Souza ACZ, Pinheiro CAM, Moreira EM (2007) Long-term load forecasting via a hierarchical neural model with time integrators. Electr Power Syst Res 77(3–4):371–378
14. Al-Hamadi HM, Soliman SA (2005) Long-term/mid-term electric load forecasting based on short-term correlation and annual growth. Electr Power Syst Res 74(3):353–361
15. Ghods L, Kalantar M (2008) Methods for long-term electric load demand forecasting: a comprehensive investigation. In: IEEE international conference on industrial technology—ICIT
16. Shi Y, Yang H, Ding Y, Pang N (2008) Research on long term load forecasting based on improved genetic neural network. In: Proceeding of Pacific-Asia workshop on computational intelligence and industrial application—PACIIA, vol 2, pp 80–84
17. Khoa TQD, Phuong LM, Binh PTT, Lien NTH (2004) Application of wavelet and neural network to long-term load forecasting. In: Proceeding of international conference on power system technology—PowerCon vol 1, pp 840–844
18. Kandil MS, El-Debeiky SM, Hasanien NE (2002) Long-term load forecasting for fast developing utility using a knowledge-based expert system. IEEE Trans Power Syst 17(2):491–496
19. Zhang S, Wang D (2009) Medium and long-term load forecasting based on PCA and BP neural network method. In: Proceeding of international conference on energy and environment technology—ICEET vol 3, pp 389–391

Chapter 5
Single-bus Generation Expansion Planning

5.1 Introduction

Generation Expansion Planning (GEP) is the first crucial step in long-term planning issues, after the load is properly forecasted for a specified future period. GEP is, in fact, the problem of determining *when*, *what* and *where* the generation plants are required so that the loads are adequately supplied for a foreseen future. This problem is dealt with in this chapter. We will see how complex the problem is, so that, we first ignore the transmission system to make the problem easy to handle. This single-bus GEP is in contrast to a multi-bus GEP which will be dealt in Chap. 6 in which transmission system effects will also be considered. Problem definition is described in Sect. 5.2. Some detailed description is provided in Sect. 5.3 through simple examples. A detailed mathematical modeling is demonstrated in Sect. 5.4. The solution procedure through Wien Automatic System Planning (WASP) package, developed by International Atomic Energy Agency (IAEA), is discussed in Sect. 5.5. Numerical results are provided in Sect. 5.6 using WASP.

5.2 Problem Definition

Generally speaking, GEP, is an optimization problem in which the aim is to determine the new generation plants in terms of *when to be available*, *what type and capacity they should be* and *where to allocate* so that an objective function is optimized and various constraints are met. It may be of static type in which the solution is found only for a specified stage (typically, year) or a dynamic type, in which, the solution is found for several stages in a specified period. The objective function consists, generally, of

$$\text{Objective function} = \text{Capital costs} + \text{Operation costs} \tag{5.1}$$

H. Seifi and M. S. Sepasian, *Electric Power System Planning*, Power Systems,
DOI: 10.1007/978-3-642-17989-1_5,

The first term is, mainly due to

- Investment costs (C_{inv})
- Salvation value of investment costs (C_{salv})
- Fuel inventory costs (C_{finv})

while, the second term, consists, mainly, of

- Fuel costs (C_{fuel})
- Non-fuel operation and maintenance costs ($C_{O\&M}$)
- Cost of energy not served (C_{ENS})

In Sect. 5.3, we will clarify some of the objective function terms through simple examples. Later on, we will develop a basic mathematical formulation.

Besides the objective function, some constraints should also be met. A simple constraint is the one which describes the available generating capacity to be greater than the load. Obviously, if a reserve margin is required, the difference should also take the reserve into account. More constraints will be required as we will discuss the problem in Sect. 5.3 through examples and in Sect. 5.4 through mathematical modeling.

5.3 Problem Description

Let us consider a case in which the aim is to determine the generation capacity for year t in which the peak load is PL_t. If PG_t denotes the available generating capacity in year t, it will be a function of K_t, where[1,2,3]

$$K_t = \text{Already committed units} + \text{New units additions} - \text{Units retired} \tag{5.2}$$

Moreover, if Res_t denotes the minimum reserve margin (in %), the following inequality should be met

$$(1 + Res_t/100)PL_t \leq PG_t \tag{5.3}$$

Moreover, suppose the available plant candidate plants are

- A: 150 MW thermal power plant (with oil fuel)
- B: 250 MW thermal power plant (with coal fuel)
- C: 100 MW gas turbine power plant (with natural gas fuel)

Let us assume that, the existing capacity is 500 MW, consisting of two already committed units (2 × 250), denoted by D. The plants specifications are provided in Table 5.1.

[1] Already committed units—from previous period.

[2] New units additions—to be determined.

[3] Units retired—due to age.

Table 5.1 Plants data

Unit name	Max capacity (MW)	Investment cost (R/kW)	Plant life (Year)	Fuel cost[a] (R/MWh)	Fixed O&M cost (R/kW month)	Variable O&M cost[b] (R/MWh)	Scheduled maintenance[b] (day/year)
A	150	300	20	20.409	1	1	10
B	250	350	30	14.000	3	3	30
C	100	250	25	25.953	2.5	2.5	50
D	250	–	–	14.355	–		–

[a] The fuel cost is considered to be independent of the operating point
[b] This data will, later on, be used in Sect. 5.6

Before going further towards examples, we define some of the terms in Table 5.1 as follows

- **Investment cost**. This term represents the cost of a power plant, in terms of R/kW. The total investment cost is the product of this value with the power plant capacity.
- **Plant life**. Two plants with the same total investment costs, but with different lives, have different values. If the plant life is say, 20 years, and the study period is say, 5 years, at the end of this period, still some values are left, defined as salvation value.[4] This value will be deducted from the capital cost so that the actual investment cost can be determined.
- **Fuel cost**. The fuel cost of a plant is, in fact, dependent on its production level (i.e. $f(PG_t)$). In other words, the cost varies with the production level. For simplicity, however, the cost (R/MWh) is considered to be fixed here. Total cost is calculated from the product of this value and the energy production of the unit.
- **O & M cost parameters**. Operation and Maintenance (O & M) is the process required for the proper operation of power plants, defined in terms of the number of days per year. Two cost parameters are also normally defined for maintenance
  - *A fixed term*, independent of energy production (in terms of R/kW month); the total value is calculated from the product of this value times the plant capacity times 12 (12 months).
  - *A variable term*, defined in terms of R/MWh. Note that the total variable cost[5] is affected by the period of maintenance, as during these days, the plant does not generate any power.

Note that for the sake of simplicity in this section, only the fixed term is considered. Moreover, note that except for the fuel cost, other parameters are not considered for the existing plants (Table 5.1).

[4] For economic decisions, some details are provided in Chap. 3. Discount rate should be given in order to calculate the salvation value. A very simple, but unrealistic choice is to consider this rate to be zero. In that case, after 5 years, 15/20 of its value is left (salvation value).

[5] The total fuel cost is also affected by the period of maintenance.

Table 5.2 Various test cases

Case name	Unit name	Investment cost	Fuel cost	Fixed O&M cost
CASE1_A1	A	✓	–	–
CASE1_AB1	A, B	✓	–	–
CASE1_ABC1	A, B, C	✓	–	–
CASE1_A2	A	✓	✓	–
CASE1_AB2	A, B	✓	✓	–
CASE1_ABC2	A, B, C	✓	✓	–
CASE1_A3	A	✓	✓	✓
CASE1_AB3	A, B	✓	✓	✓
CASE1_ABC3	A, B, C	✓	✓	✓

Now let us define some test cases as demonstrated in Table 5.2. The ✓ shows the parameters considered in each case. Nine cases are generated. The aim is to determine the generation capacity for a year with the following assumptions

- The load is 1000 MW ($PL_t = 1000$ MW), considered to be flat throughout the year.
- The reserve margin is considered to be 20%.
- The discount rate is taken to be zero.
- The results are summarized in Table 5.3.

Some explanations are given below

- In CASE1_AB1, due to unit B longer life, this unit type is selected, although its investment cost in R/kW is higher in comparison with the A type.
- Comparing CASE1_ABC2 with CASE1_ABC1 reveals the fact that B type unit is the attractive choice in meeting the energy requirement (in comparison with type C) due to its lower fuel cost. However, in meeting the reserve requirement, C type is attractive due to its lower investment cost.

Table 5.3 Results for the test cases

Case name	Selected units			Investment cost (kR/year)	Operation cost (fuel cost) (kR/year)	Fixed O&M cost (kR/year)	Total cost (kR/year)
	A	B	C				
CASE1_A1	5	–	–	11,250	–	–	11,250
CASE1_AB1	0	3	–	8750	–	–	8750
CASE1_ABC1	0	0	7	7000	–	–	7000
CASE1_A2	5	–	–	11,250	152,264	–	163,514
CASE1_AB2	0	3	–	8750	123,417	–	132,167
CASE1_ABC2	0	2	2	7833	124,195	–	132,028
CASE1_A3	5	–	–	11,250	152,264	9000	172,514
CASE1_AB3	2	2	–	10,333	124,195	21,600	156,128
CASE1_ABC3	–	2	2	7833	124,195	24,000	156,028

- Comparing CASE1_AB3 with CASE1_AB2 shows that B type unit is the attractive choice in meeting the energy requirement (in comparison with A type) due to its lower fuel cost. However, in meeting the reserve requirement, A type is attractive due to its lower O&M cost.
- The results of Table 5.3 are generated using the GEP1.m M-file [#GEP1.m; Appendix L: (L.1)]. Simple calculations may be carried out to justify the values given in that table.

Now let us make the situation more practical. Suppose we are going to observe the following points

- Our study period extends for several years. As described in Chap. 1, the planning problem may be described as a dynamic type; as opposed to static type. In that case, the capital as well as the operation costs should be minimized for the whole period. The costs have to be referred to a common reference point, so that comparisons of the plans are possible. To do so, the Net Present Values (NPV) should be calculated based on a given discount rate. It is assumed that full investment cost for a plant is made at beginning of the year in which it goes into service. The operational costs may be assumed to occur in the middle of each year. The salvation costs are assumed to occur at the end of each year.
- The load may not be constant throughout a year. Instead it can be described by a non-flat Load Duration Curve (LDC),[6] either in a continuous or discrete way. The continuous type may be in the form of a polynomial function. The discrete type may be defined as several levels, each of which by a specified period. A typical continuous type may be in the form of[7]

$$\text{Normalized load} = 1 - 3.6D + 16.6D^2 - 36.8D^3 + 36D^4 - 12.8D^5 \quad 0 \leq D \leq 1 \tag{5.4}$$

A typical discrete type is shown in Fig. 5.1.

- Besides defining a reserve margin, what happens if we also consider a reliability index for our solution, such as Loss Of Load Probability (LOLP)?[8] In fact, although power plants are maintained regularly, they may have unexpected outage due to any reason. The probability of such a failure is defined as Forced Outage Rate (FOR). If the FOR of a unit is, say, 5%, it means that the plant would be available only for 95% of the time it is anticipated to be in service. The LOLP of the overall generation resources is calculated based on the given FORs of the plants and the anticipated load. These FORs are normally known based on

---

[6] See Appendix E.

[7] Normalized load is the load divided by the maximum value. D is similarly the normalized total time.

[8] This index is expressed in terms of the average fraction of total time, the system is expected to be in a state of failure. For further details, see Appendix E.

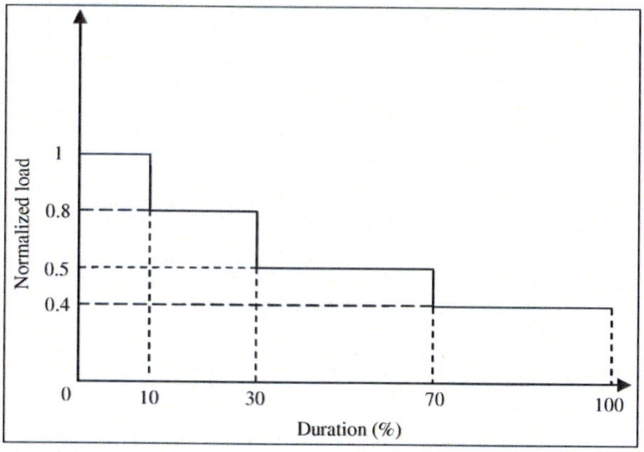

Fig. 5.1 A typical discrete LDC

the historical data of the plants.[9] Both the LOLP and the reserve margin may be simultaneously considered.

- Suppose two different plans result in acceptable performances in terms of LOLP. In other words, the resulting LOLPs are smaller than a pre-specified value, but one smaller than the other. One way to differentiate between these two plans is to consider the cost of *Energy Not Served* (ENS); as a lower LOLP implies less ENS. This cost may be calculated by ENS (which, in turn, may be calculated from LOLP) times the per unit cost of ENS (given by the user). Another way is to represent the cost as a polynomial function of ENS. If the cost of ENS is also taken into account (besides the reserve margin and the LOLP), the generation system would be expanded so far as the total cost defined in Sect. 5.2 is minimized.

We note that the problem can be quite complex if all the above mentioned points are to be considered. Moreover, still other factors may be taken into account either in terms of the objective function terms or the constraints. In Sect. 5.4, we develop a basic mathematical formulation of the problem in which some terms may be ignored or simplified. For instance, salvation value is ignored and the operation and maintenance cost is considered to be a function of only the unit capacity (the variable term, ignored). Later on and in Sect. 5.5, we introduce WASP in which nearly all terms are considered.

[9] For further details, see any book on power system reliability such as what given at the end of this chapter. Moreover, note that, by considering FORs, the total operational cost will be increased as we have to commit more expensive units once the less expensive ones are considered to be tripped out.

5.4 Mathematical Development

Based on what discussed so far, the problem is to determine from a list of available options, the number, type and capacity of each unit needed, in each year of the study period. In doing so, the total costs incurred should be minimized while various constraints, such as meeting the load, should be satisfied. If the decision variable is denoted by X_{it}, representing the number of unit type i for year t, the objective function terms and the constraints are described in the following subsections.

5.4.1 Objective Functions

Total cost, C_{total}, to be minimized may be described as[10]

$$C_{total} = C_{inv} + C_{fuel} + C_{O\&M} + C_{ENS} \tag{5.5}$$

where

C_{inv}	The investment cost
C_{fuel}	The fuel cost
$C_{O\&M}$	The operation and maintenance cost
C_{ENS}	The cost of energy not served

The details are as follows.

5.4.1.1 The Investment Cost

If X_{it} represents the number of unit type i required in year t, C_{inv} is given by

$$C_{inv} = \sum_{t=1}^{T}\sum_{i=1}^{Ng} Cost\_Inv_{it} PG_i X_{it} \tag{5.6}$$

where

$Cost\_Inv_{it}$	The cost in R/MW for unit type i in year t
PG_i	The capacity of unit i (MW)
T	The study period (in years)
Ng	The number of units types

[10] It is understood that all costs mentioned should be calculated, once referred to *base year*. This term is not repeated for convenience.

5.4.1.2 The Fuel Cost

The fuel cost of each unit is a function of its energy output,[11] normally in a nonlinear form. However, for simplicity, here we assume a linear function given by

$$C_{fuel} = \sum_{t=1}^{T} \left(\sum_{i=1}^{Ng} Cost\_Fuel_{it} Energy_{it} X_{it} + Cost\_Fuel_{et} \right) \tag{5.7}$$

where

$Cost\_Fuel_{it}$	The cost of fuel (in R/MWh) for unit type i in year t
$Energy_{it}$	Energy output for unit type i in year t
$Cost\_Fuel_{et}$	The fuel cost of existing units in year t

5.4.1.3 The Operation and Maintenance Cost

Similar to C_{inv}, the operation and maintenance cost is given as a linear function of PG_i given by

$$C_{O\&M} = \sum_{t=1}^{T} \sum_{i=1}^{Ng} Cost\_O\&M_{it} PG_i X_{it} \tag{5.8}$$

where

$Cost\_O\&M_{it}$	The operation and maintenance cost (in R/MW) for unit type i in year t

5.4.1.4 The Cost of Energy not Served

A generation unit may be tripped out in a rate given by its Forced Outage Rate (FOR). It represents the percentage of a time; the unit may be unavailable due to unexpected outages. Due to the FORs of the units and based on the demand and the available reserve, some portion of the energy demand can not be served. The so called *Energy Not Served* (ENS) can not be made zero, but should be minimized as a cost term. It is given by

$$C_{ENS} = \sum_{t=1}^{T} Cost\_ENS_t ENS_t \tag{5.9}$$

where

$Cost\_ENS_t$	The cost of the energy not served in year t (R/MWh)
ENS_t	The energy not served in year t (MWh)

---

[11] Various approaches may be used in calculating the energy outputs of the units. One simple way is to rank the units according to their fuel costs. Then, total energy requirement (as determined from LDC) is distributed among the units; based on the above ranking.

5.4.2 *Constraints*

Some constraints have to be observed during the optimization process. The ones considered here are described in the following subsections.

5.4.2.1 Technical Constraints

The generation capacity should be sufficient in meeting the load while some uncertainties are involved and the generation units may be, unexpectedly, tripped out at any time. The following two constraints may, thus, be considered

$$(1 + Res_t/100)PL_t \leq \sum_{i=1}^{Ng} PG_i X_{it} + PG_t \quad \forall t = 1,\ldots,T \tag{5.10}$$

$$LOLP_t \leq \overline{LOLP} \quad \forall t = 1,\ldots,T \tag{5.11}$$

where

Res_t	The required reserve in year t
PL_t	The load in year t
PG_t	The capacity available due to existing[12] units in year t
$LOLP_t$	The Loss Of Load Probability in year t
$\overline{LOLP}$	The maximum acceptable LOLP

The first constraint shows that the generation capacity should meet the load plus a reserve. LOLP is a reliability index normally used to represent the system robustness in response to elements contingencies.

5.4.2.2 Fuel Constraint

Fuel type j in year t may be limited to $\overline{Fuel_{jt}}$ based on its availability for the system. As a result

$$Fuel_{ejt} + \sum_{i=1}^{Ng} Fuel_{ij} Energy_{it} X_{it} \leq \overline{Fuel_{jt}} \quad \forall j \in Nf \quad \text{and} \quad \forall t = 1,\ldots,T \tag{5.12}$$

where

$Fuel_{ij}$	The fuel consumption type j for unit type i (m^3/MWh)
Nf	The number of the available fuels
$Fuel_{ejt}$	The fuel consumption type j for existing units in year t (m^3)

[12] The existing units are, in fact, the units available and justified up to that time (see (5.2)).

5.4.2.3 Pollution Constraint

Similar to fuel, the pollution generated by unit i based on pollution type j ($Pollu_{ij}$) should be limited to $\overline{Pollu_{jt}}$, so

$$Pollu_{ejt} + \sum_{i=1}^{Ng} Pollu_{ij}\, Energy_{it}\, X_{it} \leq \overline{Pollu_{jt}} \quad \forall j \in Np \text{ and } \forall t = 1, \ldots, T \qquad (5.13)$$

where

Np The number of pollution types

$Pollu_{ejt}$ The pollution type j, generated by existing units in year t

5.5 WASP, a GEP Package

WASP is a GEP package, based on single-bus modeling developed for IAEA and freely distributed to all members of this agency. It is designed to find the economically optimal generation expansion policy for an electric utility system within user-specified constraints. It utilizes probabilistic estimation of the system (production costs), unserved energy cost, reliability calculations, LP (Linear Programming) technique for determining optimal dispatch policy satisfying constraints on fuel availability, environmental emissions and electricity generation by some plants and DP (Dynamic Programming) for comparing the costs of alternative system expansion plans.

The schematic diagram of cash flows for an expansion plan is shown in Fig. 5.2.

The cost components are calculated with some details as given in Sect. 5.5.1. The WASP computer program general capabilities and characteristics are described in Sect. 5.5.2.

5.5.1 Calculation of Costs

The calculation of the various cost components is done in WASP with certain models in order to account for

a) Characteristics of the load forecast
b) Characteristics of the thermal and the nuclear plants
c) Cost of the energy not served

The load is modeled by the peak load and the energy demand for each period (up to 12) for all years (up to 30), and their corresponding inverted load duration curves. For computational convenience, the inverted load duration curves are expanded in Fourier series by the computer program.

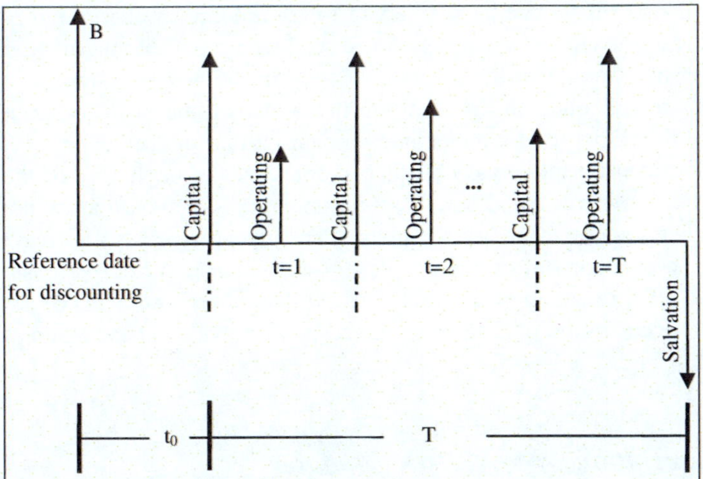

Notes

B_j	Objective function (total cost) of the expansion plan calculated in the first year of study
$Capital_t$	Sum of the investment costs of all units added in year t
$Operating_t$	Sum of all system operating costs (fuel, O & M, and energy not served) in year t
Salvation	Sum of the salvation values, at horizon, of all plants added during the study period
t_o	Number of years between the reference date for discounting and the first year of study
T	Length (in number of years) of the study period

Fig. 5.2 Schematic diagram of cash flows for an expansion plan

The model for each thermal plant is described by

- Maximum and minimum capacities
- Heat rate at minimum capacity and incremental heat rate between minimum and maximum capacities
- Maintenance requirements (scheduled outages)
- Failure probability (forced outage rate)
- Emission rates and specific energy use
- Capital investment cost (for expansion candidates)
- Variable fuel cost
- Fuel inventory cost (for expansion candidates)
- Fixed component and variable component (non-fuel) of operating and maintenance costs
- Plant life (for expansion candidates)

The cost of energy not served reflects the expected damages to the economy of the country modeled in WASP through a quadratic function relating the incremental cost of the energy not served to the amount of energy not served. In theory at least, the cost of the energy not served would permit automatic definition of the adequate amount of the reserve capacity in the power system.

In order to calculate the present-worth values of the cost components, the present-worth factors used are evaluated assuming that the full capital investment for a plant added by the expansion plan is made at the beginning of the year in which it goes into service and that its salvation value is the credit at the horizon for the remaining economic life of the plant. Fuel inventory costs are treated as the investment costs, but full credit is taken at the horizon (i.e. these costs are not depreciated). All the other costs (fuel, O & M, and energy not served) are already assumed to occur in the middle of the corresponding year. These assumptions are illustrated in Fig. 5.2.

5.5.2 Description of WASP-IV Modules

Table 5.4 summarizes the capabilities of the WASP-IV computer code.

There are seven modules in WASP. The first three can be executed independently of each other in any order. Modules 4, 5, and 6, however, must be executed in order, after execution of Modules 1, 2, and 3. There is also a seventh module, REPROBAT, which produces a summary report of the first six modules, in addition to its own results.

- **Module 1.** LOADSY (Load System Description), processes information describing period peak loads and load duration curves for the power system over the study period.

Table 5.4 Basic capabilities of WASP-IV

Parameters[a]	Maximum allowable
Years of study period	30
Periods per year	12
Load duration curves (one for each period and for each year)	360
Cosine terms in the Fourier representation of the inverted load duration curve of each period	100
Types of plants grouped by fuel types of thermal plants	10
Thermal plants of multiple units. This limit corresponds to the total number of plants in the Fixed System plus those thermal plants considered for system expansion which are described in the Variable System	88
Types of plants candidates for system expansion, of thermal plants	12
Environmental pollutants (materials)	2
Group limitations	5
System configurations in all the study period (in one single iteration involving sequential runs of modules 4–6)	5000

[a] Some of the terms are defined subsequently

- **Module 2.** FLXSYS (Fixed System Description), processes information describing the existing generation system and any pre-determined additions or retirements, as well as availability or electricity generation by some plants.
- **Module 3.** VARSYS (Variable System Description), processes information describing the various generating plants which are to be considered as candidates for expanding the generation system.
- **Module 4.** CONGEN (Configuration Generator), calculates all possible year-to-year combinations of expansion candidate additions which satisfy certain input constraints and which in combination with the fixed system can satisfy the loads. GONGEN also calculates the basic economic loading order of the combined list of FIXSYS and VARSYS plants.
- **Module 5.** MERSIM (Merge and Simulate), considers all configurations put forward by CONGEN and uses probabilistic simulation of system operation to calculate the associated production costs, energy not served and system reliability for each configuration. In the process, any limitation imposed on some groups of plants for their environmental emissions, fuel availability or electricity generation is also taken into account. The dispatching of plants is determined in such a way that plant availability, maintenance requirements, spinning reserve requirements and all the group limitations are satisfied with minimum cost. The module makes use of all previously simulated configurations. MERSIM can also be used to simulate the system operation for the best solution provided by the current DYNPRO run and in this mode of operation is called REMERSIM. In this mode of operation, detailed results of the simulation are also stored on a file that can be used for graphical representation of the results.
- **Module 6.** DYNPRO (Dynamic Programming Optimization), determines the optimum expansion plan based on previously derived operating costs along with input information on capital costs, energy not served cost, economic parameters and reliability criteria.
- **Module 7.** REPROBAT (Report Writer of WASP in a Batched Environment), writes a report summarizing the total or partial results for the optimum or near optimum power system expansion plan and for fixed expansion schedules. Some results of the calculations performed by REPROBAT are also stored on the file that can be used for graphical representation of the WASP results (see REMERSIM above).

5.6 Numerical Results

Now let us go further into some new tests, while new parameters are observed in comparison with the earlier tests. The new conditions are as follows

- The load is 500 MW for the first year, each year added by 100 MW, so that at the end of our new period of study (5 years) is 1000 MW.
- The candidate units, as well as the existing units, are the same as before (Table 5.1).

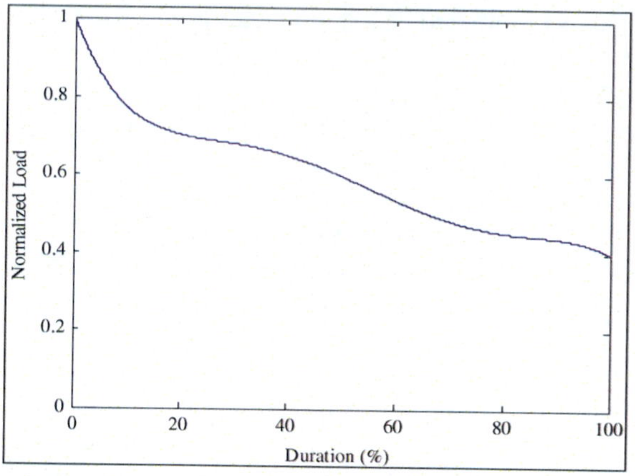

Fig. 5.3 Normalized LDC

Table 5.5 New test cases

Case name	Candidate name	Non flat LDC	FOR	LOLP constraint	ENS cost
CASE2_A1	A	–	–	–	–
CASE2_ABC1	A, B, C	–	–	–	–
CASE2_ABC2	A, B, C	✓	–	–	–
CASE2_ABC3	A, B, C	✓	✓	–	–
CASE2_ABC4	A, B, C	✓	✓	✓	–
CASE2_ABC5	A, B, C	✓	✓	✓	✓

- Discount rate is 10% per year.
- The reserve margin is considered to be 20%.
- LOLP is considered to be less than 1%.
- For the units involved $FOR_A = 15\%$, $FOR_B = 5\%$, $FOR_C = 10\%$ and $FOR_D = 9\%$.
- The cost of ENS is considered to be R 5/kWh.
- Non flat LDC may also be considered for each year. In other words, the load may be varying throughout a year. The normalized LDC is shown in Fig. 5.3. It may be described as

$$\text{Normalized load} = 1 - 3.6(\text{Duration\%}) + 16.6(\text{Duration\%})^2 - 36.8(\text{Duration\%})^3 + 36.0(\text{Duration\%})^4 - 12.8(\text{Duration\%})^5 \quad (5.14)$$

Again some test cases are generated as shown in Table 5.5. The results are demonstrated in Table 5.6. These results are generated using WASP package; for which some details are given in Sect. 5.5. The reader is encouraged to analyze the results and to see how different parameters have affected the solutions.

Table 5.6 Results for new test cases

Case name	Year 1	Year 2	Year 3	Year 4	Year 5
CASE2_A1					
Number of selected A	2	3	4	4	5
Capital investment cost (kR)	90,000	40,909	37,190	0	30,736
Salvation value (kR)	41,912	22,353	23,750	0	26,544
Operation cost (fuel and O&M costs) (kR)	81,262	91,691	99,551	103,936	107,872
Energy not served cost (kR)	–	–	–	–	–
Total annual cost (kR)	129,350	110,247	112,991	103,936	112,063
Total cumulative cost (kR)	129,350	239,597	352,588	456,524	568,587
LOLP (%)	0	0	0	0	0
CASE2_ABC1					
Number of selected A	0	0	1	1	2
Number of selected B	1	1	1	1	1
Number of selected C	0	1	1	2	2
Capital investment cost (kR)	87,500	22,727	37,190	18,783	30,736
Salvation value (kR)	45,276	13,039	23,750	14,281	26,544
Operation cost (fuel and O&M costs) (kR)	82,729	90,717	97,770	105,555	108,353
Energy not served cost (kR)	–	–	–	–	–
Total annual cost (kR)	124,953	100,405	111,209	110,057	112,545
Total cumulative cost (kR)	124,953	225,358	336,568	446,624	559,169
LOLP (%)	0	0	0	0	0
CASE2_ABC2					
Number of selected A	1	1	2	2	2
Number of selected B	0	0	0	0	0
Number of selected C	1	2	2	3	4
Capital investment cost (kR)	70,000	22,727	37,190	18,783	17,075
Salvation value (kR)	33,375	13,039	23,750	14,281	14,902

(continued)

Table 5.6 (continued)

Case name	Year 1	Year 2	Year 3	Year 4	Year 5
Operation cost (fuel and O&M costs) (kR)	47,904	53,213	56,936	61,034	64,288
Energy not served cost (kR)	–	–	–	–	–
Total annual cost (kR)	84,530	62,901	70,376	65,536	66,461
Total cumulative cost (kR)	84,530	147,431	217,806	283,342	349,803
LOLP (%)	0	0	0	0	0
CASE2_ABC3					
Number of selected A	1	1	2	2	2
Number of selected B	0	0	0	0	0
Number of selected C	1	2	2	3	4
Capital investment cost (kR)	70,000	22,727	37,190	18,783	17,075
Salvation value (kR)	33,375	13,039	23,750	14,281	14,902
Operation cost (fuel and O&M costs) (kR)	48,748	54,856	58,699	63,158	66,754
Energy not served cost (kR)	–	–	–	–	–
Total annual cost (kR)	85,373	64,544	72,139	67,660	68,927
Total cumulative cost (kR)	85,373	149,917	222,056	289,715	358,643
LOLP (%)	3.945	3.885	2.691	2.744	2.853
CASE2_ABC4					
Number of selected A	3	3	3	3	3
Number of selected B	0	0	0	0	0
Number of selected C	0	1	2	3	4
Capital investment cost (kR)	135,000	22,727	20,661	18,783	17,075
Salvation value (kR)	62,868	13,039	13,660	14,281	14,902
Operation cost (fuel and O&M costs) (kR)	49,567	55,205	60,104	64,181	67,329
Energy not served cost (kR)	–	–	–	–	–
Total annual cost (kR)	121,699	64,893	67,105	68,683	69,503
Total cumulative cost (kR)	121,699	186,592	253,697	322,380	391,882

(continued)

Table 5.6 (continued)

Case name	Year 1	Year 2	Year 3	Year 4	Year 5
LOLP (%)	0.892	0.861	0.882	0.931	0.994
CASE2_ABC5					
Number of selected A	3	4	5	6	6
Number of selected B	0	0	0	0	0
Number of selected C	1	1	1	1	2
Capital investment cost (kR)	160,000	40,909	37,190	33,809	17,075
Salvation value (kR)	75,287	22,353	23,750	25,147	14,902
Operation cost (fuel and O&M costs) (kR)	52,521	56,809	60,529	63,530	66,466
Energy not served cost (kR)	5947	5031	3934	3055	3498
Total annual cost (kR)	143,181	80,396	77,903	75,247	72,138
Total cumulative cost (kR)	143,181	223,577	301,480	376,727	448,865
LOLP (%)	0.354	0.248	0.187	0.136	0.153

Table 5.7 Generation technology data for problem 2

Technology type	Cost_Inv (investment cost) (R/kW)	T (plant life) (year)	Fuel cost (R/MWh)
A	400	30	18
B	300	20	20
C	250	25	26

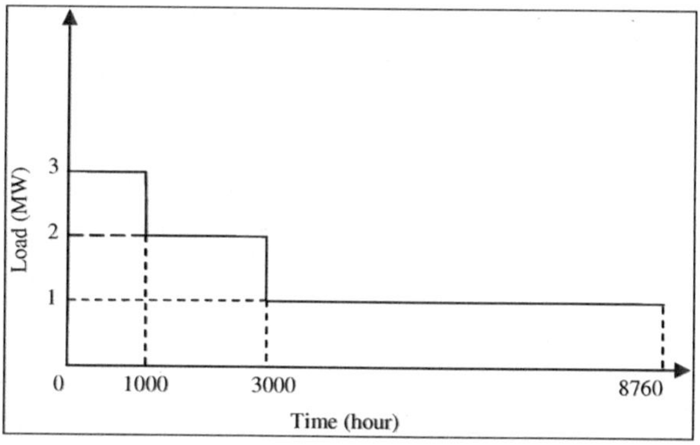

Fig. 5.4 LDC of problem 2

Problems

1. For some types of electric power generation technologies available in your area of living, find out the investment cost (in R/kW), the operational cost due to fuel (in R/kWh) and average life (in year).
2. For three generation facilities A, B and C with the details given in Table 5.7, assuming an interest rate of 10% and the possibility of choosing any generation capacity of the above mentioned technologies (A, B and C), find out the GEP results (type and capacity) for each of the following cases. In each case, calculate the investment cost (in R) as well as the operation cost (in year).

   (a) 3 MW load throughout the year (8760 h)
   (b) 3 MW load for 3000 h in a year
   (c) 3 MW load for 1000 h in a year
   (d) With LDC as shown in Fig. 5.4

3. In problem 2, if we are going to have some percentage of generation reserve, from what generation technology should it be selected? Why?
4. For supplying 1 MW load for h hours in a year, using each of the technologies outlined in problem 2, calculate and draw total cost in terms of h; as h varies

from zero to 8760 (Assume the interest rate to be 10%). Based on that, select the optimum generation technology of problem 2.

5. In Sect. 5.3, assume the operation costs of the existing units (type D) to be R18/MWh (fuel cost) due to their low efficiencies. In that case, solve CASE1_ABC3 again. Calculate the generation reserve and justify the results [#GEP1.m; Appendix L: (L.1)].

References

Reference [1] is a reference book about power system reliability evaluation. Reference [2] introduces WASP, the package developed by IAEA for GEP. Reference [3] covers some practical issues for GEP in France at the time of publication. The economic parameters affecting GEP are discussed in [4]. Some mathematical based algorithms for GEP are covered in [5–7], while some non-mathematical based ones are introduced in [8–10]. Review and comparison of these algorithms are given in [11] and [12]. If GEP and TEP are to be analyzed together, the problem becomes highly complex. Some algorithms are covered in [13–19].

1. Billinton R, Allan RN (1996) Reliability evaluation of power systems, second edition. Plenum Press, New York
2. International Atomic Energy Agency (IAEA) (2001) Wien Automatic System Planning (WASP), User Manual. www-pub.iaea.org/MTCD/publications/PDF/CMS-16.pdf
3. Montfort B, Lederer P (1986) Generation planning at Électricité de France—a sharper focus for the coming decades. Int J Electr Power Energy Syst 8(2):75–92
4. Sevilgen SH, Erdem HH, Cetin B, Akkaya AV, Dağdaş A (2005) Effect of economic parameters on power generation expansion planning. Energy Convers Manag 46(11–12):1780–1789
5. Park YM, Park JB, Won JR (1998) A hybrid genetic algorithm/dynamic programming approach to optimal long-term generation expansion planning. Int J Electr Power Energy Syst 20(4):295–303
6. Antunes CH, Martins AG, Brito IS (2004) A multiple objective mixed integer linear programming model for power generation expansion planning. Energy 29(4):613–627
7. Ramos A, Perez-Arriaga IJ, Bogas J (1989) A nonlinear programming approach to optimal static generation expansion planning. IEEE Trans Power Syst 4(3):1140–1146
8. Murugan P, Kannan S, Baskar S (2009) NSGA-II algorithm for multi-objective generation expansion planning problem. Electr Power Syst Res 79(4):622–628
9. Kannan S, Slochanal SMR, Subbaraj P, Padhy NP (2004) Application of particle swarm optimization technique and its variants to generation expansion planning problem. Electr Power Syst Res 70(3):203–210
10. Kandil MS, Farghal SA, Abdel-Aziz MR (1992) Knowledge base of an expert system for generation expansion planning. Electr Power Syst Res 23(1):59–70
11. Zhu J, Chow M (1997) A review of emerging techniques on generation expansion planning. IEEE Trans Power Syst 12(4):1722–1728
12. Kannan S, Slochanal SMR, Padhy NP (2005) Application and comparison of metaheuristic techniques to generation expansion planning problem. IEEE Trans Power Syst 20(1):466–475
13. Pereira MVF, Pinto LMVG, Cunha SHF, Oliveira GC (1985) A decomposition approach to automated generation/transmission expansion planning. IEEE Trans Power Apparatus Syst PAS-104(11):3074–3083
14. Li W, Billinton R (1993) A minimum cost assessment method for composite generation and transmission system expansion planning. IEEE Trans Power Syst 8(2):628–635

15. Samarakoon HMDRH, Shrestha RM, Fujiwara O (2001) A mixed integer linear programming model for transmission expansion planning with generation location selection. Int J Electr Power Energy Syst 23(4):285–293
16. Liu G, Sasaki H, Yorino N (2001) Application of network topology to long range composite expansion planning of generation and transmission lines. Electr Power Syst Res 57(3):157–162
17. Ramachandran K, Sharma JD (1978) A method for generation and transmission planning. Comput Electr Eng 5(2):171–178
18. Murugan P, Kannan S, Baskar S (2009) Application of NSGA-II algorithm to single-objective transmission constrained generation expansion planning. IEEE Trans Power Syst 24(4):1790–1797
19. Sepasian MS, Seifi H, Foroud AA, Hatami AR (2009) A multiyear security constrained hybrid generation-transmission expansion planning algorithm including fuel supply costs. IEEE Trans Power Syst 24(3):1609–1618

Chapter 6
Multi-bus Generation Expansion Planning

6.1 Introduction

As detailed in Chap. 5, GEP is, in fact, the process of determining the generation requirements for a system so that the loads can be satisfied in an efficient (typically the most economical) manner while various technical or non-technical constraints are met. The approach presented in Chap. 5 was based on single bus representation of the system. In other words, we basically ignored the transmission system and found out the total generation requirements based on an optimization model.

In a practical life, we are, however, confronted with determining the nodal generation requirements. In other words, we should, somehow, allocate the total generation requirements among system buses. The solution may be simple if the transmission system strength was infinite, the fuel costs were the same for all buses, the cost of land was also similar and there were no other practical limitations. In that case, we can arbitrarily allocate the total generation requirements among the buses according to our wishes.

The assumptions cited above are not valid in practice. We should, somehow, find a solution, while easy to solve, has a sound engineering basis. If we are going to consider all details, the problem ends up with a model which may be impossible or very difficult to solve. Instead, we are going to develop a model with the following observations

- We assume that the total generation requirements as well as the types and the capacities of the generation units are known from Chap. 5.
- We assume that some practical limitations and data are available for system buses. For instance, some types of generations (for example, steam generations) may be allocated in some specific buses or the maximum generation which can be installed in a specific bus is known.
- The aim is to allocate the generations among the buses in such a way that transmission enhancement requirements are minimized.

We again emphasize the point that the transmission system modeling used here is approximate in the sense as outlined in this chapter. Detailed transmission

H. Seifi and M. S. Sepasian, *Electric Power System Planning*, Power Systems, DOI: 10.1007/978-3-642-17989-1_6, system planning algorithms are described in Chaps. 8 and 9, once substation requirements are known from Chap. 7.

Problem description is given in Sect. 6.2 through one simple example. A linear programming solution approach is provided in Sects. 6.3 and 6.4. There, we ignore some practical aspects of the problem. A simple, yet practical, Genetic Algorithm (GA) based solution is described in Sects. 6.5 and 6.6.

6.2 Problem Description

The problem is more readily described through one simple example as detailed below.

Assume that the total generation requirement of a system is known to be 500 MW (1 × 150, 1 × 250 and 1 × 100 MW units), through the approach outlined in Chap. 5. The system is the Garver test case (Fig. 6.1) with the details given in Appendix F. However, assume that the loads of the buses are increased each by 100 MW (total 500 MW) so that 500 MW new generation is required. In terms of new generation, three scenarios are assumed as follows

- *Scenario 1*. All generations are to be installed at bus 1.
- *Scenario 2*. 250 MW, 150 MW and 100 MW are to be installed at buses 1, 3 and 4, respectively.
- *Scenario 3*. 400 MW (1 × 250 and 1 × 150) and 100 MW are to be installed at buses 2 and 4, respectively.

A summary of some load flow results is shown in Table 6.1 [#DCLF.m; Appendix L: (L.5)]. For our purposes, we have included a sum of lines over-loadings (in normal condition) both in absolute values and multiplied by respective lines lengths. As seen, if either absolute values or the values multiplied by lengths

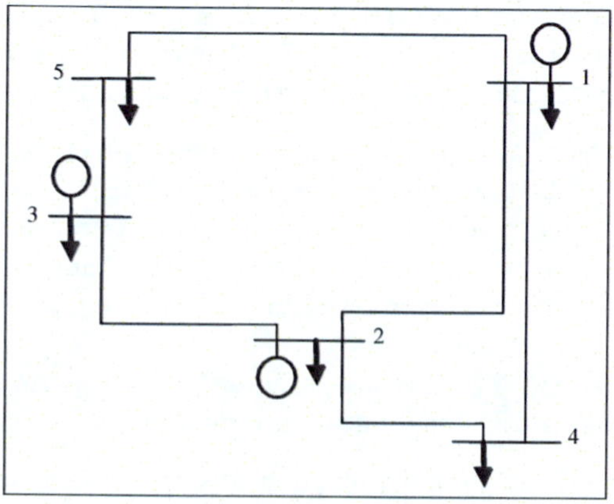

Fig. 6.1 Garver test system

Table 6.1 A summary of load flow results

Scenario	Overloading	
	Sum (absolute values)	Sum (multiplied by lengths)
1	2.089	652.2
2	0.253	50.6
3	0.404	80.6

(somehow proportional to enhancement requirements) are used, scenario 2 is the best choice. However, as in scenario 3, only two locations are justified for new generations, this may be more attractive in comparison with scenario 2.[1]

This simple example shows the fact that although the result of the approach in Chap. 5 is a necessity; an effort should be followed to allocate, somehow, the generation requirements among the buses. If the system is small and the number of alternatives (scenarios) is limited, the approach presented above may suffice. In a practical life, in which the system and the number of alternatives are large, some advanced algorithms should be followed. It is worth mentioning here that our main emphasis in this chapter is GEP and not the actual transmission enhancement requirements. In other words, the approximation mentioned above that the transmission enhancement requirement is proportional to the length-based overloads does not result in determining the actual transmission enhancement routes. In Chaps. 8 and 9 we come back to this important step, once the generations are allocated according to the approach presented in this chapter and substation expansion requirements are known according to the algorithms discussed in Chap. 7. The proportionality of the transmission enhancement requirements to the length-based overloads is not the only way to observe this point and other criteria may be proposed and employed.

A linear programming approach for problem solving is described in Sects. 6.3 and 6.4. There, we would discuss that although the proposed approach is robust in terms of the mathematical formulation, it has drawbacks in terms of some practical issues. That is why a Genetic Algorithm (GA) based approach is presented in Sects. 6.5 and 6.6 in which while practical considerations are observed, any more extension may be readily applied.

6.3 A Linear Programming (LP) Based GEP

6.3.1 Basic Principles

The flows through transmission lines are functions of both the loads in the load buses and the generations in the generation buses. The loads are assumed to be

[1] Think of an alternative index in which the number of generation units is, somehow, accounted for.

known and distributed among the load buses. Total generation is assumed to be known but its distribution among the generation buses is assumed to be unknown. If DCLF is used to model the system behavior (see Appendix A), the line flows would be a linear function of the loads and the generations. In that case, as optimization problem may be formulated as follows in which the aim is to allocate the total generation requirements among the buses.

For an N-bus, M-line network, DCLF equations are

$$\mathbf{P_G} - \mathbf{P_D} = \mathbf{B}\boldsymbol{\theta} \tag{6.1}$$

where

- $\mathbf{P_G}$ A vector of generations $(N \times 1)$
- $\mathbf{P_D}$ A vector of loads (or demands) $(N \times 1)$
- $\boldsymbol{\theta}$ A vector of bus angles $(N \times 1)$
- $\mathbf{B}$ The admittance matrix with $R = 0$ $(N \times N)$

The line flows are calculated as follows

$$\mathbf{P_L} = \mathbf{b}\,\mathbf{A}\,\boldsymbol{\theta} \tag{6.2}$$

where

- $\mathbf{P_L}$ A vector of line flows $(M \times 1)$
- $\mathbf{b}$ A matrix $(M \times M)$ in which b_{ii} is the admittance of line i and non-diagonal elements are zero
- $\mathbf{A}$ The connection matrix $(M \times N)$ in which a_{ij} is 1, if a line exists from bus i to bus j; otherwise zero. Moreover, for the starting and the ending buses, the elements are 1 and -1, respectively

From (6.1) and (6.2), we have

$$\mathbf{P_L} = \mathbf{bAB}^{-1}(\mathbf{P_G} - \mathbf{P_D}) \tag{6.3}$$

For a specific line i, the line flow (P_{Li}) is

$$P_{Li} = \sum_{j=1}^{N} s_{ij}(P_{Gj} - P_{Dj}) \tag{6.4}$$

where P_{Gj} and P_{Dj} are the generation and the demand (load) of bus j, respectively. s_{ij} is, in fact, the ijth element of $\mathbf{bAB}^{-1}$ matrix, describing the ith line flow sensitivity with respect to the generation and the load difference of bus j.

Now let us make the situation more practical by assuming that there are some areas, each composed of some generation and load buses. Assume that the load and the generation of bus j in an area k, as represented by P_{Gj} and P_{Dj}, respectively, are some portion of the total load and generation of area k (PD^k and PG^k, respectively). In other words

$$P_{Dj} = \alpha_{Dj} PD^k \quad j \in \text{Area}(k) \quad k = 1, \ldots, Na \tag{6.5}$$

$$P_{Gj} = \alpha_{Gj} PG^k \quad j \in \text{Area}(k) \quad k = 1, \ldots, Na \tag{6.6}$$

where

$$\sum_{j \in \text{Area}(k)} \alpha_{Dj} = 1.0 \quad k = 1, \ldots, Na \tag{6.7}$$

$$\sum_{j \in \text{Area}(k)} \alpha_{Gj} = 1.0 \quad k = 1, \ldots, Na \tag{6.8}$$

where Na is the number of areas while α_{Dj} and α_{Gj} are the jth load and generation participation factors in an area, respectively.

Assume that we are mainly interested in the generation allocations among the areas and not the buses. Moreover, the flows through the lines between the areas are of interest.

Combining (6.4) through (6.8) results in

$$P_{Li} = \sum_{k=1}^{Na} \left(A^k_{Gi} PG^k - A^k_{Di} PD^k \right) \tag{6.9}$$

where

$$A^k_{Gi} = \sum_{j \in \text{Area}(k)} s_{ij} \alpha_{Gj} \tag{6.10}$$

$$A^k_{Di} = \sum_{j \in \text{Area}(k)} s_{ij} \alpha_{Dj} \tag{6.11}$$

where A^k_{Gi} and A^k_{Di} are the ith line flow sensitivity with respect to the generation and the load of area k, respectively. As for the planning horizon, the load allocation is assumed to be fixed, we have

$$P_{Li} = \left(\sum_{k=1}^{Na} A^k_{Gi} PG^k \right) + c_i \tag{6.12}$$

where c_i is a constant.

The flow through a transmission line i (P_{Li}) should be within its thermal capacity limits ($\bar{P}_{Li}$), i.e.

$$-\bar{P}_{Li} \le P_{Li} \le \bar{P}_{Li} \tag{6.13}$$

Moreover, the kth area generation should be within its maximum ($\overline{PG}^k$) and minimum ($\underline{PG}^k$) limits, i.e.

$$\underline{PG}^k \le PG^k \le \overline{PG}^k \tag{6.14}$$

where these two limits are specified by the user according to any technical or non-technical observations.

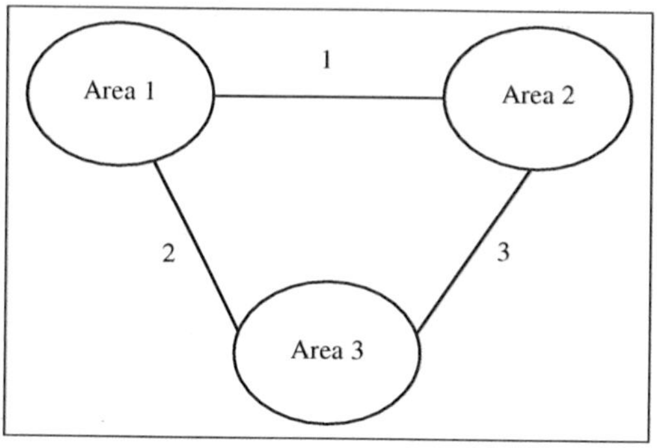

Fig. 6.2 A three-area case

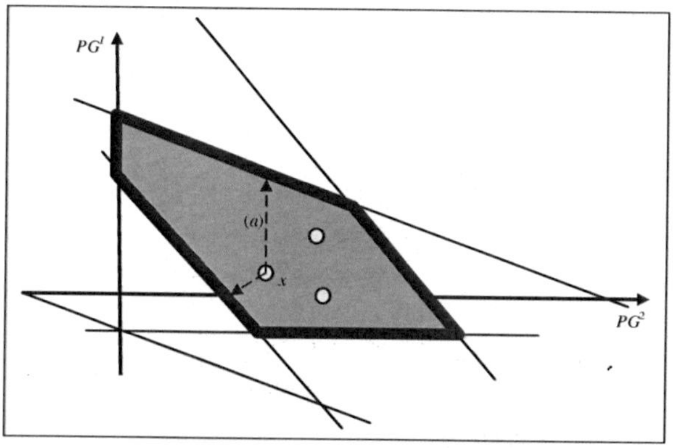

Fig. 6.3 Feasible zone

Before going through the mathematical formulation of the problem, let us describe a graphical observation of the above problem in a three-area case as depicted in Fig. 6.2.

As the total generation is assumed to be known, the generations of all three areas cannot be independent. If PG^1 and PG^2 are assumed to be independent variables, PG^3 is then determined as a dependent variable. Coming back to (6.12), for three inter-area lines denoted by 1, 2 and 3, we would have

$$\begin{aligned} P_{L1} &= a_1 PG^1 + b_1 PG^2 + c_1 \\ P_{L2} &= a_2 PG^1 + b_2 PG^2 + c_2 \\ P_{L3} &= a_3 PG^1 + b_3 PG^2 + c_3 \end{aligned} \tag{6.15}$$

If these equations are drawn as shown in Fig. 6.3 along with the limitations imposed by (6.13) and (6.14), a dark zone appears, showing the feasible points.

Any point within the zone shows a feasible point in terms of meeting the generation limitations as well as the line flow constraints. From a current operating point shown as x, it is evident that if an increase of "a" is applied to the generation of area 1 (i.e. PG^1), while PG^2 is fixed (PG^3 should be accordingly reduced), one line reaches its thermal limit and should be reinforced. The graphical representation above cannot be applied to larger test cases. A mathematical formulation is given in Sect. 6.3.2.

6.3.2 Mathematical Formulation

In a practical situation, the investment cost of a generation unit, besides the actual cost of equipment, depends also on some technical or non-technical factors such as the cost of land, the fuel supply piping cost, the interconnection cost to the main grid, etc. It is assumed that the effect of all terms can be reflected into β^k (R/MW) showing the generation cost in area k. A mathematical optimization problem is then developed with the details given below.

6.3.2.1 Objective Function

As we discussed earlier, the investment cost of a generation unit is area dependent, reflected as β^k. Moreover, once a generation unit is installed at a bus, any of the existing lines may be needed to be enhanced to a higher capacity. As a result, the objective function considered in this chapter is

$$F = \sum_{k=1}^{Na} \beta^k PG^k + \sum_{i=1}^{M} \gamma L_i (b_i - 1) \tag{6.16}$$

where the first term is the generation investment cost and the second term is the transmission enhancement cost (L_i is the length of the line i). Note that γ is the investment cost (R/km)[2] of a line and b_i is loading of line i, if the line is overloaded.[3] Note that if line is not overloaded, b_i is set to 1.0.

The decision variables are PG^ks and b_is. It is worth mentioning that in an extreme case, an area may consist of a single bus so that, instead of area-based, the problem may be solved bus-based.

6.3.2.2 Constraints

The constraints to be observed during the optimization process are as follows

$$-b_i \bar{P}_{Li} \leq \left(\sum_{k=1}^{Na} A_{Gi}^k PG^k + c_i \right) \leq b_i \bar{P}_{Li} \quad i = 1, \ldots, M \tag{6.17}$$

[2] γ is the average cost per unit length of a line.

[3] b_i is expressed in terms of loading of an overloaded line. If for instance, the capacity of a line is 200 MVA and its loading is 240 MVA, b_i is 1.2.

$$1 \leq b_i \leq \bar{b} \quad i = 1, \ldots, M \tag{6.18}$$

$$\underline{PG^k} \leq PG^k \leq \overline{PG^k} \quad k = 1, \ldots, Na \tag{6.19}$$

$$\sum_{k=1}^{Na} PG^k = PG^0 \tag{6.20}$$

(6.17) is derived from (6.12) and (6.13) except for that the inequality is checked for the overloaded lines (M is the sum of the number of the lines between the areas.). $\bar{b}$ is the maximum capacity that a line may be expanded (to be specified by the user). (6.19) is the same as (6.14) repeated here for convenience. PG^0 is the total generation capacity as determined from the approach presented in Chap. 5.

6.3.2.3 Final Model

The optimization problem to be solved is as follows

$$\begin{array}{c} \text{Minimize (6.16)} \\ \text{Subject to (6.17) through (6.20)} \end{array} \tag{6.21}$$

6.4 Numerical Results

The algorithm proposed above is tested on the test grid already shown in Fig. 6.1. The total generation requirement is assumed to be 5.0 p.u. Five scenarios are assumed as follows (Table 6.2) [#GEP2.m; Appendix L: (L.2)].

- *Scenario 1*. Assume that the extra generation required (5.0 p.u.) is distributed among the existing units and in proportion to their existing generations.
- *Scenario 2*. Assume the generation allocation is possible, with equal (and negligible) geographical investment cost for generation. Moreover, assume the transmission enhancement cost is proportional to the line length.
- *Scenario 3*. The same as scenario 2 except considering the generation investment cost of bus 1 to be higher than those of buses 2 and 3.
- *Scenario 4*. The same as scenario 3 except considering the transmission enhancement cost to be zero.
- *Scenario 5*. The same as scenario 3 except ignoring the maximum generation capacities for the buses.

In scenario 1, the way the generations is distributed results in 0.13 p.u overload. If generation allocation with the aid of optimization modeling is permitted, this overload is readily removed as shown in scenario 2. In scenario 3, the generations are shifted towards less expensive buses (2 and 3) with no transmission enhancement cost. If the transmission enhancement cost is considered to be zero

Table 6.2 Numerical results for various scenarios

Scenario	Description	PG^1 (p.u.)	PG^2 (p.u.)	PG^3 (p.u.)	Overloading (p.u.)	Enhanced lines	Enhancement required (%)
1	Base case	3.61	1.60	2.08	0.13	–	–
2	$\beta^1 = 0$ $\beta^2 = 0$ $\beta^3 = 0$ $\gamma = 20$ $\overline{PG}^1 = 10.0$ $\overline{PG}^2 = 1.5$ $\overline{PG}^3 = 3.6$ $\underline{PG}^1 = 1.13$ $\underline{PG}^2 = 0.50$ $\underline{PG}^3 = 0.65$	3.22	1.27	2.79	0.0	–	–
3	$\beta^1 = 100$ $\beta^2 = 0$ $\beta^3 = 0$ $\gamma = 20$ $\overline{PG}^1 = 10.0$ $\overline{PG}^2 = 1.5$ $\overline{PG}^3 = 3.6$ $\underline{PG}^1 = 1.13$ $\underline{PG}^2 = 0.50$ $\underline{PG}^3 = 0.65$	2.89	1.5	2.89	0.0	–	–
4	$\beta^1 = 100.0$ $\beta^2 = 0$ $\beta^3 = 0$ $\gamma = 0$[a] $\overline{PG}^1 = 10.0$ $\overline{PG}^2 = 1.5$ $\overline{PG}^3 = 3.6$ $\underline{PG}^1 = 1.13$ $\underline{PG}^2 = 0.50$ $\underline{PG}^3 = 0.65$	2.18	1.50	3.60	0.48	Bus2–Bus3 Bus3–Bus5	9 39
5	$\beta^1 = 100.0$ $\beta^2 = 0$ $\beta^3 = 0$ $\gamma = 0$[a] $\overline{PG}^1 = 10.0$ $\overline{PG}^2 = 10.0$ $\overline{PG}^3 = 10.0$ $\underline{PG}^1 = 1.13$ $\underline{PG}^2 = 0.50$ $\underline{PG}^3 = 0.65$	1.13	4.81	1.34	0.45	Bus2–Bus4 Bus3–Bus5	23 22

[a] Set γ to a very low value in the software

(scenario 4), the maximum generations possible are installed at buses 2 and 3, while lines 2–3 and 3–5 are enhanced 9 and 39%, respectively. In scenario 5, where no limit is imposed on generation capacities, the minimum generation is installed at bus 1 (the most expensive bus) while two lines as shown have to be enhanced sufficiently.

6.5 A Genetic Algorithm (GA) Based GEP

In Sect. 6.3, the area or the bus generations as the decision variables were assumed to be continuous. This assumption is not valid in practice, as the generation capacities available are of discrete nature. Moreover, the installation of some specific power plants may be impractical in some specific buses/areas. The reasons may be technical and/or non-technical (such as environmental considerations). That is why a modified algorithm is proposed in this section for which GA is used as the solution tool.

Assume that Ng power plants with the given capacities and types are justified based on the algorithms discussed in Chap. 5. The aim is to allocate the plants among the buses in such a way that the transmission enhancement requirements are minimum.

If X_m is introduced as the decision variable for which the mth element shows the bus number in which the mth power plant is to be installed, the objective function (see (6.16)) and the constraints (see (6.17–6.20)) are modified as follows

$$\min \sum_{m=1}^{Ng} \beta_m(X_m) + \sum_{i=1}^{M} \gamma L_i(b_i - 1)$$

$$\text{s.t.} \; -b_i\overline{P}_{Li} \leq \sum_{j=1}^{N} s_{ij} \sum_{m=1}^{Ng} Z_m^j PG^m + c_i \leq b_i\overline{P}_{Li} \quad i = 1, 2, \ldots, M \tag{6.22}$$

$$1 \leq b_i \leq \overline{b} \quad i = 1, 2, \ldots, M$$

$$\underline{PG^i} \leq \sum_{j=1}^{N} \sum_{m=1}^{Ng} Z_m^j PG^m + PG^{i0} \leq \overline{PG}^i \quad i = 1, \ldots, N$$

$$1 \leq X_k \leq Nc \quad k = 1, 2, \ldots, Ng$$

where

$\beta_m(X_m)$ The installation cost of the mth power plant in bus number X_m

Ng The number of power plants, justified from Chap. 5

Z_m^j An auxiliary variable; 1 if the mth power plant is installed at bus j; otherwise zero

Nc The number of candidate buses for the power plants

Table 6.3 GA-based algorithm (with 1.0 p.u. plants capacities)

Scenario	Description	PG^1 (p.u.)	PG^2 (p.u.)	PG^3 (p.u.)	Overloadings (p.u.)	Enhanced lines	Enhancement required
3	$\beta^1 = 100.0$ $\beta^2 = 0$ $\beta^3 = 0$ $\gamma = 20$ $\overline{PG}^1 = 10.0$ $\overline{PG}^2 = 1.5$ $\overline{PG}^3 = 3.6$ $\underline{PG}^1 = 1.13$ $\underline{PG}^2 = 0.50$ $\underline{PG}^3 = 0.65$	2.0 + 1.13[a]	1.0 + 0.5	2.0 + 0.65	0.0	–	–
4	$\beta^1 = 100.0$ $\beta^2 = 0$ $\beta^3 = 0$ $\gamma = 0$ $\overline{PG}^1 = 10.0$ $\overline{PG}^2 = 1.5$ $\overline{PG}^3 = 3.6$ $\underline{PG}^1 = 1.13$ $\underline{PG}^2 = 0.50$ $\underline{PG}^3 = 0.65$	2.0 + 1.13	1.0 + 0.5	2.0 + 0.65	0.0	–	–

[a] 1.13 p.u. existing and 2.0 p.u. new

Table 6.4 The details of power plants

Description	Gas turbines (GT)	Steam turbines (ST)	Hydraulic turbine (HT)
Number of units required	3	2	1
The capacity of each unit (p.u.)	0.5	1.0	1.5
Base cost (R)	250	400	1000

PG^m The generation capacity of the mth generation unit candidate
PG^{i0} The existing generation at bus i

The proposed model is of non-linear type for which GA is used as the solution tool.

6.6 Numerical Results for GA-based Algorithm

As discussed in Sect. 6.5, in the proposed algorithm, it is possible to define the standard capacities available along with their bus-dependent installation costs. For

Table 6.5 The cost factors

Bus	GT	ST	HT
1	1.0	1.0	100.0
2	1.5	1.0	100.0
3	1.0	1.0	100.0
4	1.5	1.0	1.0
5	1.0	1.0	1.0

The cost factor is the factor multiplied by the base cost shown in Table 6.4 for various buses

Table 6.6 Final installation results

Bus no.	GT (p.u.)	ST (p.u.)	HT (p.u.)
1	0.5	0.0	0.0
2	0.0	1.0	0.0
3	0.5	0.0	0.0
4	0.0	1.0	0.0
5	0.5	0.0	1.5

instance, the installation cost of a specific power plant may be different if the plant is installed in bus 2 instead of bus 3.

To verify the algorithm, initially it is assumed that the power plants are identical with a 0.1 p.u. capacity (similar to Sect. 6.4). Scenarios 3 and 4 of Table 6.2 are repeated. As expected, the results are the same as before. Now, repeat the same tests; however assuming 1.0 p.u. capacity for the plants. The results are shown in Table 6.3. Comparing the results with those of Table 6.2, it is evident that the capacities allocated for the buses are rounded off to higher or lower values.

Let us consider a more realistic case. Assume that six power plants are justified based on the algorithm of Chap. 5; the details are given in Table 6.4.

Moreover, assume that five buses of Fig. 6.1 are considered as the generation candidate buses with the details given in Table 6.5.

As the hydraulic turbines may only be installed in buses 4 and 5, the cost factors of buses 1–3 are assumed high values. Moreover, the cost factors for steam turbines are assumed to be identical for all buses. In terms of gas turbines, the cost factors for buses 2 and 4 are assumed to be higher due to gas piping cost requirements.

Assuming the transmission enhancement cost to be R 20/km, the GA-based algorithm results are shown in Table 6.6. The power plants are so allocated that no transmission enhancement is required while no overloading is also observed.

Problems

1. For the Garver base test system, assume that the load has a 10% annual increase for all buses. If after 15 years, new generations are required and the generation

installation cost is assumed to be identical for all buses, find out the generation expansion plans for the following three cases [#DCLF.m; Appendix L: (L.5)]

(a) The generations of existing buses are uniformly increased.
(b) The new generation requirement is applied at the southern part of the system (bus 4).
(c) The new generation requirement is uniformly distributed among all buses.

For all cases, report the DCLF results as well as overloads time lengths. Compare the results.

2. Repeat problem 1, if the generation installation cost is R 55/p.u. for bus 2, R 65/p.u. for bus 4, R 50/p.u. for the remaining buses and the transmission construction cost is R 0.05/km [#DCLF.m; Appendix L: (L.5)].
3. In problem 2, find out the generation expansion plan for the following three cases [#GEP2.m; Appendix L: (L.2)]

   (a) Ignoring any limit on the generation level of each bus.
   (b) Assuming the generation limits of 2.0 and 3.0 p.u. on buses 1 and 5, respectively.
   (c) Repeat (a), assuming the generation installation costs for buses 2 and 4 are R 60/p.u. and R 75/p.u., respectively.

4. With the software provided and for the test system of problem 1, find out the generation expansion plans for the following limiting cases [#GEP2.m; Appendix L: (L.2)]

   (a) Very high transmission enhancement cost, very low and uniform generation installation cost and ignoring any generation limit for each bus.
   (b) Very high transmission enhancement cost, very low but non-uniform generation installation cost (much higher for buses 2 and 4) and ignoring any generation limit for each bus.
   (c) Very low transmission enhancement cost, very high and uniform generation installation cost and ignoring any generation limit for each bus.
   (d) Very low transmission enhancement cost, very high and non-uniform generation installation cost (much higher for buses 2 and 4) and ignoring any generation limit for each bus.
   (e) Repeat (b) and (d), provided the generation limit for each bus is considered to be twice of its load.

5. For the Garver base test system of problem 1, draw a figure (similar to Fig. 6.3) if bus 1 is located in area 1, buses 2 and 4 are located in area 2 and buses 3 and 5 are located in area 3. Assume the maximum generation limits are ignored and the generation of bus 1 is a dependent variable [#DCLF.m; Appendix L: (L.5)].
6. Investigate and discuss in some details the geographical characteristics affecting both the generation installation costs and generation capabilities of various types of units.

7. In the modeling introduced by (6.21), γ was selected to be an average value of a transmission line. In practice, due to various voltage levels and geographical conditions, this assumption is not strictly correct. Modify (6.21) appropriately and also in the Matlab code [#GEP2.m; Appendix L: (L.2)], generated so that this point is observed. Devise and solve some new exercises with the new development.
8. In the modeling introduced by (6.21), a multi-area system is assumed where in each area, some generation buses exist. However, in the Matlab code generated, it is assumed that only one bus is available in each area. Modify the code appropriately so that multi-bus multi-area cases may be considered. Devise and solve some new exercises with the new development.

References

The references addressed for this chapter are the same as those introduced in Chap. 5. [1] is a reference book about power system reliability evaluation. [2] introduces WASP, the package developed by IAEA for GEP. [3] covers some practical issues for GEP in France at the time of publication. The economic parameters affecting GEP are discussed in [4]. Some mathematical based algorithms for GEP are covered in [5–7], while some non-mathematical based ones are introduced in [8–10]. Review and comparison of these algorithms are given in [11, 12]. If GEP and TEP are to be analyzed together, the problem becomes highly complex. Some algorithms are covered in [13–19].

1. Billinton R, Allan RN (1996) Reliability evaluation of power systems, 2nd edn. Plenum Press, New York
2. International Atomic Energy Agency (IAEA) (2001) Wien automatic system planning (WASP), user manual. www-pub.iaea.org/MTCD/publications/PDF/CMS-16.pdf
3. Montfort B, Lederer P (1986) Generation planning at Électricité de France—a sharper focus for the coming decades. Int J Electr Power Energy Syst 8(2):75–92
4. Sevilgen SH, Erdem HH, Cetin B, Akkaya AV, Dağdaş A (2005) Effect of economic parameters on power generation expansion planning. Energy Convers Manage 46(11–12): 1780–1789
5. Park YM, Park JB, Won JR (1998) A hybrid genetic algorithm/dynamic programming approach to optimal long-term generation expansion planning. Int J Electr Power Energy Syst 20(4):295–303
6. Antunes CH, Martins AG, Brito IS (2004) A multiple objective mixed integer linear programming model for power generation expansion planning. Energy 29(4):613–627
7. Ramos A, Perez-Arriaga IJ, Bogas J (1989) A nonlinear programming approach to optimal static generation expansion planning. IEEE Trans Power Syst 4(3):1140–1146
8. Murugan P, Kannan S, Baskar S (2009) NSGA-II algorithm for multi-objective generation expansion planning problem. Electr Power Syst Res 79(4):622–628
9. Kannan S, Slochanal SMR, Subbaraj P, Padhy NP (2004) Application of particle swarm optimization technique and its variants to generation expansion planning problem. Electr Power Syst Res 70(3):203–210
10. Kandil MS, Farghal SA, Abdel-Aziz MR (1992) Knowledge base of an expert system for generation expansion planning. Electr Power Syst Res 23(1):59–70
11. Zhu J, Chow M (1997) A review of emerging techniques on generation expansion planning. IEEE Trans Power Syst 12(4):1722–1728

12. Kannan S, Slochanal SMR, Padhy NP (2005) Application and comparison of metaheuristic techniques to generation expansion planning problem. IEEE Trans Power Syst 20(1):466–475
13. Pereira MVF, Pinto LMVG, Cunha SHF, Oliveira GC (1985) A decomposition approach to automated generation/transmission expansion planning. IEEE Trans Power Apparatus and Syst PAS-104(11):3074–3083
14. Li W, Billinton R (1993) A minimum cost assessment method for composite generation and transmission system expansion planning. IEEE Trans Power Syst 8(2):628–635
15. Samarakoon HMDRH, Shrestha RM, Fujiwara O (2001) A mixed integer linear programming model for transmission expansion planning with generation location selection. Int J Electr Power Energy Syst 23(4):285–293
16. Liu G, Sasaki H, Yorino N (2001) Application of network topology to long range composite expansion planning of generation and transmission lines. Electr Power Syst Res 57(3): 157–162
17. Ramachandran K, Sharma JD (1978) A method for generation and transmission planning. Comput Electr Eng 5(2):171–178
18. Murugan P, Kannan S, Baskar S (2009) Application of NSGA-II algorithm to single-objective transmission constrained generation expansion planning. IEEE Trans Power Syst 24(4):1790–1797
19. Sepasian MS, Seifi H, Foroud AA, Hatami AR (2009) A multiyear security constrained hybrid generation-transmission expansion planning algorithm including fuel supply costs. IEEE Trans Power Syst 24(3):1609–1618

Chapter 7
Substation Expansion Planning

7.1 Introduction

With electric power consumption growth, desired new transmission system elements are needed to overcome the possible lack of adequacy problems so that with the least costs, various operational constraints are met. In the so-called Substation Expansion Planning (SEP), the problem is to determine the required expansion capacities of the existing substations as well as the locations and the sizes of new substations together with the required availability times, so that the loads can be adequately supplied.

The loads to be supplied are widely geographically distributed. For substation planning, the normal procedure is to initially determine the distribution substation requirements and moving upward, to finally determine the transmission substation requirements. This approach, although accurate and practical for short and midterm plannings, may prove impractical for long-term studies (say, 5 years onward) of transmission substations, as the transmission owner (developer) may wish to determine the possible allocations and sizes of the substations (either new or expansion of existing) without involving in much details of the downward grids (sub-transmission and distribution). One way to overcome this problem is to propose an algorithm in which the geographically distributed loads are somehow assigned to transmission substations. Although this does not happen in practice, the final transmission substations allocations and capacities can prove appropriate, provided various constraints are properly observed. The assigning procedure is, however, crucial as an unsuitable procedure can result in improper solutions.

In this chapter, the problem of SEP is described for transmission and sub-transmission levels. The approaches described are, however, general enough to be applied for distribution level, too, with minor modifications. The SEP problem is defined in Sect. 7.2. A basic case is covered, in Sect. 7.3, so that the reader can readily follow up some basic objective function terms and constraints. An optimization based view is then covered (Sect. 7.4) in which some practical objective function terms and constraints are defined. An advanced case is then followed, in

H. Seifi and M. S. Sepasian, *Electric Power System Planning*, Power Systems,
DOI: 10.1007/978-3-642-17989-1_7,

Sect. 7.5 in which a complex optimization problem is also defined together with its specific solution methodology. Numerical results are demonstrated in Sect. 7.6.

7.2 Problem Definition

The SEP may be defined as an optimization problem in which all the investment costs as well as the operational costs have to be minimized, while various constraints are met. The final solution should determine

(a) The expansion capacity of any existing substation (provided feasible),
(b) The allocation and the size of any new substation,
(c) The investment costs.

In mathematical terms, the problem may be defined as

$$\text{Minimize } C_{total} = C_{inv} + C_{opt} \tag{7.1}$$

$$\text{Subject to Constraints} \tag{7.2}$$

where C_{inv} refers to all investment costs and C_{opt} denotes the operational costs. A typical investment cost is the cost of constructing a new substation, whereas the cost of providing the losses is a typical operational cost.

Various constraints should also be observed during the optimization process. For instance, the capacity of a single substation should not violate a specified limit, or a feeder (line) loading should not violate its thermal capacity.

Before proceeding any further in terms of mathematical formulation, we will proceed with a basic case in Sect. 7.3 to apprehend some basic issues. We will come back to mathematical aspects in Sect. 7.4.

7.3 A Basic Case

7.3.1 Problem Description

Consider an 11-load node case as depicted in Fig. 7.1, fed in 33 kV and to be supplied through high-voltage (HV) substation(s) (say 230 kV:33 kV).

The aim is to determine the HV substation(s) required so that the loads are completely supplied. A high voltage line (bold) is assumed as the supplying grid for the HV substation(s).

Some simple and feasible solutions are

(a) Allocate a HV substation at each load node (Fig. 7.2). Feed each HV substation by the HV grid.

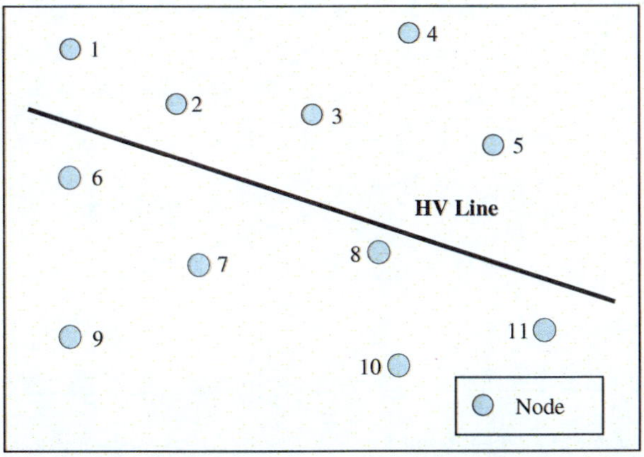

Fig. 7.1 A simple 11-load node case

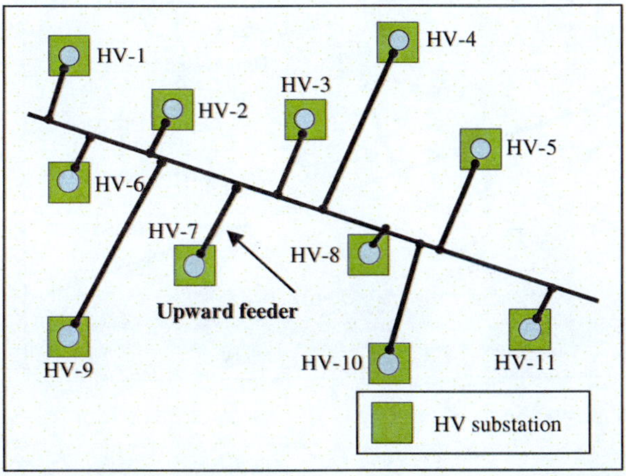

Fig. 7.2 HV substation at each load node

(b) Allocate one HV substation and feed all load nodes through it (Fig. 7.3). Supply the HV substation by the HV grid.
(c) Allocate more than one HV substation and distribute loads among them. A two-HV substation case is shown in Fig. 7.4.

In case (a), the capacity of each HV substation can be equal to its respective load magnitude. There is no cost for the downward grid, while there are some costs for the upward grid (i.e., the grid for supplying the substations). In case (b), the HV substation capacity should be equal to the sum of the loads. There are some costs for both the downward and the upward grids. The problem is, however, where to allocate the HV substation. In case (c), again, there are some costs for both the downward and the upward grids. However, the decision is more complicated as we should determine the number, allocations and sizes of the HV substations.

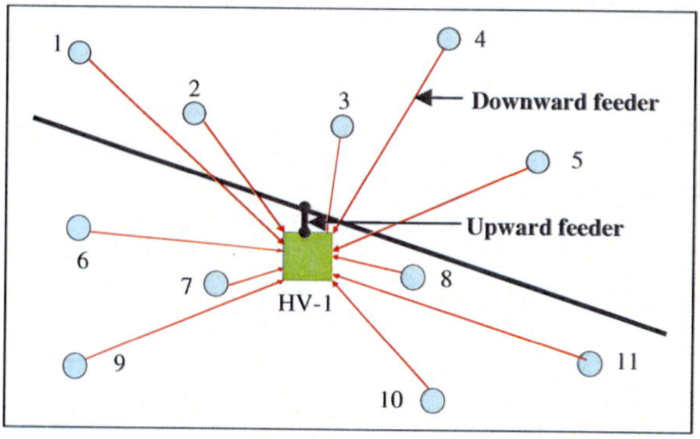

Fig. 7.3 HV substation at a single point

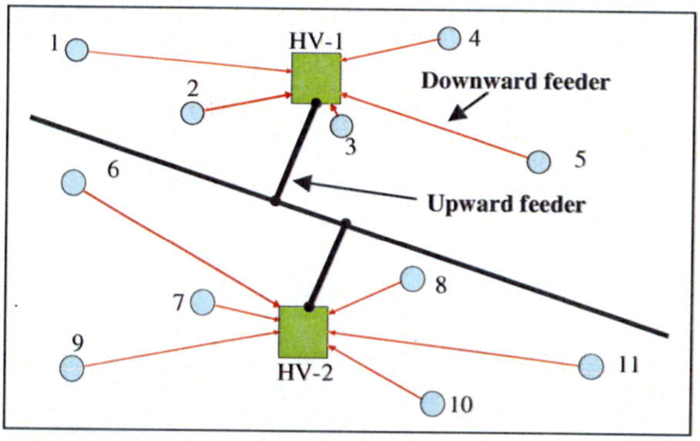

Fig. 7.4 A two HV substation case

A planner should decide on the best choice. The *best* implies the *lowest cost* choice. The overall costs (see (7.1)) may be divided into three main terms

1. *The cost associated with the HV substations*. This cost term, is divided into three main terms

   - *Land cost*. Normally, the land cost near the load nodes is higher. Moreover, although the overall HV capacities of solutions (a), (b) and (c) are the same, the lands required are not the same.
   - *Equipment cost*. This is due to transformers, switchgears, etc. for each substation. This is proportional to the substation capacity. However, it is not linearly proportional, i.e., a 2 × 30[1] MVA substation is not necessarily two-times (but lower) more costly than a 1 × 30 MVA substation.

[1] i.e. a substation with two 30 MW transformers.

- *The cost of losses*. While the former two costs refer to the investment costs, another cost to be observed is the cost of substation losses, as an operational cost (see (7.1)).

As a result, the cost associated with the HV substation is

$$\begin{aligned}\text{HV substation cost} = &\ \text{MVA independent term (due to land)}\\ &+ \text{MVA dependent term (equipment)}\\ &+ \text{Cost of substation losses}\end{aligned} \tag{7.3}$$

2. *The cost associated with the downward grid*. This cost term primarily depends on the feeder cost itself, i.e., the cost per unit length (depending on the type and the cross sectional area) and the length. Later on, we will talk why we should choose an appropriate downward feeder. In terms of this cost term, solution (a) is the best, as there is no downward grid cost. Solution (c) can be better than (b) as more lengthy feeders are used in (b).
3. *The cost associated with the upward grid*. The discussion here is similar to the discussion for the cost of the downward grid. In terms of this cost term, solutions (b), (c) and (a) may be regarded as the prior choices, respectively, based on upward grid lengths.

For the downward and the upward grids costs, another cost term of interest is the operational cost, mainly due to the feeder losses (C_{opt} in (7.1)). More lengthy, lower cross sectional area feeders result in higher losses. The cost of losses should be observed for the feeder life (say 30 years).

As discussed so far, even for this primitive simple case, if the planner is to observe the *lowest* cost choice, the decision is not so easy. However, the decision is even more complicated as some constraints, which are more of technical nature, should also be observed. At this stage, we only consider the following two constraints regarding the upward and the downward grids feeders

- *Thermal capacity of a feeder*. Thermal capacity of a feeder should not be violated upon feeding a load node. The lowest thermal capacity feeder (appropriate for feeding a specific load) should be selected, as it is normally the lowest cost choice.
- *Acceptable voltage drop along a feeder*. The voltage drop along a feeder should be less than a prespecified value (say 5%). A higher cross sectional area feeder (i.e., a feeder with lower resistance) results in lower voltage drop, however higher in terms of feeder cost.

Normally, for low-length feeders, the thermal capacity is the limiting constraint while for the high-length feeders; voltage drop is the limiting one. In terms of these two constraints, upward and downward grids should be appropriately selected for solutions (a), (b) and (c); otherwise, a lower cost solution may be justified, while technical requirements are not met. Moreover, in terms of the HV substation, its capacity should be observed as a constraint.

Let us make the situation as in solutions (a), (b) and (c), even more complex. Assume that there are already two existing substations supplying the loads in current year. Moreover, assume that the loads shown in Fig. 7.1 are the amount of load increments (in comparison with the current year) for a target year. The aim is, again, to supply the load increments via both *new substations* (similar to Figs. 7.2, 7.3 and 7.4) and *existing substations* (if they can be expanded). A typical combination with *two existing substations* is shown in Fig. 7.5.

Note that the costs associated with expanding an existing substation is normally lower than constructing a new one with a similar capacity. While there may be opportunities for supplying some parts of the loads via the existing substations (by their expansions, if feasible), the rest should be supplied through new substations; properly, allocated and sized.

7.3.2 Typical Results for a Simple Case

Let us assume a simple case in which the cost of the upward grid is totally ignored. Moreover, assume that the downward grid cost is directly proportional to the length of the feeder, supplying a load, via a substation. With this assumption, it is implied that only one feeder type is used for supplying the loads. As already noted, the cost of each substation can be mainly divided into a fixed (independent from the capacity) cost (due to the land required) and a variable (dependent on the capacity) cost (due to the equipment). The cases to be considered are shown in Table 7.1.

Detailed descriptions of the cases, as well as the overall results are followed. The system under study is shown in Fig. 7.6, showing a 37-load node case with no existing substations. Let us assume that 25 candidate substations are assumed as shown in this figure.

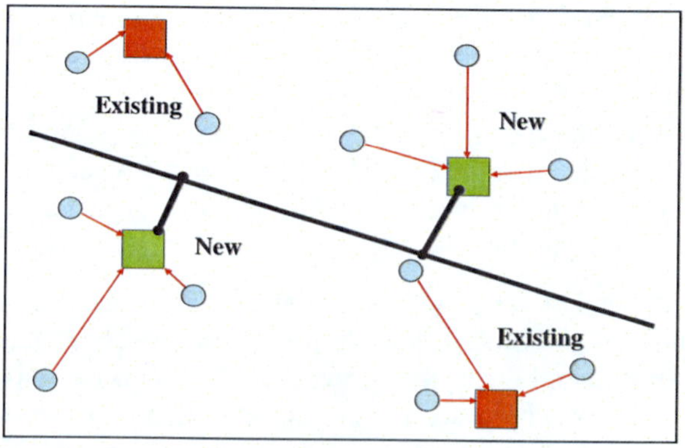

Fig. 7.5 Solution with existing substations

Table 7.1 Test cases

Case no.	Descriptions
1	Prevailing substation cost while ignoring substation capacity limits
2	Prevailing substation cost while considering substation capacity limits
3	The same as case 2, however, with prevailing cost of land for some specific areas
4	Prevailing downward grid cost

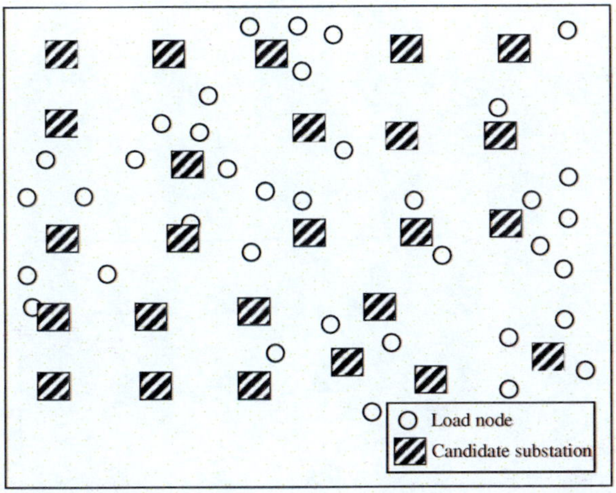

Fig. 7.6 System under study

7.3.2.1 Case 1

If the substation cost is the main term of the total cost (in comparison with other terms) and moreover, there is no limit on the capacity of each substation, it is expected that only one substation is justified for supplying all loads (with enough capacity, Fig. 7.7). The reason for justifying only one substation is that the cost of the land required is assumed to be independent of the capacity of the substation. Besides, the substation would be justified at the load center of gravity of all load nodes, to make sure that the overall downward grid length is the lowest and the downward grid cost is at minimum. More details are provided in Sect. 7.4.

7.3.2.2 Case 2

Now assume that each substation has a specified capacity limit so that more than one substation is required to supply the loads. It is expected that more substations to be justified; however, so allocated that the overall downward grid lengths are again minimum. The results are shown is Fig. 7.8.

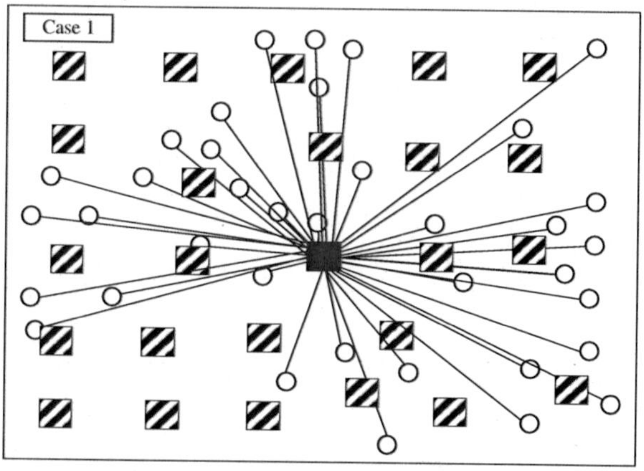

Fig. 7.7 Results for case 1

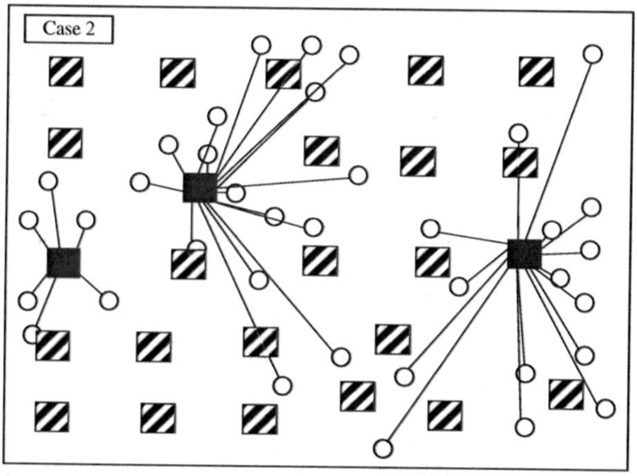

Fig. 7.8 Results for case 2

7.3.2.3 Case 3

Now assume that the conditions are the same as in case 2, except that the cost of the land required is different for each point. In fact, normally for high density load centers, the land cost is much higher than the others. The results obtained are shown in Fig. 7.9. As expected, the substations are justified more towards the areas with lower land costs.

7.3.2.4 Case 4

In this case, it is assumed that the downward grid cost is much higher than the substation cost. The results are shown in Fig. 7.10. As expected, each load point is

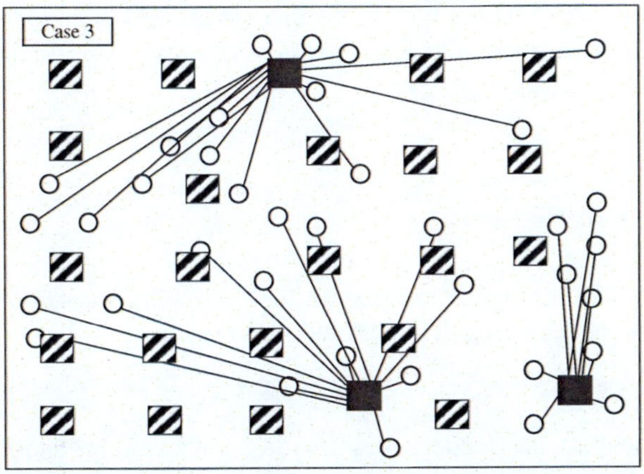

Fig. 7.9 Results for case 3

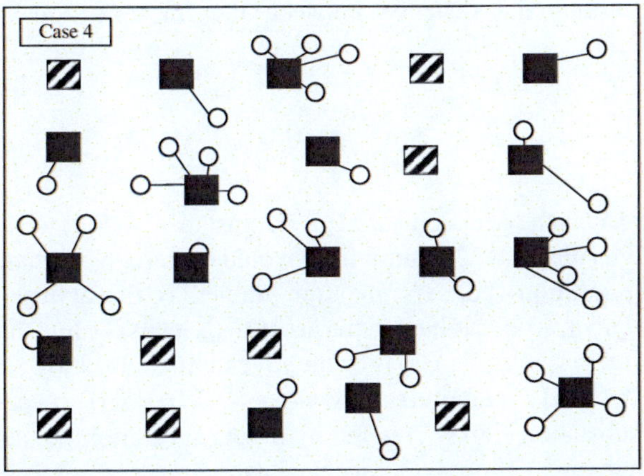

Fig. 7.10 Results for case 4

connected to its closest substation so that the overall downward grid cost is at minimum.

7.4 A Mathematical View

In this section, we try to formulate the problem of Sect. 7.3 as a mathematical optimization problem; however, in a simplified form. From the three cost terms addressed in Sect. 7.3.1, only the first two, i.e., the cost associated with the HV substations and the cost associated with the downward grid are considered.

Moreover, it is assumed that the cost of the downward grid is merely proportional to the distance of the load node to the feeding substation. In the following subsections, more details are presented.

7.4.1 Objective Function

The objective function, C_{total}, consists of the following two terms; $C_{down-line}$ (downward grid cost) and C_{stat} (HV substation cost), i.e.

$$C_{total} = C_{down-line} + C_{stat} \tag{7.4}$$

Let us assume that the feeder used for the downward grid, is a type with $g_L(i)$ (for the ith load) as the cost of its unit length (say 1 km) per one unit power transfer capability (say 1 MVA). For instance, if 25 MVA in position i is to be transmitted over 10 km, the cost would be $250g_L(i)$. As a result, if Ns and Nl represent the number of supply points (substations) and load nodes, respectively, and $D(i, j)$ represents the distance between the ith load node from the jth substation, we have

$$C_{down-line} = \sum_{i=1}^{Nl}\sum_{j=1}^{Ns} g_L(i)X(i,j)D(i,j)S_L(i) \tag{7.5}$$

where $X(i,j)$ represents the decision variable. For instance, $X(5,2)$ is 1, if load node 5 is supplied through substation 2; otherwise it would be zero. Note that $X(i,j)$ will be obtained upon the solution of the optimization problem so that at the end, the supply point of each load node is determined. In terms of C_{stat}, let us assume that the variable cost of a substation per MVA is $g_s^v(j)$ for the jth candidate location.[2] As a result, if $S_L(i)$ represents the load i magnitude in MVA, $g_s^v(j)X(i,j)S_L(i)$ represents the cost associated with substation j, if $S_L(i)$ is fed by the jth substation (i.e., $X(i,j) = 1.0$). As, in general, there are Ns supply points, we have

$$C_{stat} = C_{stat-fix} + C_{stat-var} \tag{7.6}$$

where

$$C_{stat-fix} = \sum_{j=1}^{Ns} g_s^f(j)X_s(j) \tag{7.6a}$$

$$C_{stat-var} = \sum_{j=1}^{Ns}\left(g_s^v(j)\left(\sum_{i=1}^{Nl} X(i,j)S_L(i) - C_{exis}(j)\right)\right) \tag{7.6b}$$

[2] Depending on the type of a substation (normal, underground, GIS, etc.), the variable cost may vary.

Note that for a new substation, the existing capacity (C_{exis}) is zero. This term is added to represent the fact that if the capacity required to supply the loads is less than the capacity of an existing substation, no cost is required in terms of the substations. $X_s(j)$ is 1 if the jth substation is selected; otherwise zero. $g_s^f(j)$ represents the fixed cost of a substation (land cost) and assuming to be zero for the existing ones.

7.4.2 Constraints

If a load is supplied through a substation far from the load node, the voltage drop along the feeder may be larger than a permissible value (say 5%). In fact, we can define this constraint as follows[3]

$$X(i,j)D(i,j) \leq \bar{D} \quad \forall\, i = 1,\ldots,Nl, \quad \forall\, j = 1,\ldots,Ns \tag{7.7}$$

where $\overline{D}$ shows the maximum distance a load can be supplied through a substation. For instance, if $\overline{D}$ is 10 km, it means that any load can be supplied through a substation with a distance not greater than 10 km. Otherwise, the voltage drop constraint would not be satisfied.

A second constraint to be met is the substation capacity as follows

$$\sum_{i=1}^{Nl} X(i,j)S_L(i) \leq \bar{S}_j \quad \forall\, j = 1,\ldots,Ns \tag{7.8}$$

where the $\sum$ term represents the burden on substation j. $\overline{S}_j$ represents the maximum capacity of the jth substation.

7.4.3 Problem Formulation

Considering the objective function (Sect. 7.4.1) and the constraints (Sect. 7.4.2), the optimization problem may be summarized as follows

$$\text{Min} \sum_{i=1}^{Nl}\sum_{j=1}^{Ns} g_L(i)X(i,j)\, D(i,j)S_L(i) + \sum_{j=1}^{Ns}\left(g_s^v(j)\left(\sum_{i=1}^{Nl} X(i,j)\, S_L(i) - C_{exis}(j)\right)\right) + \sum_{j=1}^{Ns} g_s^f(j)X_s(j) \tag{7.9}$$

[3] See the problems at the end of the chapter.

Subject to

$$
\begin{aligned}
& X(i,j)D(i,j) \le \overline{D} \quad \forall i = 1,\ldots,Nl \quad \forall j = 1,\ldots,Ns \\
& \sum_{i=1}^{Nl} X(i,j)S_L(i) \le \overline{S}_j \quad \forall j = 1,\ldots,Ns \\
& \sum_{j=1}^{Ns} X(i,j) = 1.0 \quad \forall i = 1,\ldots,Nl
\end{aligned}
\tag{7.10}
$$

(Expressing the requirement of feeding a load node through only one substation)

$$\sum_{i=1}^{Nl} X(i,j) \le X_S(j)Nl \quad \forall j = 1,\ldots,Ns$$

(Determining the value of X_S (j) to be either zero or one)

$$X(i,j),\ X_S(j) : \text{Binary integer (zero or 1)}$$

7.4.4 Required Data

The problem as outlined in Sect. 7.4.3 should be solved based on some available (input) data. The required information is as follows.

7.4.4.1 Load Data

The load of each load node should be known in terms of its magnitude (in MVA) as well as its geographical location (i.e., geographical *X* and *Y*). The load is normally predicted based on some forecasting algorithms (see Chap. 4). Its value should be less than the thermal capacity of an available supplying feeder. If the load magnitude is greater than the thermal capacity of an available feeder, it may be decomposed into two or more parts (equal or unequal), at the same geographical point so that more than one feeder may be justified for its supplying.

7.4.4.2 Distances Between the Load Nodes and the Substations

Several substations (both expandable existing substations and some new ones) should be initially selected as feasible feeding (supplying) points. Once these are known, *D*(*i*, *j*), can be easily calculated. Note that in its simplest case, one substation may be allocated as candidate at each load point.

7.4.4.3 Cost Terms

As outlined in Sect. 7.4.1, g_L, g_s^v and g_s^f should be known in advance. At this stage, we assume an average value for g_L. In terms of g_s^v and g_s^f, they may be determined, substation by substation, as, for instance, the cost of the land required is different with attention to its location.

7.4.4.4 Solution Methodology

As there are both binary integer and non-integer variables in (7.9), the problem is a Binary Integer Linear Programming (BILP) one which can be solved by any existing optimization package.

7.5 An Advanced Case

In Sect. 7.3, an overall view of SEP was covered. It was discussed how the upward grid, the downward grid and the substations may affect the solution. A mathematical formulation of the problem was demonstrated in Sect. 7.4. Although some objective function terms and constraints were considered in the problem formulation as defined in (7.9) and (7.10), some were ignored as follows

(a) *Objective function*. The cost of the upward grid (the investment cost as well as the cost of losses) was ignored.
(b) *Constraints*. Acceptable voltage drop and thermal capacity of the upward feeders and some reserve capacity for the substations should be considered in the problem formulation.
(c) *Modeling*. Besides adding new objective function terms and constraints, the investment costs of the downward as well as the upward feeders should be properly improved; as a very simplified approach was already used. Moreover, the substation and the feeders should be selected from a set of available options.
(d) *Solution Methodology*. If the formulation is modified and improved, the resulting problem would be nonlinear so that new solution techniques are required, especially for large scale systems.

These points are considered in this section. Some other practical issues are also covered.

7.5.1 General Formulation

7.5.1.1 Objective Functions

The aim is to supply the loads through all transmission (transmission to subtransmission) substations so that

$$C_{total} = C_{down-line} + C_{stat} + C_{up-line} + C^{LL}_{loss} + C^{S}_{loss} \tag{7.11}$$

is minimized. C_{total} is the overall plan cost. Other terms are described below. Note that

- Each load is represented with its magnitude (in MVA) and geographical characteristics (X and Y) for the horizon year.
- Each load is assumed to be radially supplied by an upward substation (to ignore the downward grids[4]). Although not accurate in practical terms, this approach facilitates the planning procedure with due attention to some practical considerations (The explanation will be given afterwards).

(a) $C_{down-line}$

A load may be supplied by several nearby substations. The cost is dependent on the distance between the load center and the substation as follows

$$C_{down-Line} = \sum_{j \in S} \sum_{i \in L(j)} C_L(A^{LL}_i) D^{LL}_{ij} \tag{7.12}$$

in which

$C_{down-Line}$	The cost of all downward feeders
$C_L(A^{LL}_i)$	The cost of the feeder for supplying the ith load by conductor A^{LL}_i (see Sect. 7.6.3 for details)
D^{LL}_{ij}	The distance between the ith load and the jth substation
S	The set of all new and expanded substations
$L(j)$	The set of all loads connected to the jth substation

(b) C_{stat}

A major cost is the investment cost for all substations defined as

$$C_{stat} = \sum_{j \in S} \left(\alpha^S_j + \beta^S_j S^S_{cap\,j} \right) - \sum_{j \in SE} AF_j \left(\alpha^S_j + \beta^S_j S^{ES}_{cap\,j} \right) \tag{7.13}$$

α^S_j	Fixed cost for the jth substation (mainly due to the land cost required)
β^S_j	Variable cost factor for the jth substation (dependent on the capacity)
$S^S_{cap\,j}$	Capacity of a new substation j
SE	The set of all expanded substations

[4] For instance, for SEP of sub-transmission substations, the loads are assigned according to medium voltage feeders (say 33 kV). For SEP of transmission substations, the loads are assigned according to HV (sub-transmission voltage) feeders (say 63 kV).

AF_j Amortizing coefficient for the existing substation j
$S^{ES}_{cap\,j}$ Capacity of the existing substation j

The second $\sum$ term in (7.13) denotes the cost associated with the expansions of the existing substations. If, for instance, $AF_j = 0.2$, it means that 80% of the practical life of the substation is expired (20% remaining), so that this amount (20%) is considered as a negative cost term.

(c) $C_{up\text{–}line}$

Obviously, the closer a substation is to an existing transmission grid, the more attractive it is, in terms of the general costs. To consider this effect, a term $C_{up\text{–}line}$ is included in (7.11), as defined in (7.14) consisting of two terms; namely, a fixed cost for the right of way, tower, etc. (dependent on the voltage level) and a variable cost (dominantly conductor cost) dependent on the line capacity. Therefore

$$C_{up-line} = \sum_{j \in S} \left(\alpha_j^{HL} + \beta_j^{HL} S^{HL}_{cap\,j} \right) D_j^{HL} \tag{7.14}$$

where
α_j^{HL} Fixed cost of the upward grid for supplying substation j
β_j^{HL} Variable cost factor of the upward grid for supplying substation j
$S^{HL}_{cap\,j}$ Upward grid capacity for supplying substation j[5]
D_j^{HL} The distance between substation j to the nearest feeding point of HV transmission network

It is evident that D_j^{HL} does not show the exact distance for a practical situation. It somehow considers the upward grid in problem formulation so that substations far from the existing network are not justified.

(d) C^{LL}_{loss}

The losses of the downward grid as operational losses should also be minimized. That is why C^{LL}_{loss} is introduced as

$$C^{LL}_{loss} = P^{LL}_{loss} \sum_{j \in S} \sum_{i \in L(j)} R\left(A_i^{LL}\right) D_{ij}^{LL} \left(S^i_{load}\right)^2 \tag{7.15}$$

[5] $S^{HL}_{cap\,j}$ is, at least, equal to the required substation capacity, as determined by the algorithm. In practice, it may be higher due to system security aspects.

where[6]

P_{loss}^{LL} The cost of the downward grid losses calculated as in base year (for 30 years operational period)

$R(A_i^{LL})$ The conductor resistance of the feeder supplying the ith load (For details, see Sect. 7.6.3)

S_{load}^{i} The change of MVA of the ith load with respect to the base value (current year) (for details, see Sect. 7.6.3

D_{ij}^{LL} As before

(e) C_{loss}^{S}

Another term to be considered is the cost of transformer losses (operational losses), denoted by C_{loss}^{S}, and defined as

$$C_{loss}^{S} = P_{loss}^{S} \sum_{j \in S} \left(\alpha_{loss\,j}^{S} + \beta_{loss\,j}^{S} \left(\frac{S_j^S}{S_{cap\,j}^{S}} \right)^2 \right) \tag{7.16}$$

where

$\alpha_{loss\,j}^{S}$ The fixed losses of the jth substation

$\beta_{loss\,j}^{S}$ The variable losses of substation j for full load conditions

P_{loss}^{S} The cost of transformer losses calculated as in base year (for 30 years operational period)

S_j^S The actual loading of the jth substation in MVA

$S_{cap\,j}^{S}$ As before

7.5.1.2 Constraints

The following constraints are considered in the optimization problem

- *For the downward grid.* Thermal capacity of the feeder for supplying the load (see (a) below) and with acceptable voltage drop (see (b) below).
- *For the substations.* Maximum and minimum installation capacities (see (c) below) as well as standard capacities (see (d) below).
- *For the upward grid.* Thermal capacity of the upward transmission line (see (e) below).

(a) *Thermal capacity of the downward feeder*

$$S_{load}^{i} \leq S_i^{LL} \quad \forall i \subset L \tag{7.17}$$

---

[6] In (7.15), R and S are, in terms of p.u./unit length and p.u., respectively; while P_{loss}^{LL} is defined in terms of R/p.u. If actual values are going to be used, (7.15) should, appropriately, be modified.

where

L Set of loads

S_i^{LL} The required capacity of selected feeder for supplying the ith load

(b) *Voltage drop*

$$\Delta U^i \leq \Delta U - \Delta U^S \quad \forall i \subset L \tag{7.18}$$

where

ΔU^i Actual voltage drop for load i

ΔU Acceptable voltage drop

ΔU^S A factor for considering the fact that an already existing substation may have some voltage problems and the least amount of extra load may be applied to this substation

(c) *Maximum and minimum installation capacities*

$$\underline{S}_j \leq S^S_{cap\,j} \leq (1.0 - res_j)\overline{S}_j \tag{7.19}$$

where res_j refers to the required reserve capacity for the jth substation. For the existing substations, $\overline{S}_j$ refers to the maximum expansion capacity of the substation. For this type of substation, $\underline{S}_j$ may be set at a value less than its existing capacity (or even zero). In that case, the substation may be de-rated (the extra capacity is considered as a benefit) or even totally removed, provided the optimization procedure finds it economical.

(d) *Standard capacities*

$$S^S_{cap\,j} \subset S_{stand} \tag{7.20}$$

shows that the substations should be selected from a set of standard list (available from the planning departments).

(e) *Thermal capacity of the upward lines*

Similar to (7.17), (7.21) applies to upward transmission lines.

$$S^{HL}_{cap\,j} \leq S^{HH}_j \quad \forall j \subset S \tag{7.21}$$

where S is defined before.

7.5.2 Solution Algorithm

The problem defined so for is similar to (7.9) and (7.10); however, with added and improved objective function terms and constraints. It is a non-linear optimization problem which can not readily be solved by existing packages. Metaheuristic algorithms; such as Genetic Algorithm (GA), Simulated Annealing (SA), Tabu Search (TS), etc.; are powerful enough to be applied for these types of the problems, even for large scale systems. In the following subsections, the authors experiences in using GA are demonstrated. For some details on GA, the reader is encouraged to, initially; follow the materials covered in Chap. 2. GA is a meta-heuristic approach used for optimization problems. Some chromosomes are initially generated. Two operators, namely, crossover and mutation, are thereafter applied and new chromosomes are then generated. In what follows, the crossover and mutation operators, in improved forms, are described.

The decision variables considered in the chromosomes are in fact the supplying substations as

$$W_i = [X_1, X_2, \ldots, X_N] \tag{7.22}$$

where

W_i The ith chromosome

X_j The supplying substation number for feeding the jth load

Two crossover operators, namely, normal and mathematical, are applied as shown in (7.23) and (7.24).

Random point

$$W_1 = \left[X_1^1, X_2^1, X_3^1, \ldots, X_N^1\right]$$
$$W_2 = \left[X_1^2, X_2^2, X_3^2, \ldots, X_N^2\right]$$

Normal crossover (7.23)

$$W_1' = \left[X_1^1, X_2^1, X_3^2, \ldots, X_N^2\right]$$
$$W_2' = \left[X_1^2, X_2^2, X_3^1, \ldots, X_N^1\right]$$

$$W_1 = \left[X_1^1, X_2^1, X_3^1, \ldots, X_N^1\right] \qquad W_1' = \alpha\, W_1 + (1-\alpha)\, W_2$$

$$\xrightarrow{\text{Mathematical crossover}} \tag{7.24}$$

$$W_2 = \left[X_1^2, X_2^2, X_3^2, \ldots, X_N^2\right] \qquad W_2' = (1-\alpha)\, W_1 + \alpha\, W_2$$

where α is a random number [0, 1].

Regarding mutation operator, four options are proposed to improve the optimization procedure

- *Normal mutation* as shown in (7.25)

$$\begin{gathered} \downarrow \text{Random point} \\ W_1 = \left[X_1^1, X_2^1, \ldots, X_j^1, \ldots, X_N^1\right] \\ \Big\downarrow \text{Normal mutation} \\ W_1' = \left[X_1^1, X_2^1, \ldots, x_j^1, \ldots, X_N^1\right] \end{gathered} \tag{7.25}$$

where x_j^1 is a random number in the *j*th variable range.

- *The most suitable mutation* as shown in (7.26)

$$\begin{gathered} \downarrow \text{Random point} \\ W_1 = \left[X_1^1, X_2^1, \ldots, X_j^1, \ldots, X_N^1\right] \\ \Big\downarrow \text{Most suitable mutation} \\ W_1' = \left[X_1^1, X_2^1, \ldots, x_j^{1*}, \ldots, X_N^1\right] \end{gathered} \tag{7.26}$$

where x_j^{1*} is the most suitable substation for supplying the *j*th load.

- *Substation elimination mutation* in which a substation is randomly selected and all of its connecting loads are disconnected and then connected to its closest substation.
- *Dual displacement mutation* as shown in (7.27)

Random point ↓ ↓ Random point

$$W_I = \left[X_1^I, X_2^I, \ldots, X_i^I, \ldots, X_j^I, \ldots, X_N^I\right]$$

Dual displacement mutation (7.27)

$$W_I' = \left[X_1^I, X_2^I, \ldots, X_j^I, \ldots, X_i^I, \ldots, X_N^I\right]$$

In each stage, a fitness value is calculated for each population with assigning a penalty factor to the infeasible solutions (i.e., the ones violating the constraints). To speed up the convergence properties of the algorithm and at the same time, to use the information which may still be useful in rejected chromosomes, this penalty factor is linearly increased (through iterations) from zero toward a very high value. The fitness function is in fact the cost as detailed in (7.11).

7.6 Numerical Results

Following what we have covered in Sects. 7.4 and 7.5, in this section we present the numerical results on a typical system so that the algorithm capabilities may be assessed.

7.6.1 System Under Study

The pictorial representation of the system is already depicted in Fig. 7.6. It shows a 37 load-node system, each with 30 MVA (0.3 p.u.) consumption. The system has four existing substations (1 through 4). Twenty-one more substations are considered as new candidates. The geographical distributions of the substations are shown in Table 7.2 in terms of *X* and *Y*. Moreover, the distance of each candidate substation to the upward grid is shown and defined as *S*. Note that in practical conditions, *X* and *Y* should be determined using GIS[7] (Geographical Information System).[8]

---

[7] For more details, see Appendix G.

[8] In this section, the distance between two substations is calculated using $\sqrt{(X_1 - X_2)^2 + (Y_1 - Y_2)^2}$. If, however, *X* and *Y* are defined using GIS, the distance calculation is different (see the problems at the end of the chapter).

Table 7.2 Geographical distributions of substations (existing and candidate)

No.	*X*	*Y*	*S*	No.	*X*	*Y*	*S*
1	15	33	–	14	55	58	30
2	35	50	–	15	75	55	28
3	85	33	–	16	33	33	49
4	55	33	–	17	70	33	43
5	48	70	65.3	18	12	19	41
6	60	14	56.6	19	28	19	55
7	65	55	85.6	20	44	21	46
8	92	15	118	21	60	22	34
9	15	70	81.5	22	12	10	66
10	32	70	58.3	23	28	10	76
11	68	70	44	24	44	10	73
12	88	70	79	25	70	10	79
13	15	50	48				

7.6.2 Load Model

It is assumed that the existing network supplies the base load of each *load node* and the new downward grid should be planned for the load increase. As a result, each load (as denoted by its magnitude and geographical properties, i.e., *X* and *Y*) is divided into a *basevalue* and an *increase*. For planning the downward grid, only the increase part is considered, while for substation loadings and the upward grid design, both parts (the base and the increase) are considered. In this example, we assume the base values to be zero. The geographical distributions of the load nodes are shown in Table 7.3 in terms of *X* and *Y*.

7.6.3 Downward Grid

The downward grid of the system under study is in fact the sub-transmission level of the system. It comprises 63 kV elements. For cost analysis of the downward grid, four curves are used as shown in Fig. 7.11, where the horizontal axis shows the typical standard conductors available, while the vertical axes are *thermal capacity*, *voltage drop*, *investment cost* and *resistance*, respectively.

A linear approximation is assumed between the points, as indicated. The way these curves are used is as follows. Initially, based on an acceptable voltage drop for the specified load (*b*), a conductor size is selected (b'). Also, based on the load magnitude in MVA, from the line thermal capacity curve, a conductor size is selected, too (a'). Max (a',b') is selected as the final choice (for the case demonstrated, b'). For the selected conductor, the cost and the resistance are then determined (c and d, respectively). Right-of-way sitting difficulties are also

Table 7.3 Geographical distributions of load nodes

No.	X	Y	No.	X	Y
1	2	40	20	54	66
2	20	40	21	60	53
3	2	28	22	58	19
4	5	21	23	64	17
5	20	28	24	60	6
6	10	50	25	67	40
7	25	50	26	85	57
8	30	57	27	95	73
9	40	61	28	95	44
10	37	55	29	90	40
11	43	45	30	94	33
12	35	35	31	91	32
13	46	42	32	93	28
14	52	40	33	93	19
15	44	32	34	98	12
16	46	15	35	75	32
17	44	76	36	85	17
18	52	76	37	85	10
19	58	74			

observed by a large factor in the investment cost. Moreover, specific loads[9] (to a specific substation) are also considered. For the example to be tested here, for the downward grid, two options are considered as shown in Table 7.4.

7.6.4 Upward Grid

The upward grid of the system under study comprises 230 kV elements. The line capacity and the type are determined based on capacity, voltage level and the type of the supplying substation. For the test system, a 4.0 p.u. capacity line with a cost of R 214 × 10^3/km is considered as the only option available.

7.6.5 Transmission Substation

The transmission substation considered would be of 230 kV:63 kV type. Various capacities are available and may be considered. Alternatives and classifications, as required for each region, can also be considered. The standard capacities available

[9] The loads which have to be supplied by a specific substation.

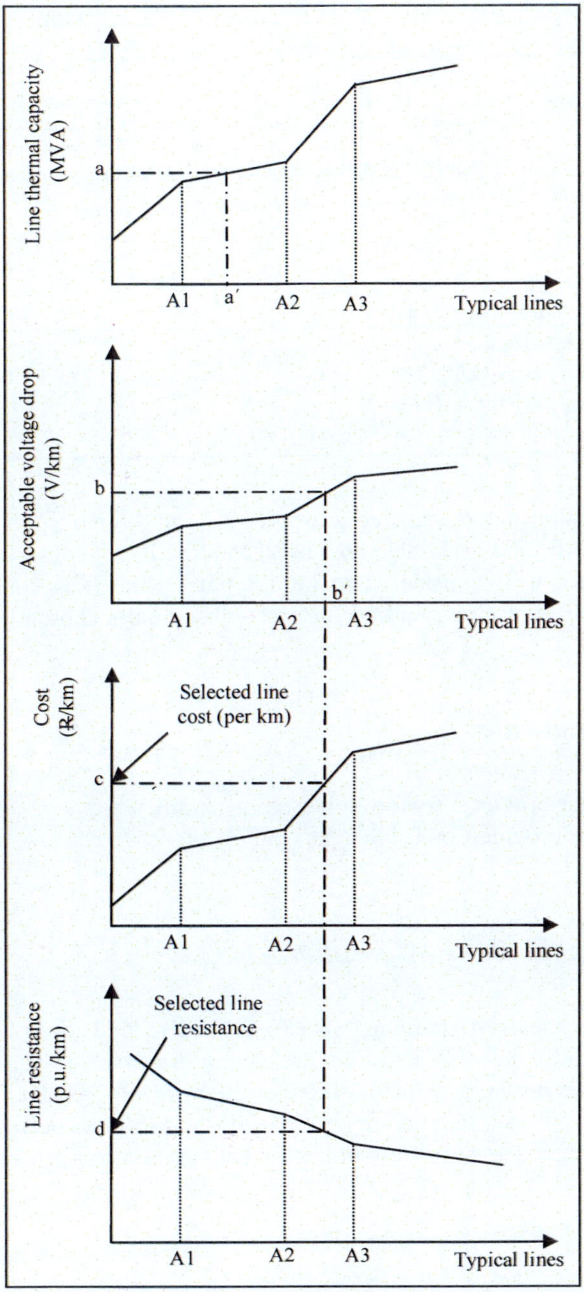

Fig. 7.11 Line selection curves

Table 7.4 Available downward grid feeders

Option	Capacity (p.u.)[a]	Cost (10^3 R/km)	*R* (p.u./km)	*X* (p.u./km)
1	0.5	65	0.004	0.010
2	1.0	95	0.002	0.005

The information provided here is used in Sect. 7.6.7. For the BILP solution presented in this section, an average cost of R 80 × 10^3/km is used for the cost of the downward grid, while the maximum permissible length is 50 km

[a] 1 p.u. = 100 MVA

Table 7.5 Miscellaneous data

Loads power factors	1.0
The cost of losses (R/p.u.)	3500 × 10^3
Amortizing coefficient (%)	10
Downward grid acceptable voltage drop (%)	5

for these substations are considered to be 1.8, 2.7 and 4.8 p.u.,[10] with a fixed cost of R 17000 × 10^3 and a variable cost of R 2500 × 10^3/p.u. The capacity of the existing substations is considered to be 1.8 p.u., while it is considered to be unexpandable. The reserve required for each substation is chosen to be 15%.

7.6.6 Miscellaneous

Besides the data provided so for, some other parameters are required for Sect. 7.6.8. These are provided in Table 7.5.

7.6.7 Results for BILP Algorithm

Binary Integer Linear Programming (BILP) is used to find the solution. For the parameters already shown, the results are demonstrated in Table 7.6 and as in Fig. 7.12. The results are generated using the SEP.m M-file [#SEP.m; Appendix L: (L.3)]. As shown, five substations are justified with the capacities noted. The way that the loads are assigned to each substation is so that various constraints are satisfied.

[10] The information provided here is used in Sect. 7.6.7. For the BILP solution presented in this section, the substation capacity is considered to be a continuous parameter with the fixed and the variable cost values, shown. The maximum capacities of existing and candidate substations are 1.8 and 4.8 p.u., respectively.

Table 7.6 Results of substation expansion planning for the system

No.	X	Y	Required capacity (p.u.)
1	15	33	1.8
2	35	50	1.8
3	85	33	1.8
4	55	33	1.8
7	65	55	3.9

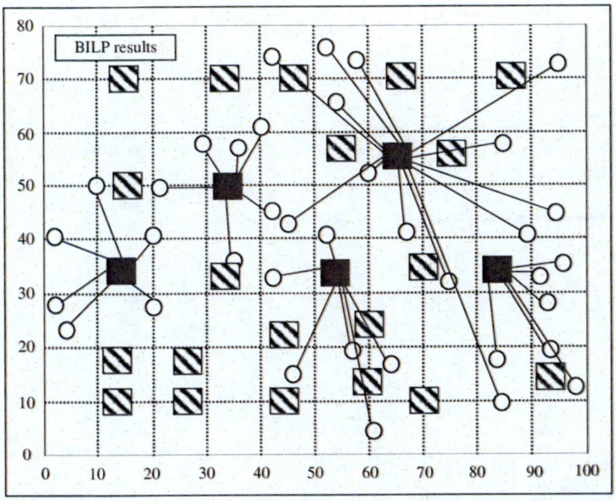

Fig. 7.12 BILP results

7.6.8 *Results for GA*

The problem formulation was fully discussed in Sect. 7.5. Results for a practical system are covered in this section.

Besides those parameters already noted, for the GA to be used in this section, the parameters are shown in Table 7.7. The results are shown in Table 7.8. As demonstrated, the ways the loads are assigned to the substations are in some instances different from those shown in Fig. 7.12. The reason is that in using GA, in general, the global optimality can not be guaranteed. However, the constraints are still satisfied.

Problems

1. In your area of living, find out the fixed and variable costs of substations for various voltage levels.

Table 7.7 GA parameters

Crossover probability (normal)	0.8	Mutation probability (normal)	0.0
Crossover probability (mathematical)	0.0	Mutation probability (the most suitable)	0.5
Mutation probability (elimination)	0.7	Mutation probability (dual displacement)	0.5
Population size	500	No. of converging iterations[a]	20

[a] If after 20 consecutive iterations, the objective function value does not change significantly, the algorithm is over

Table 7.8 GA results

Selected substation[a]	Initial capacity (p.u.)	Maximum capacity	Load number[b]							Capacity requirement plus reserve (p.u.)	Capacity requirement (p.u.)
1	1.8	1.8	1	3	4	5	6	–	–	1.8	1.5
2	1.8	1.8	2	7	8	10	12	–	–	1.8	1.5
3	1.8	1.8	21	25	26	27	29	–	–	1.8	1.5
4	1.8	1.8	11	13	14	15	35	–	–	1.8	1.5
5	0	4.8	9	17	18	19	20	–	–	1.8	1.5
7	0	4.8	28	30	31	32	33	34	36	2.7	2.1
21	0	4.8	16	22	23	24	37	–	–	1.8	1.5

[a] The number shown is the substation number from Table 7.2
[b] The number shown is the load number from Table 7.3

2. Prove (7.7).
3. An optimization problem is dependent upon its input parameters such as economical factors. As these parameters may exhibit some uncertainties, the planner should investigate the sensitivity of the solution with respect to the uncertainties of these parameters. Analyze the robustness of the solution reported in Sect. 7.6 with respect to the changes in lines and transformer costs, employed there [#SEP.m; Appendix L: (L.3)].
4. In the example reported in Sect. 7.6, find the solution and analyze the result, if the maximum feeder length is 50 km [#SEP.m; Appendix L: (L.3)].
5. Repeat the example reported in Sect. 7.6, assuming no initial substation exists and all given existing substations are considered as candidates [#SEP.m; Appendix L: (L.3)].
6. If (X_1, Y_1) and (X_2, Y_2) are the geographical properties of two points, prove the distance between these two points (D) to be as follows[11]

[11] For the calculations of the distances referred to in this book and by using the relationship given in this problem, the distance is set = 1.0 km, if it is calculated to be less than 1.0 km.

$$D = 20000 \times \cos^{-1}[(1 - A/2)/\pi]$$

where

$$A = (\cos(Y_1)\cos(X_1) - \cos(Y_2)\cos(X_2))^2 + (\cos(Y_1)\sin(X_1) - \cos(Y_2)\sin(X_2))^2 + (\sin(Y_1) - \sin(Y_2))^2.$$

References

As we discussed earlier in the chapter, SEP is the process of finding the allocation and sizes of both the expandable and the new substations. As the normal practice is to move from distribution substations towards the transmission substations, most of the research reported in literature are devoted to distribution substations. However, in [1], the problem is discussed from a transmission view. Some other general aspects are reviewed in [2].
Distribution substation planning is considered as a part of distribution planning. Distribution planning models (both SEP and feeder routing) are reviewed in some references such as [3, 4]. The frameworks for large scale systems are presented in [5–7]. Some practical and/or mathematical issues of the problem are covered in [8–10].
Distribution substation planning has also been received attention, separately, in literature such as [11–13]. Some other issues of the problem are covered in [14–16].

1. Sepasian MS, Seifi H, Foroud AA, Hosseini SH, Kabir EM (2006) A new approach for substation expansion planning. IEEE Trans Power Syst 21(2):997–1004
2. Peng W, Liu W (2009) A new method for substation planning problem based on weighted K-means. In: Yu W, He H, Zhang N (eds) Advances in neural networks—ISNN'09, part II. Springer, Berlin, pp 647–654
3. Gönen T, Ramirez-Rosado IJ (1986) Review of distribution system planning models: a model for optimal multistage planning. IEE Proc Gener Transm Distrib 133(7):397–408
4. Khator SK, Leung LC (1997) Power distribution planning: a review of models and issues. IEEE Trans Power Syst 12(3):1151–1159
5. Quintana VH, Temraz HK, Hipel KW (1993) Two-stage power-system-distribution-planning algorithm. IEE Proc Gener Transm Distrib 140(1):17–29
6. Navarro A, Rudnick H (2009) Large-scale distribution planning—part I: simultaneous network and transformer optimization. IEEE Trans Power Syst 24(2):744–751
7. Najafi S, Hosseinian SH, Abedi M, Vahidnia A, Abachezadeh S (2009) A framework for optimal planning in large distribution networks. IEEE Trans Power Syst 24(2):1019–1028
8. Ramirez-Rosado IJ, Bernal-Agustin JL (1998) Genetic algorithms applied to the design of large power distribution systems. IEEE Trans Power Syst 13(2):696–703
9. Hsu Y-Y, Chen J-L (1990) Distribution planning using a knowledge-based expert system. IEEE Trans Power Deliv 5(3):1514–1519
10. Willis HL, Tram H, Engel MV, Finley L (1995) Optimization applications to power distribution. Comput Appl Power IEEE 8(4):12–17
11. El-Fouly THM, Zeineldin HH, El-Saadany EF, Salama MMA (2008) A new optimization model for distribution substation siting, sizing, and timing. Int J Electr Power Energy Syst 30(5):308–315
12. Temraz HK, Salama MMA (2002) A planning model for siting, sizing and timing of distribution substations and defining the associated service area. Electr Power Syst Res 62(2):145–151

13. Hongwei D, Yixin Y, Chunhua H, Chengshan W, Shaoyun G, Jim X, Yi Z, Rui X (1996) Optimal planning of distribution substation locations and sizes—model and algorithm. Int J Electr Power Energy Syst 18(6):353–357
14. Khodr HM, Melian JA, Quiroz AJ, Picado DC, Yusta JM, Urdaneta AJ (2003) A probabilistic methodology for distribution substation location. IEEE Trans Power Syst 18(1):388–393
15. Zifa L, Jianhua Z (2006) Optimal planning of substation of locating and sizing based on GIS and adaptive mutation PSO algorithm. In: Proceeding of international conference on power system technology—PowerCon
16. Haghifam M-R, Shahabi M (2002) Optimal location and sizing of HV/MV substations in uncertainty load environment using genetic algorithm. Electr Power Syst Res 63(1):37–50

Chapter 8
Network Expansion Planning, a Basic Approach

8.1 Introduction

In previous chapters, we looked at the expansion planning of generations and substations. Although in both GEP and SEP (Chaps. 6 and 7), the network conditions were, somehow, accounted for, the modeling was very approximate and needs much further investigations. The so called *Network Expansion Planning* (NEP) process tries to find the optimum routes between the generation buses (determined in GEP phase) and the load centers (determined from load forecasting) via substations (determined in SEP phase), in such a way that

- Loads are completely supplied during both
  - Normal conditions
  - Once some types of contingencies occur on some system elements[1]
- Least costs are incurred

In fact, NEP is an optimization process in which the allocation (the sending and the receiving ends) and class (voltage level, number of conductors, conductor type) of new transmission elements, together with their required availability times are specified.

In this chapter, a basic case is analyzed. A more complex situation is dealt with in Chap. 9. Some of the aspects (such as voltage level) are considered there.

8.2 Problem Definition

Generally speaking, in NEP, the problem is to determine the transmission paths between substations (both existing and new) as well as their characteristics (voltage level, number of circuits, conductor type, and so on).

[1] We will see what an *outage* or *element* means in practical terms.

H. Seifi and M. S. Sepasian, *Electric Power System Planning*, Power Systems,
DOI: 10.1007/978-3-642-17989-1_8,

In doing so

- The investment cost should be minimized
- The operational cost should be minimized
- Various constraints should be met during
  - Normal conditions
  - Contingency conditions

We will see shortly that in its simplest form, the investment cost involves the cost of adding new transmission elements. Moreover, the operational cost would be the cost of power losses during the element life. Modeling the operational cost as well as some other new terms will be defined and added in Chap. 9.

In terms of the constraints, an obvious case is the limiting transfer capability of an element, which should not be violated. The contingency is, in fact, an outage occurring on a single element (such as a line, a transformer, a power generation unit) or some elements. The single element case is commonly referred to N − 1 conditions.[2] Simultaneous contingencies on two elements (for instance one line and one transformer, two lines, etc.) are referred to N − 2 conditions and so on. By contingency conditions (say N − 1), we mean that the network should be so planned that with every single element, out, the load is completely satisfied and no violation happens.

Before proceeding any further on mathematical modeling, we will talk a little bit more in Sect. 8.3 to understand the NEP problem with some more details.

8.3 Problem Description

To understand some basic concepts outlined in Sect. 8.2, a simple test case as depicted in Fig. 8.1 is used. This system is the one normally used for basic network planning issues, proposed by Garver.[3] The relevant data are provided in Appendix F. A normal load flow solution procedure may be used to determine the power transfer of each line. However, a simplified type of load flow, the so called DCLF,[4] is normally used in power system planning problems, by which the power transfers may be calculated very fast.[5] Whatever the calculation procedure is, the normal flow conditions are shown in Fig. 8.2, in which the numbers within the arrows show the per unit power transfers of lines [#DCLF.m; Appendix L: (L.5)]. Based on the lines capacities provided in Appendix F, it is evident that no violation happens in this condition.

---

[2] Which means *normal minus one* element.

[3] See the list of the references at the end of the chapter.

[4] Direct Current Load Flow.

[5] In a later section and Appendix A, DCLF is discussed more.

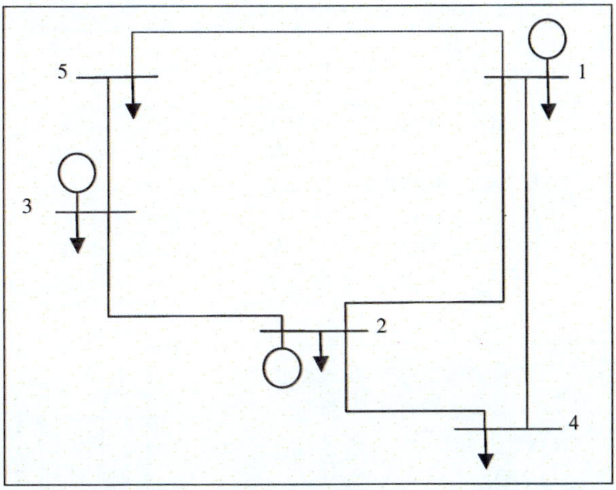

Fig. 8.1 Garver test system

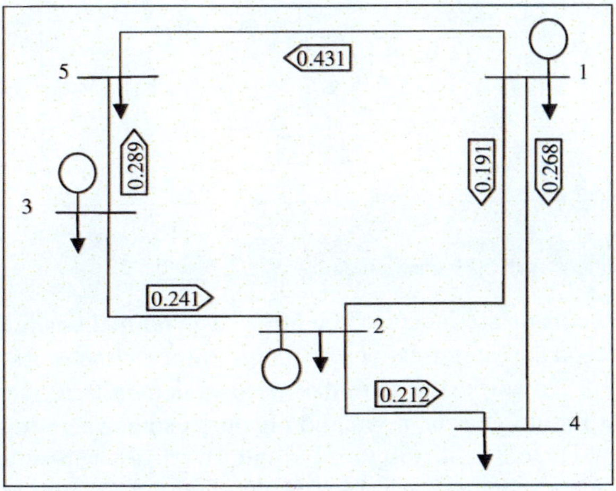

Fig. 8.2 Flow conditions for the base case

Now assume that a single contingency occurs on each line. In other words, assuming each line to be out, one-by-one, we are going to find out how the powers are distributed throughout the network. Again DCLF is employed for each case. Table 8.1 shows a summary of the results [#DCLF.m; Appendix L: (L.5)]. As shown, again there is no line capacity violation, even for N – 1 conditions. Briefly

> The network condition is acceptable for both normal and N – 1 conditions

Table 8.1 N − 1 results (base case)

Contingency on line	Flow on line (p.u.)					
	1–2	1–4	1–5	2–3	2–4	3–5
1–2	0.000	0.340	0.550	−0.360	0.140	0.170
1–4	0.352	0.000	0.538	−0.348	0.480	0.182
1–5	0.499	0.391	0.000	0.190	0.089	0.720
2–3	0.363	0.337	0.190	0.000	0.143	0.530
2–4	0.064	0.480	0.346	−0.156	0.000	0.374
3–5	−0.016	0.186	0.720	−0.530	0.294	0.000

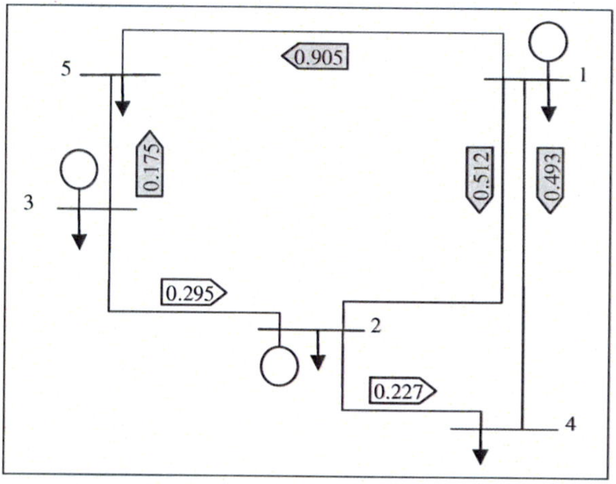

Fig. 8.3 Flow conditions with 50% load increase

Now suppose the loads are increased by 50%.[6] The normal condition as shown in Fig. 8.3 is acceptable in terms of the flows through the lines [#DCLF.m; Appendix L: (L.5)]. However, the semi-dark arrows demonstrate the lines that if any of them is out for any reason, a violation happens somewhere in the network. For instance, from Table 8.2, if line (1–5) is out, overloads happen on lines 1–2 and 3–5 [#DCLF.m; Appendix L: (L.5)]. Briefly, with 50% higher loads

The network condition is acceptable for normal but not for N − 1 conditions[7]

Now if the loads are increased by 116.5% (in comparison with the base case shown in Appendix F), the results shown in Fig. 8.4 demonstrate the fact that even in

[6] For any test involving load change or generation contingencies, it is assumed that the generation change is compensated by the generation of bus 1 (slack bus).

[7] Even if for a specific line to be out, a violation happens, the design is considered to be unacceptable.

Table 8.2 N − 1 results (50% load increase)

Contingency on line	Flow on line (p.u.)					
	1–2	1–4	1–5	2–3	2–4	3–5
1–2	0.000	0.685	1.225	−0.615	0.035	−0.145
1–4	0.808	0.000	1.102	−0.492	0.720	−0.022
1–5	1.159	0.751	0.000	0.610	−0.031	1.080
2–3	0.723	0.577	0.610	0.000	0.143	0.470
2–4	0.376	0.720	0.814	−0.204	0.000	0.266
3–5	0.387	0.443	1.080	−0.470	0.277	0.000

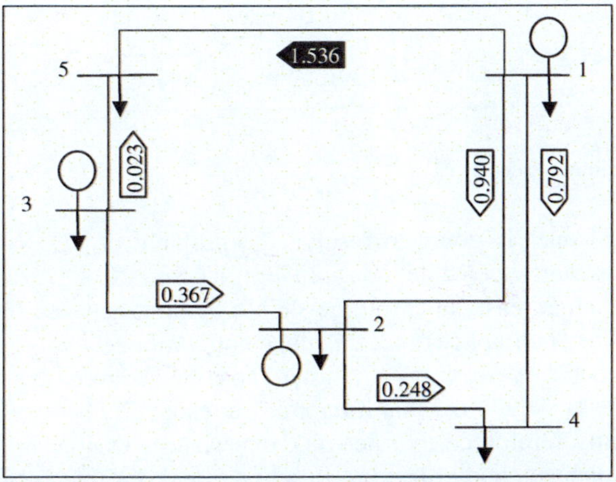

Fig. 8.4 Flow conditions with 116.5% load increase

Table 8.3 N − 1 results (116.5% load increase)

Contingency on line	Flow on line (p.u.)					
	1–2	1–4	1–5	2–3	2–4	3–5
1–2	0.000	1.145	2.125	−0.955	−0.105	−0.565
1–4	1.416	0.000	1.854	−0.684	1.040	−0.294
1–5	2.039	1.231	0.000	1.170	−0.191	1.560
2–3	1.203	0.897	1.170	0.000	0.143	0.390
2–4	0.792	1.040	1.438	−0.268	0.000	0.122
3–5	0.924	0.786	1.560	−0.390	0.254	0.000

normal conditions, line (1–5) is overloaded (dark arrow). For N − 1 conditions, there are still some more violations (Table 8.3) [#DCLF.m; Appendix L: (L.5)]. Briefly

The network condition is not acceptable for both normal and N − 1 conditions

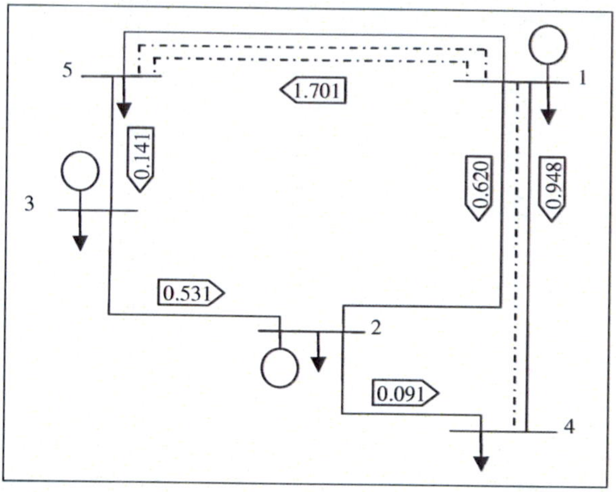

Fig. 8.5 Flow conditions (scenario 1)

Suppose the planner is going to resolve the problem of Fig. 8.4 so that the system conditions are acceptable for both normal and N – 1 conditions. The planner notices that, at least, an extra line should be built between buses 1 and 5. However, he or she soon notices that this does not completely solve the problem. So, the planner considers two lines of similar capacities between those two buses. Following that, he or she notices that if the transfer capability between buses 1 and 4 is not sufficiently reinforced, the network experiences trouble in some N – 1 conditions. The planner finally decides on adding an extra line between buses 1 and 4, too. His or her final choice, as depicted in Fig. 8.5 (scenario 1), completely solves the problem in both normal and N – 1 conditions.[8] Table 8.4 shows the results for N – 1 conditions [#DCLF.m; Appendix L: (L.5)].

Suppose another planner suggests a new topology to resolve the same problem, based on his or her own engineering judgments. His or her plan of Fig. 8.6 (scenario 2) results in acceptable conditions for both normal and N – 1 conditions (Table 8.5) [#DCLF.m; Appendix L: (L.5)].

Now a simple question is that what happens if another planner comes into play? How many solutions are there for this specific problem? How should we select the best solution?

Suppose the number of existing lines to be N (6 in our case), the number of candidate corridors to be M (If between any two buses, extra lines may be considered, M is 10 in our case) and the number of extra candidate lines to be feasible in each corridor is K (2 for our case). As a result, it can be shown[9]

[8] The arrows for lines 1–5 and 1–4 are for the total transfers. These numbers reflect the figures for the normal conditions.

[9] Left as an exercise for the reader.

Table 8.4 N − 1 results (scenario 1)

Contingency on line	Flow on line (p.u.)					
	1–2	1–4	1–5	2–3	2–4	3–5
1–2	0.000	1.196	2.073	−0.903	−0.157	−0.513
1–4	0.746	0.714[a]	1.809	−0.639	0.326	−0.249
1–5	0.668	0.976	1.626[a]	−0.456	0.064	−0.066
2–3	0.958	1.142	1.170	0.000	−0.102	0.390
2–4	0.571	1.040	1.659	−0.489	0.000	−0.099
3–5	0.710	1.00	1.560	−0.390	0.040	0.000

[a] Note that the flows are for the remaining line(s) on the route

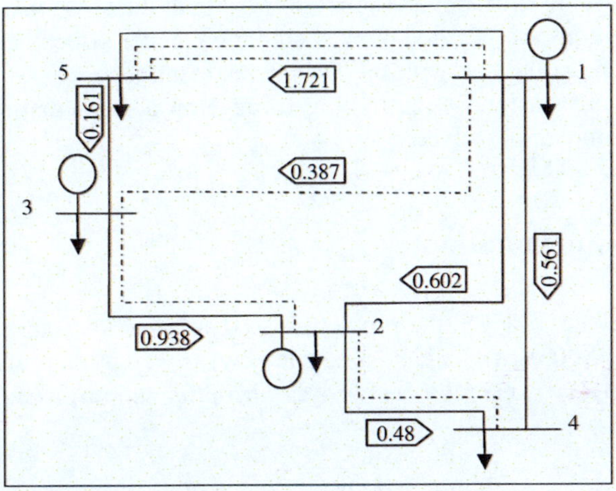

Fig. 8.6 Flow conditions (scenario 2)

Table 8.5 N − 1 results (scenario 2)

Contingency on line	Flow on line (p.u.)						
	1–2	1–3	1–4	1–5	2–3	2–4	3–5
1–2	0.000	0.575	0.707	1.989	−1.394	0.334	−0.429
1–3	0.710	0.000	0.615	1.944	−0.774	0.424	−0.385
1–4	0.821	0.527	0.000	1.923	−1.280	1.040	−0.362
1–5	0.642	0.444	0.581	1.604	−0.878	0.460	−0.043
2–3	0.702	0.325	0.611	1.632	−0.787	0.430	−0.073
2–4	0.569	0.366	0.644	1.692	−0.888	0.396	−0.131
3–5	0.656	0.466	0.588	1.560	−0.856	0.452	0.000

$(K+1)^M$	All possible system topologies
$\left[\frac{K \times M}{K+1} + N\right]$	Average number of contingencies for each topology
$(K+1)^M\left[1 + N + \frac{K \times M}{K+1}\right]$	Total number of load flows required for all topologies and in normal and contingency conditions

In fact, there may be, at most 59×10^3 topologies, for this specific problem. Obviously many of them are not feasible, as there may be some types of violations in either normal or N − 1 conditions. For each of the cases cited above, a DCLF should be run (807×10^3 DCLF for our case). Find out the running time to be nearly 2.24 h, if a single load flow for this specific case takes 0.01 s. What happens for large scale systems? If many of them are feasible in the sense that both normal and N − 1 conditions are satisfied, how should the planner select the best choice? An obvious choice is the one with the least investment cost.

In the following section, we are going to develop a mathematical model by which the problem can be solved.

8.4 Problem Formulation

As already described, in NEP, the problem is to determine the transmission paths between substations (buses); both existing and new; as well as their characteristics. The problem may be, generally, viewed as an optimization problem as shown below

$$\begin{gathered}\text{Minimize (Objective Function)}\\ \text{s.t. (Constraints)}\end{gathered} \tag{8.1}$$

In its simplest form, the objective function consists of the investment cost for new transmission lines, while the constraint terms consist of load-generation balance and transmission limits. The terms are described below.

8.4.1 Objective Function

The aim is to minimize the total cost (C_{total}), consisting of the investment cost for new transmission lines[10] ($C_{new\text{-}line}$), i.e.

$$C_{total} = C_{new\text{-}line} \tag{8.2}$$

[10] In Chap. 9, new terms will be added, as more practical cases are considered.

where

$$C_{new\text{-}line} = \sum_{i \in Lc} C_L(x_i) L_i \tag{8.3}$$

where L_i is the transmission length (km) of the candidate, Lc is the set of candidates, x_i is the transmission type of the candidate (set of various types such as number of bundles, conductor types and number of circuits) and $C_L(x_i)$ is the investment cost per km for type x_i.

8.4.2 Constraints

As mentioned before, the load-generation balance should be observed during the optimization process. Moreover, the capacities of transmission lines should not be violated, too. These constraints are described below.

8.4.2.1 Load Flow Equations

For most basic planning studies, it is of normal practice to use DCLF equations, as the planner avoids any anxiety about voltage problems and possible convergence difficulties. Moreover, especially for large-scale power systems, the solution time may be exceptionally high (see Sect. 8.3), if ACLF is employed. It is obvious that in the final stage, ACLF should be performed to have an acceptable voltage profile during normal as well as contingency conditions (Chap. 10). Appendix A provides more details on DCLF.

The DCLF equations are in the form of (8.4)

$$\begin{aligned} &\sum_{j=1}^{N} B_{ij}(\theta_i - \theta_j) = P_{Gi} - P_{Di} \quad \forall\, i \subset n \\ &\sum_{j=1}^{N} B_{ij}^{m}\left(\theta_i^{m} - \theta_j^{m}\right) = P_{Gi}^{m} - P_{Di} \quad \forall\, i \subset n \cap m \subset C \end{aligned} \tag{8.4}$$

where θ_i and θ_j are the voltage phase angles of buses i and j, respectively; B_{ij} is the imaginary part of the element ij of the admittance matrix, P_{Gi} is the power generation at bus i, P_{Di} is the power demand at bus i, and n is the set of system buses. The index m shows the contingency parameters and variables. C is the set of contingencies. N is the system number of buses.

8.4.2.2 Transmission Limits

For each of the transmission lines, the power transfer should not violate its rating during both normal and contingency conditions (N − 1, in our examples[11]), so

$$\begin{aligned} b_k(\theta_i - \theta_j) &\leq \overline{P}_k^{No} \quad \forall\, k \in (Lc + Le) \\ b_k^m\left(\theta_i^m - \theta_j^m\right) &\leq \overline{P}_k^{Co} \quad \forall\, k \in (Lc + Le) \cap m \in C \end{aligned} \tag{8.5}$$

where $\overline{P}_k^{No}$ and $\overline{P}_k^{Co}$ are the line k ratings during normal and contingency conditions, respectively; θ_i and θ_j are the voltage phase angles of line k during normal conditions; θ_i^m and θ_j^m are the voltage phase angles of line k following contingency m; and Le is the set of existing lines. Lc is defined earlier. b_k and b_k^m represent the line k admittances in normal and contingency conditions, respectively.[12]

8.5 Solution Methodologies

The problem formulated in Sect. 8.4 may be solved by available optimization techniques. Both mathematical based options and heuristic types may be tried, each with its own capabilities and drawbacks. For a practical specially large scale system, the approach employed should be robust and flexible enough to be applied.

Two methods are proposed here to solve the NEP problem. The solution methodologies are demonstrated through observations on the Garver test system.

8.5.1 Enumeration Method

If the system is not large, the search space can be completely checked to find the best solution. In other words, various topologies may be checked to find out the solutions which are feasible; in other words resulting in acceptable normal and N − 1 conditions. From those feasible, the one resulting in the least investment cost would be the final solution.

Obviously, for large scale systems, the enumeration method fails to find a solution as the search space is exceptionally large.

---

[11] It should be mentioned the approaches presented may be employed for other cases, such as N − 2, etc., too.

[12] Generally for most *i–j* pairs, b_k^m is zero. If a line is tripped out, b_k^m for that specific line would be zero, but for any other line, its value would be identical to the value of the normal case. If for a double circuit line between bus i and bus j, one circuit is tripped out, b_k^m would be equal to that of the remaining circuit.

8.5.2 Heuristic Methods

One way to solve such a problem is to choose the methods based, somehow, on engineering judgments. For instance in the so called *forward* method, the candidates are added one-by-one. We proceed so far as the system conditions are acceptable for both normal and N − 1 conditions. The so called *backward* approach works vice versa in such a way that, all candidates are initially added to the network and the candidates are removed, one-by-one so far as a violation happens in either normal or N − 1 conditions. As a matter of fact, the *backward* approach may start from a point within the feasible region while the *forward* approach may start from outside such a region.[13]

As the number of candidates may be much higher than the real number justified and required, the execution time of the backward approach is normally higher than that of the forward approach. However, as it starts within the feasible region, the solutions will remain feasible through the solution process. As a result, the solutions may be more favorable in comparison with the forward approach especially when some feasible solutions are to be compared.

On the other hand, once there are some new substations with no initial connections to the rest of the network, the calculation of the performance index[14] encounters difficulty during the initial stages of the algorithm so that during this time and until all such new substations are somehow connected, the search does not follow any specific route towards the solution so that the algorithm may even fail in reaching a solution at all.

In fact, as in the *backward* approach, we remain in the feasible region throughout the solution process, the most costly candidates are, normally, removed first. However, in the *forward* approach, as we start from a point outside the feasible region, the most effective candidates are initially selected. As a result, typically, the backward process ends up with more justified candidates in comparison with the forward process; however with less costly paths. There is no guarantee that either of the approaches ends up at the same results or one makes sure that the solution of one is better than the other.

The trajectories are shown in Fig. 8.7 for a typical two-variable case. The dark area shows the feasible region with the boundary conditions dictated by various constraints. The *backward* search starts from a point within the feasible region while the *forward* search follows a trajectory from a point outside that region. Both move towards the optimum solution.

---

[13] Indeed, if adding all candidates does not result in a non-problematic network, the backward stage starts from a point outside the feasible region. Same reasoning applies to the forward stage.

[14] The performance index is in fact, the evaluation function defined, later on, in (8.6). There, we have a constraint violation term. During the initial stages where the new substations are not connected to the rest of the network, we come across difficulty in calculating this term of the evaluation function.

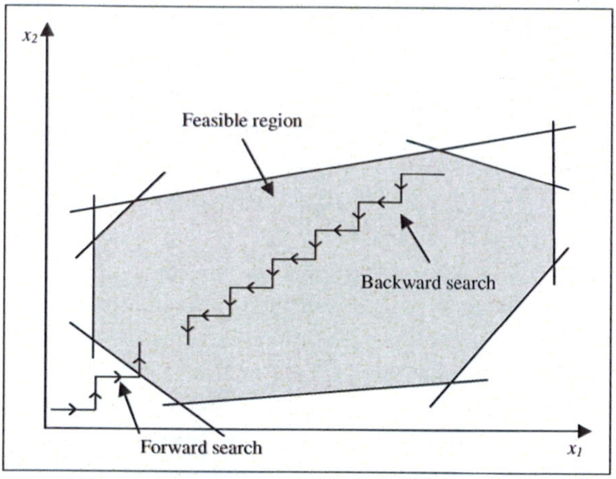

Fig. 8.7 Backward and forward approaches

Initially we will discuss the *backward* and the *forward* methods as the basic approaches. Following that, we will talk about the so called *decrease* method as an improved algorithm to the above. A *hybrid* method is finally demonstrated.

8.5.2.1 Backward Method

Let us propose a simple method in which, all candidates are initially added to the network. Thereafter, the candidates are removed, one-by-one, and an evaluation function is calculated in each case.[15] The one (i.e. with one of the candidates removed) with the lowest evaluation function is chosen as the starting point and the procedure is repeated until we come to a point that a violation happens in either normal or N − 1 conditions.

To make the points clear, first, recall that the best solution is the one with the lowest investment cost (see (8.2)) while there is no violation in both normal and N − 1 conditions. As an evaluation function, let

$$\text{Evaluation Function} = C_{total}\ (\text{see (8.2)}) + \alpha\ (\text{Constraints violations}) \quad (8.6)$$

where α is a large number and the constraints violations are calculated as the sum of the absolute values of all violations.

As a result, the solution will end up with the least cost choice and with no constraints violations.

Let us now again consider the case of Fig. 8.1. Assume that there are six corridors as shown in Fig. 8.8. If the combination 111111 denotes the case in which all (corridors) candidates are assumed *in*, the backward approach is best illustrated as in Fig. 8.9.

[15] We will see shortly what the evaluation function is.

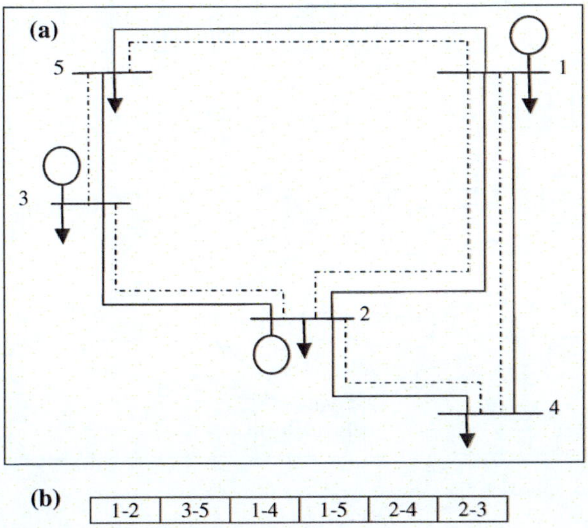

Fig. 8.8 Six candidates for the test case. **a** One-line diagram representation and **b** block representation (binary coded in Fig. 8.9)

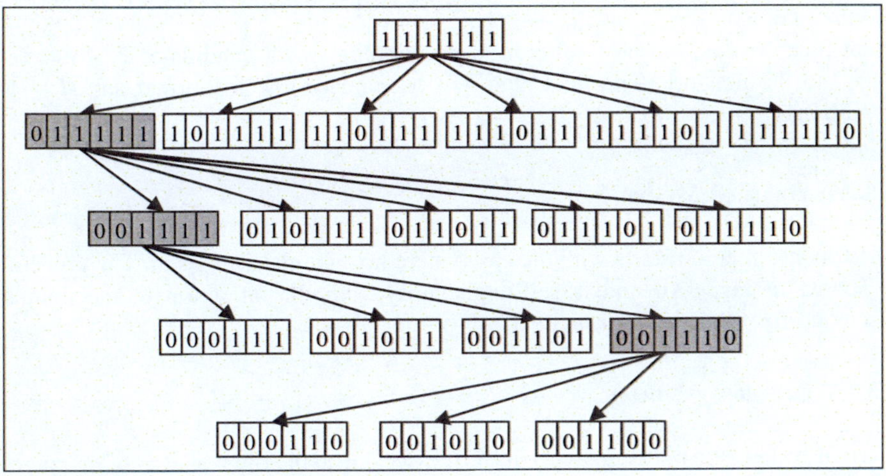

Fig. 8.9 Backward approach for the test case

Initially all candidates are added to the network (block 111111). Thereafter, each candidate is removed (blocks 011111–111110), one-by-one and the evaluation function (8.6) is calculated in each case.[16] If, for instance, 011111 results in

[16] Note that the term of constraints violations has to be calculated as the sum for both normal and all N − 1 conditions.

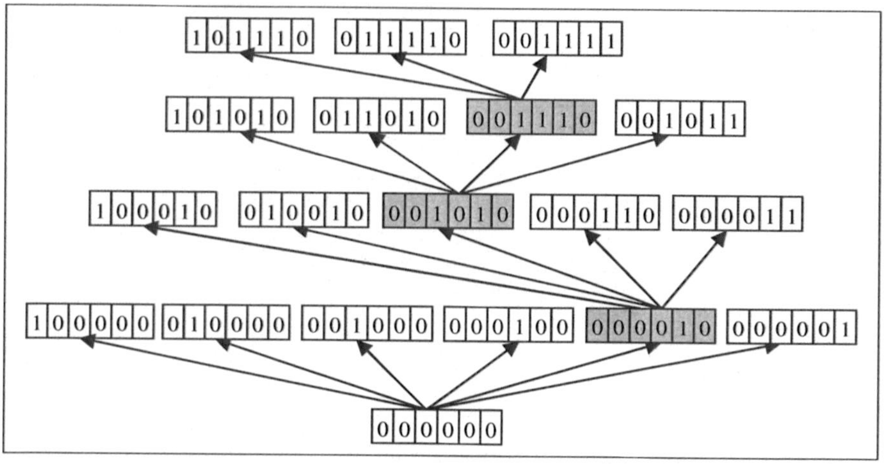

Fig. 8.10 Forward approach for the test case

the least evaluation function, the method continues, with candidate 2 removed (blocks 001111–011110). The algorithm is repeated until we reach to a condition with the lowest evaluation function (i.e. the lowest cost and no constraints violations). For instance we may reach to 001111 at the next stage and to 001110 as the final choice.[17]

In other words, the best solution is the topology with candidates 3, 4 and 5, added to the network. The final plan will be robust for both normal and N − 1 conditions.

8.5.2.2 Forward Method

The forward method starts with the case where no candidate line is *in*. The process is shown in Fig. 8.10 and is self-explanatory with due attention to the points discussed for the backward method.

8.5.2.3 Decrease Method

In a real system, the major cost of a line is the one due to the right-of-way of the route or the corridor. Once this right-of-way is acquired, there may be some alternatives of building various capacities or types of transmission lines within that corridor.

In both the backward and the forward approaches, a single line is assumed *in* or *out* in each stage. As the right-of-way acquiring cost is a major cost for a line, the optimal solution approach should initially search for the least cost corridors. Once

[17] Block 001110 is the final choice as moving further (blocks 000110 through 001100) results in some types of violations.

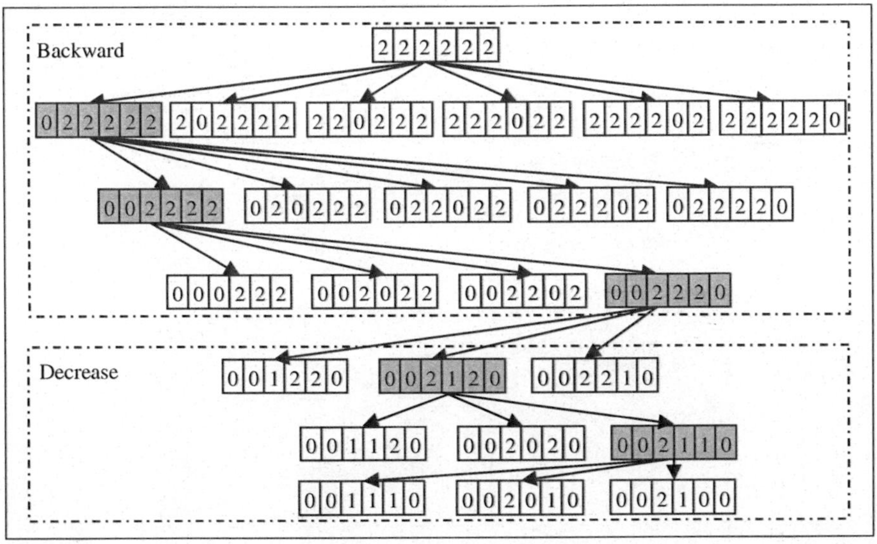

Fig. 8.11 Backward-decrease approach for the test case

these corridors are selected, the types and the capacities of the required transmission lines may be chosen.

Now assume that in our earlier example, two alternatives are possible for each corridor. For instance, either a single-circuit line or a double-circuit line is possible. As another example, two single-circuit lines with capacities A and B may be assumed with A > B.

The *decrease* method may be explained as follows.

In either the *backward* or the *forward* approaches, the solution process proceeds with the highest capacity available candidate for each corridor. Once done, in a *decrease* stage, the lower capacity (cost) options for each corridor are tried to see if they can perform the job.

For a *backward* approach, the process is shown in Fig. 8.11. Number 2 demonstrates the higher capacity option for each corridor in the *backward* stage.

In the *decrease* stage, the lower capacity for each corridor is shown by number 1. 002110 is the final choice in which the higher capacity is selected for candidate 3 while the lower ones are selected for candidates 4 and 5. Moving further results in some types of violations.

8.5.2.4 Backward–Forward-Decrease Method, a Hybrid Approach

As already described, the use of forward approach is undesirable if a new substation is to be supplied from nearby buses. On the other hand, the search space is enormous for large scale systems, if backward approach is tried for both normal and contingency conditions. So, what do we have to do for a large scale system?

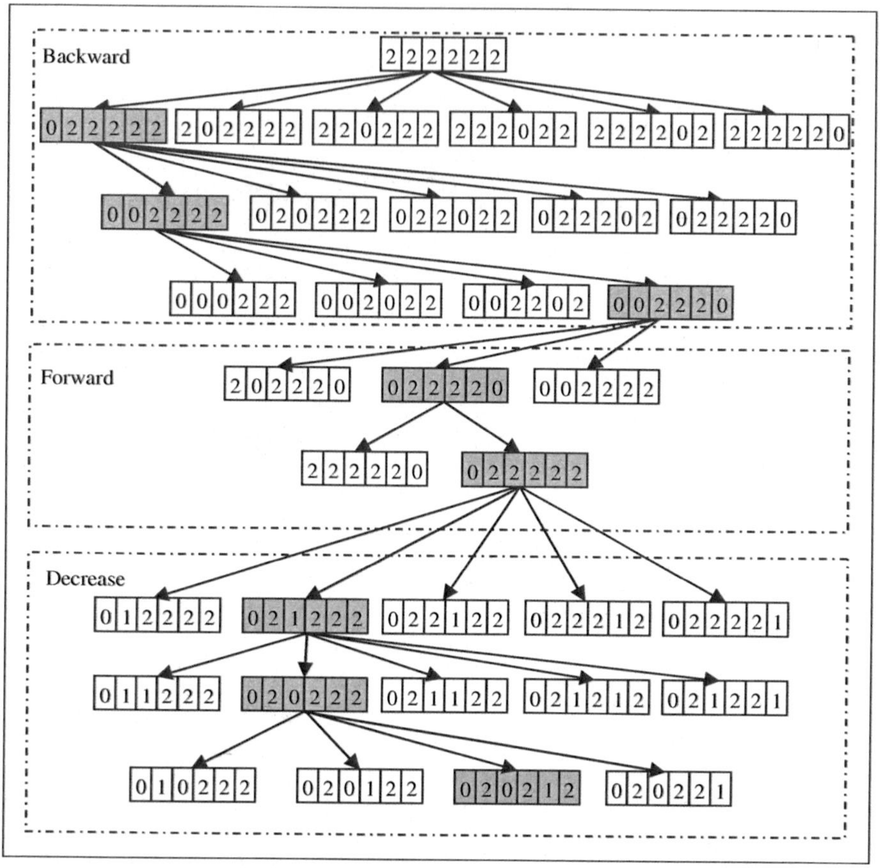

Fig. 8.12 Hybrid approach

One way to overcome the difficulties, is to plan, initially the network for normal conditions (no contingency) using the backward approach. As no contingency is considered at this stage, the solution speed will be high and acceptable. Thereafter, the forward approach is employed to find the solution in the presence of all foreseen contingencies (N − 1).

To illustrate how this hybrid method works, suppose that the backward approach is used only for the normal conditions.[18] If 002220 is the final choice, we proceed with the forward approach from 002220, as shown in Fig. 8.12, now for N − 1 conditions. Each time, a candidate is added (blocks 202220–002222) and the evaluation function (8.6) is calculated, considering all contingences. For instance, assume 022220 results in the least evaluation function, while still there are some constraints violations (for some contingencies). Thereafter, blocks

[18] i.e. the violations are calculated only for the normal conditions.

222220 and 022222 are tried to check the one which results in the least evaluation function, while no constraints violations happen (for instance block 022222). As a result, the final solution which results in the least investment cost and no violations in both normal and N − 1 conditions is the one with candidates 2, 3, 4, 5 and 6 to be *in*. Then the *decrease* stage is tried to reach at the final solution of 020212 based on the process already described.

Obviously, for a large scale system, the number of candidates would be high and the solution normally ends up with limited number of choices. Moreover, although the optimality of the solution cannot be guaranteed, the solution speed and accuracy would be quite acceptable.

Before proceeding for some numerical results, let us add a new term to the evaluation function (8.6) which makes it more practical. Suppose that in an intermediate stage, in either normal or any of contingency conditions, a situation happens that an isolated substation (bus) appears. This condition is referred to an *island* and should be avoided during normal and contingency conditions, so

$$N_{island} = 0 \tag{8.7}$$

If a line contingency is modeled in the algorithm by choosing a very high value for the line reactance, an island is detected by checking the phase angle difference across the line to be a large number. This happens due to the fact that the far end of the line terminates at a load bus.

To avoid any islanding, let us expand (8.6) as

$$\text{Evaluation function} = (8.6) + \beta(8.7) \quad \alpha \gg 1,\ \beta \gg \alpha \tag{8.8}$$

Provided α and β are arbitrarily chosen very high, the final solution will end up with no islanding conditions as well as with no constraints violations. β is chosen to be much higher than α for the following two reasons

- An islanding removal is considered to be more important that removal of a constraint violation. In fact, if $\beta \approx \alpha$, the algorithm may attempt to remove violations, while still there may be islands which are not removed.
- The quantity of the term representing the constraints violations is normally much higher than the term representing the number of islands.

8.6 Numerical Results

Two test cases are used for evaluating the proposed algorithms. One, is a small test system for which various algorithms are checked. The other is a large test system for which the hybrid algorithm is tried.

Table 8.6 Results for the Garver test system

Description	Enumeration method	Forward method	Backward method	Hybrid method
Selected lines			1–5	
	1–5	1–5	1–5	1–5
	1–5	1–5	2–3	1–5
	1–4	1–4	2–4	1–4
			3–5	
Number of load flows	17,825,800	897	4,067	252
Lines lengths justified (km)	1,000	1,000	1,200	1,000

8.6.1 Garver Test System

Let us first perform the approaches presented in Sect. 8.5 on the Garver test system[19] (Appendix F), as follows

- Case I: Enumeration method (Sect. 8.5.1) [#DCLF.m; Appendix L: (L.5)]
- Case II: Backward method (Sect. 8.5.2.1) [#Backwardsearch.m; Appendix L: (L.4)]
- Case III: Forward method (Sect. 8.5.2.2) [#Forwardsearch.m; Appendix L: (L.4)]
- Case IV: Hybrid method (Sect. 8.5.2.4) [#Hybridsearch.m; Appendix L: (L.4)]

The candidate lines are assumed to be a set of all possible connections between any two buses.[20] The plans summarized in Table 8.6 show that all methods, except for the backward case, result in the same optimal configuration. The fact that the backward approach fails in reaching the same solution was already explained in Sect. 8.5.2.

8.6.2 A Large Test System

To assess the capability of the proposed hybrid approach for a large scale system, an 84-bus test system as depicted in Fig. 8.13 is employed.

This is a single level voltage network with detailed information as outlined in Appendix H. The general data are provided in Table 8.7.

DCLF results in observing some violations in both normal and contingency conditions, a summary of which is provided in Table 8.8. The contingencies are assumed to be tripping of any single transmission line or 10% reduction in the generation level of any generation bus. As shown, the system is confronted by some violations for which NEP has to provide some types of solutions.

In order to perform NEP, some candidate lines should be initially selected. To do so, in our test example, we have assumed that any route between any two of the

[19] With modified load (116.5% of base values; Fig. 8.4).

[20] Two lines are considered for each connection.

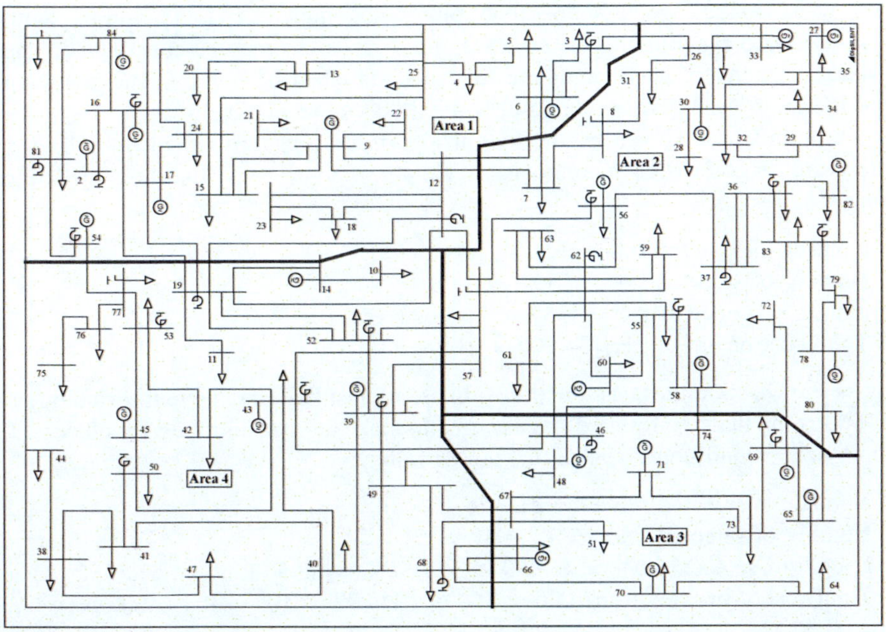

Fig. 8.13 84-Bus test system

Table 8.7 General data of the 84-bus network

No of buses	84	Total load (MW)	22,253
No of generation buses	25	No of lines	128

Table 8.8 A summary of violations with no expansion

The number of contingencies resulting in islands	16
Sum of lines overloads in normal conditions (per unit)	0.328
Sum of lines overloads in generation contingencies (10% reductions) (per unit)	0.0
Sum of lines overloads (for all contingencies) in contingency conditions (per unit)	62.6

buses with a straight length of less than 500 km may be a choice. The geographical information of the buses is provided in Appendix H. As a result, 1,287 candidate routes are selected. With due attention to the existing overloads (in either normal or contingency conditions), an extra four candidate routes are added by the planner. The list of total 1,291 candidate routes is provided in Appendix H.[21]

[21] In practice, the candidate lines as well as their specifications are selected by the planner based on some technical/environmental/engineering judgments; the number of which is quite lower than what shown here. In Chap. 9, we present an approach by which the planner may use some indices in rational choices of candidate lines.

For each route, we assume that either a single or two circuits with $R = 0.000015$ p.u./km and $X = 0.00025$ p.u./km (each) may be constructed. The loading capacity of each candidate line is assumed to be 3.3 p.u. The costs are considered to be R 250,000/km and R 400,000/km, respectively.

The problem is solved by the hybrid algorithm already outlined. Initially the backward stage is tried to remove all violations in normal condition, with the following results

- Number of selected routes = 6
- Number of selected lines = 12
- Length of selected lines = 337 × 2 = 674 km
- Total cost of lines = R 134,800 × 10^3

As shown, six routes, each with two lines, are justified to remove the violations.

Following that, the forward stage is tried to make the system robust for all N − 1 contingency conditions. The results are as follows

- Number of selected routes = 21 + 6
- Number of selected lines = 42 + 12 = 54
- Length of selected lines = (3,880 + 337) × 2 = 8,434 km
- Total cost of lines = 1,552,000 × 10^3 + 134,800 × 10^3 = R 1,686,800 × 10^3

21 extra routes (42 extra lines), are now justified for reaching at a robust network for contingency conditions.

As the final stage, the decrease stage is tried to check that if lower capacity solutions may be used. The results shown below demonstrate the fact that number of the routes would not decrease, but the number of the lines is reduced by five.

- Number of selected routes = 21 + 6 = 27
- Number of selected lines = 42 + 12 − 5 = 49
- Length of selected lines (single circuits) = 532 km
- Length of selected lines = (3,880 + 337) × 2 − 532 = 7,902 km
- Total cost of lines = 1,552,000 × 10^3 + 134,800 × 10^3 – 79,800 × 10^3 = R 1,607,000 × 10^3

Some details of the results for the large test system are provided in Appendix I.

Problems

1. Assuming the generation cost to be observed, modify the model of Sect. 8.4, appropriately.
2. For Fig. 8.4, suggest an alternative expansion plan.
3. Assuming that the generation in bus 1 can be reduced and those of buses 2 and 3 can be increased, find out the optimum generation plan for problem 2 from various generation plans so that the transmission enhancement requirements are minimized [#DCLF.m; Appendix L: (L.5)].

4. Find out the optimum expansion plan for problem 3 [#DCLF.m; Appendix L: (L.5)].
5. Calculate the number of load flows and the topologies to be considered in each of the hybrid stages, namely, backward, forward and decrease. From the results, calculate the number of load flows for the test network in Sect. 8.6.2 and analyze the relative execution times of various stages.
6. For the network planned in Sect. 8.6.1 (resulting from the hybrid method)
   (a) Find out the system load increase (in %) for which the plan remains robust.
   (b) Assuming an annual load increase of 3%, perform NEP for the system in 6-year time [#Hybridsearch.m; Appendix L: (L.4)].
   (c) Assuming two 3-year periods, repeat part (b) using a quasi-dynamic approach[22] [#Hybridsearch.m; Appendix L: (L.4)].
7. Referring to recent published literature, find out alternative objective functions and constraints for the model developed in Sect. 8.4.

References

Reference [1] is the typical one of most transmission planning studies, whereas [2] reviews the research up to 2003. Test models are covered in [3]. Some mathematical based approaches are given in [4–8]. Non-mathematical based algorithms are quite a few. Some are introduced in [9–12]. Some of these are compared in [13]. If the transmission system comprises of several voltages; the substation configuration is to be determined in combination with transmission network; GEP and TEP are to be analyzed together or a multi-year approach is to carried out, the problem becomes more complex. Some of these issues are covered in [14–17].

1. Villasana R, Garver LL, Salon SJ (1985) Transmission network planning using linear programming. IEEE Trans Power Appar Syst PAS-104(2):349–356
2. Latorre G, Cruz RD, Areiza JM, Villegas A (2003) Classification of publications and models on transmission expansion planning. IEEE Trans Power Syst 18(2):938–946
3. Romero R, Monticelli A, Garcia A, Haffner S (2002) Test systems and mathematical models for transmission network expansion planning. IEE Proc Gener Transm Distrib 149(1):27–36
4. Rider MJ, Garcia AV, Romero R (2008) Transmission system expansion planning by a branch-and-bound algorithm. IET Gener Transm Distrib 2(1):90–99
5. Haffner S, Monticelli A, Garcia A, Mantovani J, Romero R (2000) Branch and bound algorithm for transmission system expansion planning using a transportation model. IEE Proc Gener Transm Distrib 147(3):149–156
6. Alguacil N, Motto AL, Conejo AJ (2003) Transmission expansion planning: a mixed-integer LP approach. IEEE Trans Power Syst 18(3):1070–1077

[22] The quasi dynamic approach is checked for both planning-ahead and planning-back algorithms. In the planning-ahead algorithm, the network is planned from the first year towards the final year. For each year, the planned network of the last year would be considered as the base plan. For the planning-back algorithm, the network is initially planned for the final year. Coming back towards the first year, the lines added for each year would be considered as candidates for the previous year. From the available candidates, the optimal plan is found.

7. Romero R, Monticelli A (1994) A hierarchical decomposition approach for transmission network expansion planning. IEEE Trans Power Syst 9(1):373–380
8. Romero R, Rocha C, Mantovani M, Mantovani JRS (2003) Analysis of heuristic algorithms for the transportation model in static and multistage planning in network expansion systems. IEE Proc Gener Transm Distrib 150(5):521–526
9. Oliveira GC, Costa APC, Binato S (1995) Large scale transmission network planning using optimization and heuristic techniques. IEEE Trans Power Syst 10(4):1828–1834
10. Sum-Im T, Taylor GA, Irving MR, Song YH (2009) Differential evolution algorithm for static and multistage transmission expansion planning. IET Gener Transm Distrib 3(4):365–384
11. Romero R, Rider MJ, Silva IdeJ (2007) Metaheuristic to solve the transmission expansion planning. IEEE Trans Power Syst 22(4):2289–2291
12. Galiana FD, McGillis DT, Marin MA (1992) Expert systems in transmission planning. Proc IEEE 80(5):712–726
13. Gallego RA, Monticelli A, Romero R (1998) Comparative studies on nonconvex optimization methods for transmission network expansion planning. IEEE Trans Power Syst 13(3):822–828
14. Seifi H, Sepasian MS, Haghighat H, Foroud AA, Yousefi GR, Rae S (2007) Multi-voltage approach to long-term network expansion planning. IET Gener Transm Distrib 1(5):826–835
15. Li W, Lu J (2005) Risk evaluation of combinative transmission network and substation configurations and its application in substation planning. IEEE Trans Power Syst 20(2):1144–1150
16. Sepasian MS, Seifi H, Foroud AA, Hatami AR (2009) A multiyear security constrained hybrid generation-transmission expansion planning algorithm including fuel supply costs. IEEE Trans Power Syst 24(3):1609–1618
17. Xie M, Zhong J, Wu FF (2007) Multiyear transmission expansion planning using ordinal optimization. IEEE Trans Power Syst 22(4):1420–1428

Chapter 9
Network Expansion Planning, an Advanced Approach

9.1 Introduction

In Chap. 8, we discussed in some details the basic network expansion problem. We learnt how the problem may be formulated in a case that only one voltage level is involved and our aim is to minimize the total investment cost on transmission lines. In this chapter, we deal with a practical case in which several transmission voltages are involved. We will see how the problem may become more complicated and how it may be solved. Section 9.2 deals with the problem description. The way it can be formulated is shown in Sect. 9.3. The solution algorithm is discussed in Sect. 9.4. A process for reducing the number of candidates is shown in Sect. 9.5. Numerical results are provided in Sect. 9.6.

9.2 Problem Description

As described in Chap. 1, a power grid consists of various voltages, namely, EHV (or UHV), HV, MW and LV; the interconnection between these voltages is established through transformers, grouped in substations (Chap. 7). The grid is normally structured and classified into transmission, sub-transmission and distribution networks. As noted in Chap. 1, in a practical case, various voltages may exist in a level; such as 400 and 230 kV in the transmission level. We will see how the NEP problem may become complicated in such a case.

Figure 9.1 depicts a typical EHV transmission system, which is a modified Graver test system of Chap. 8; its parameters are provided in Appendix F. As shown, substations 1 and 3 are dual voltages (400 and 230 kV) while substations 2, 4 and 5 are shown as being single voltage (230, 230 and 400 kV, respectively). Both the generations and/or the loads may be connected to either voltage; as demonstrated. To make notations easier to follow, we have used a two character figure to represent each voltage level. For instance 32 and 34 represent the 230 and

H. Seifi and M. S. Sepasian, *Electric Power System Planning*, Power Systems,
DOI: 10.1007/978-3-642-17989-1_9, © Springer-Verlag Berlin Heidelberg 2011

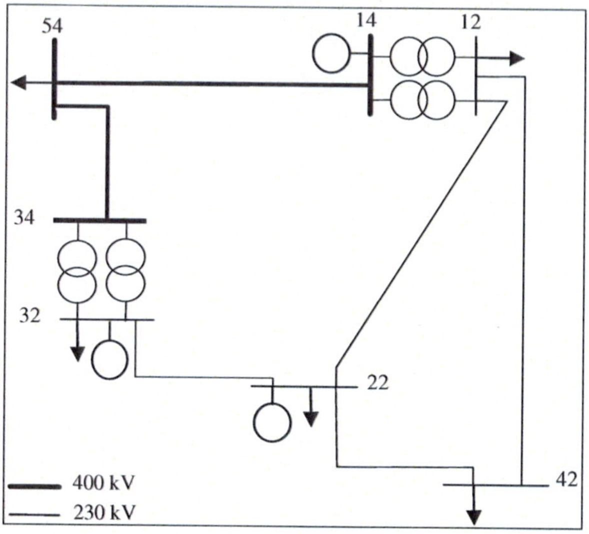

Fig. 9.1 Modified Garver test system

400 kV buses of substation 3, respectively. 54 represents the 400 kV bus of substation 5, etc. Note that in practice, the generations and/or the loads are not directly connected to EHV level buses, but connected through some transformers; their details are not of interest here.

Moreover, the four existing transformers are modeled by their impedances each equal to $R = 0.002$ p.u. and $X = 0.04$ p.u. However, in this section we have not considered their respective capacity limits and contingencies. In other words, we have assumed that they are, always, in service and any contingency on any other elements does not result in a transformer overload. Later on, in our detailed modeling, we consider, transformers, too.

We solved the NEP problem in Chap. 8 for the case in which the loads were increased by 116.5%. The network was designed to be robust both in normal as well as contingency conditions; provided a double circuit transmission line was built from bus 1 to bus 5 and a single circuit one from bus 1 to bus 4 (see Fig. 8.5). If the EHV grid consists of only one voltage level, the solution presented there would be fine. Now what happens, if the presented network there, is in fact the per unit representation of the network in Fig. 9.1? Does it mean that we have to build a double circuit 400 kV line from bus 1 to bus 5 and a 230 kV line from bus 1 to bus 4? Is this solution optimal based on the indices defined in Chap. 8?

Suppose one of our choices is to use a 400 kV line with 1/5 per unit reactance and 1/16 per unit resistance (per km) of the respective values of an available 230 kV line. In terms of the susceptance, it is assumed that B of the 400 kV line is 2.5 times that of the 230 kV line. Moreover assume that its thermal capacity is three times higher than that of the 230 kV line. An alternative solution is shown in

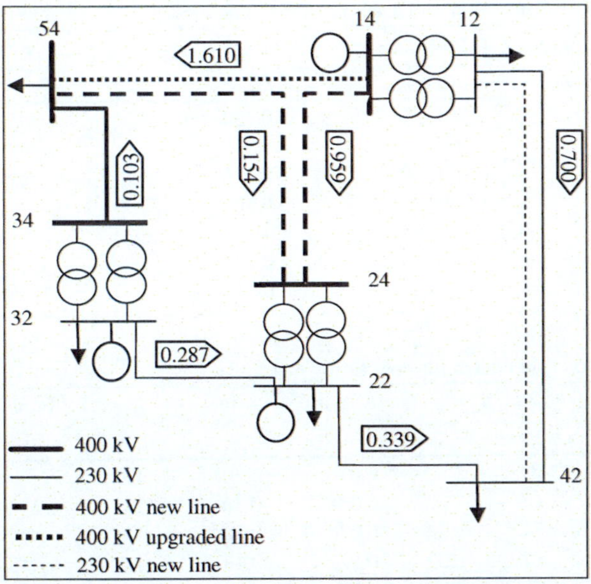

Fig. 9.2 An alternative solution

Table 9.1 N − 1 results

Contingency on line	Flow on line (p.u.)						
	12–42	14–24	14–54	22–32	22–42	24–54	34–54
12–42	0.507[a]	1.067	1.696	−0.307	0.533	−0.219	0.083
14–24	0.820	0.000	2.450	−0.411	0.220	−0.869	−0.021
14–54	0.864	2.406	0.000	−0.033	0.176	1.203	0.357
22–32	0.720	1.062	1.488	0.000	0.320	−0.318	0.390
22–42	1.040	0.770	1.460	−0.250	0.000	−0.040	0.140
24–54	0.712	1.070	1.487	−0.317	0.328	0.000	0.073
34–54	0.694	0.922	1.654	−0.390	0.346	−0.094	0.000

[a] Note that the flows are for the remaining line(s) on the route

Fig. 9.2 in which, the first circuit (14–54) is an upgraded 400 kV line[1] and instead of the second circuit from bus 14 to bus 54, the circuit is drawn to substation 2 (bus 22) through two 400 kV:230 kV transformers. Line 54–24 is a new 400 kV line, while line 14–24 is, in fact, the upgraded (from 230 to 400 kV) earlier line (12–22 in Fig. 9.1), now reconfigured between buses 14 and 24. The results shown in Table 9.1 [#DCLF.m; Appendix L: (L.5)] demonstrate the fact that this alternative is attractive from technical viewpoint. However, it is demonstrated, too, that substation 2 has to be upgraded to a higher voltage (400 kV).

[1] Upgraded from the earlier 400 kV line with a lower capacity to a higher capacity type.

Table 9.2 Generation and load data for the new case

Bus	Load		Generation
	P_D (p.u.)	Q_D (p.u.)	P_G (p.u.)
12	0.520	0.252	–
22	1.560	0.755	0.500
32	0.260	0.126	1.650
42	1.040	0.503	–
14	–	–	1.440
54	1.560	0.755	–
6	0.650	0.315	2.000

Table 9.3 Candidates for connecting bus 6

Line no.	Bus		R (p.u.)	X (p.u.)	Capacity limit (p.u.)
	From	To			
7	62	22	0.075	0.30	1.0
8	62	22	0.075	0.30	1.0
9	62	24	0.075	0.30	1.0
10	62	32	0.12	0.48	1.0

Next, suppose that from the SEP analysis, a new substation 6 would be added to the system; with unknown voltage. Moreover, assume that some generations are retired and some new added; their details are shown in Table 9.2. With due attention to the candidates available for connecting bus 6 to the rest of the network (Table 9.3), some solutions are possible. Two alternatives are shown in Figs. 9.3 and 9.4. The results are summarized in Tables 9.4 and 9.5, respectively [#DCLF.m; Appendix L: (L.5)].

Still there are other points to be considered or have to be observed. For instance, from Fig. 9.4, the solution ends up with 6 connected lines to bus 2.

Normally there may be some limitations on the number of possible connections. Moreover, suppose the planner considers the possible use of an intermediate 400 kV:230 kV substation. In that case, the NEP problem should find its optimal connections to its nearby substations, through either 230 kV and/or 400 kV transmission lines. Even an intermediate substation may be justified, while no voltage conversion happens.[2]

Based on the limitations involved and the possible candidates,[3] the NEP problem should be so formulated that from all technically acceptable solutions, the most economical one is selected as the final choice. The problem formulation is described in Sect. 9.3.

[2] This type of substation is normally referred to a *switching substation*.

[3] In terms of connecting lines in various voltages, voltage upgrading of existing substations, expansions of existing substations by adding new transformers, etc.

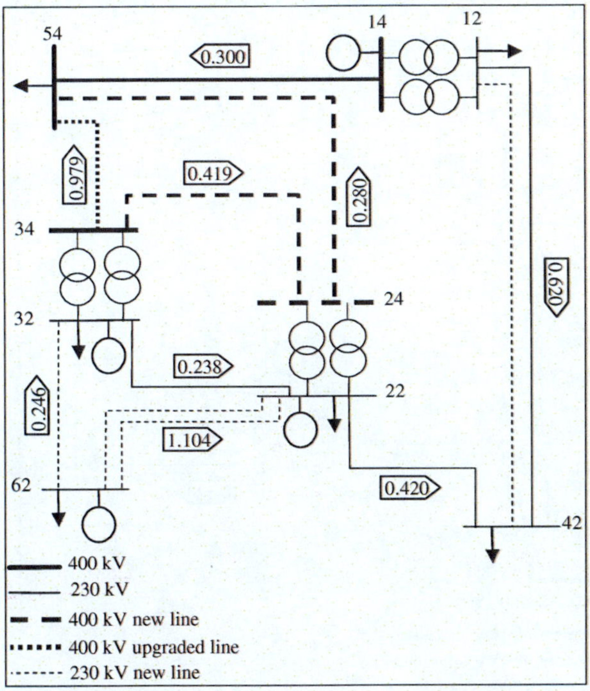

Fig. 9.3 The new test case (scenario 1)

9.3 Problem Formulation

As discussed in Chap. 8, in NEP, the problem is the determination of transmission paths for the system so that the loads are adequately supplied in both normal and N − 1 conditions. For optimum results, the objective function terms as well as the constraints are defined in the following subsections. Before that, however, in Sect. 9.3.1, we briefly review the basic requirements. Following that, the objective function terms and the constraints are described.

9.3.1 Basic Requirements

In this subsection, the basic requirements are discussed.

9.3.1.1 Various Voltage Levels

In a practical case, various transmission voltages are simultaneously in use. For instance, if for a grid, both 400 and 230 kV are available as the transmission media,

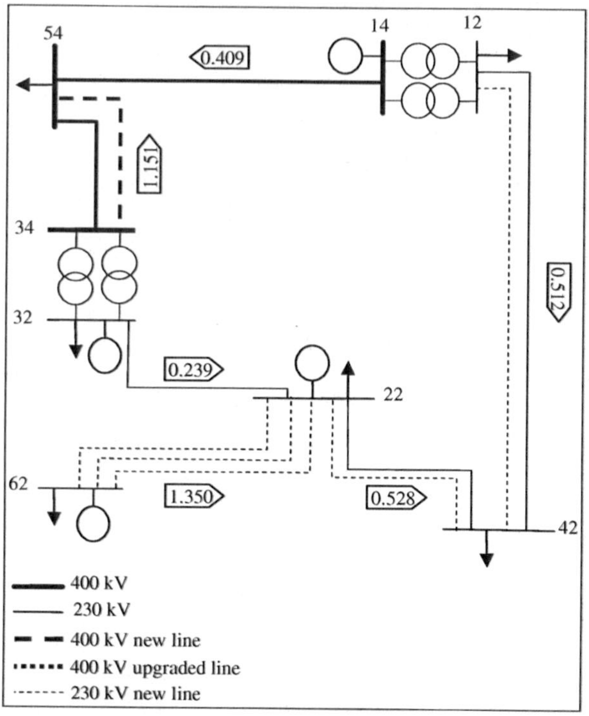

Fig. 9.4 The new test case (scenario 2)

Table 9.4 N − 1 results for the new test case (scenario 1)

Contingency on line	Flow on line (p.u.)								
	14–54	24–54	34–54	34–24	12–42	22–32	22–42	32–62	22–62
14–54	0.000	0.417	1.143	0.309	0.920	−0.196	0.120	−0.259	−1.090
24–54	0.320	0.000	1.240	0.202	0.600	−0.205	0.440	−0.256	−1.094
34–54	0.381	1.179	0.000	1.223	0.540	−0.371	0.501	−0.204	−1.146
34–24	0.283	0.034	1.243	0.000	0.638	−0.356	0.403	−0.209	−1.142
12–42	0.446	0.214	0.900	0.472	0.474[a]	−0.258	0.566	−0.240	−1.110
22–32	0.292	0.231	1.037	0.575	0.628	0.000	0.412	−0.223	−1.128
22–42	−0.120	0.471	1.209	0.266	1.040	−0.180	0.000	−0.264	−1.086
32–62	0.308	0.322	0.930	0.285	0.612	−0.175	0.428	0.000	−1.350
22–62	0.294	0.246	1.020	0.527	0.626	−0.288	0.414	−0.455	−0.959[a]

[a] Note that the flows are for the remaining line(s) on the route

400, 230 and 400 kV:230 kV substations exist in the system (The sub-transmission voltage may be either 63 or 132 kV or similar). The transmission voltage level of a new substation is initially unknown. The proposed NEP algorithm should be able to determine

Table 9.5 N − 1 results for the new test case (scenario 2)

Contingency on line	Flow on line (p.u.)						
	14–54	34–54	34–54[b]	12–42	22–32	22–42	22–62
14–54	0.000	1.300	0.260	0.920	0.170	0.120	−1.350
34–54	0.577	0.000	0.983	0.342	−0.407	0.698	−1.350
34–54[b]	0.417	1.143[a]	0.000	0.504	−0.247	0.536	−1.350
12–42	0.529	0.859	0.172	0.391[a]	−0.359	0.650	−1.350
22–32	0.170	1.158	0.232	0.750	0.000	0.290	−1.350
22–42	0.319	1.034	0.207	0.602	−0.149	0.439[a]	−1.350
22–62	0.409	0.959	0.192	0.512	−0.239	0.528	−1.350[a]

[a] Note that the flows are for the remaining line(s) on the route
[b] Line with lower capacity

- The best voltage level (e.g. 400 or 230 kV) of a new transmission line,
- The voltage level of the new substations [e.g. 400 to 230 kV, 400 kV to (sub-transmission voltage) or 230 kV to (sub-transmission voltage)],
- The possible upgrading of an existing substation [e.g. 230 kV to (sub-transmission voltage)] to a higher voltage level (e.g. 400 to 230 kV).

9.3.1.2 Switching Substations

In practice, switching substations may be economically and technically justified in which no local load is supplied. The algorithm should be so formulated that these possibilities are determined.

9.3.1.3 Line Splitting

Sometimes, it may be economically and technically justified that a substation is fed through a nearby line by splitting the line and connecting its two parts as an *in-feeder* and an *out-feeder* to the substation. These cases should also be observed.

9.3.1.4 System Losses

The system losses were ignored in Chap. 8. Although DCLF formulation is based on considering only the lines reactances (i.e. losses ignored), we have to, somehow, observe the losses on our modeling; as various network plans result in different system losses (or in fact, costs).

9.3.1.5 Substation Limitations

Any substation may have some limitations in terms of the number of connecting lines. This point has to be observed in our modeling, as a more economical solution with connecting lines (to a substation) more than the allowable limit is actually impractical.

9.3.2 Objective Functions

The aim is to minimize the total cost (C_{total}) as shown in (9.1)

$$C_{total} = C_{new\text{-}line} + C_{exp\text{-}sub} + C_{chn\text{-}sub} + C_{up\text{-}sub} + C_{sw\text{-}sub} + C_{sp\text{-}line} + C_{loss} \quad (9.1)$$

where each term is defined below.

(a) $C_{new\text{-}line}$

It is the investment cost for new transmission lines defined as

$$C_{new\text{-}line} = \sum_{i \in Lc} C_L(x_i) L_i \quad (9.2)$$

where L_i is the transmission line length (km) of the *i*th candidate, *Lc* is the set of candidates, x_i is the transmission line type of the *i*th candidate (set of various types such as voltage level, number of bundles and number of circuits) and C_L (x_i) is the investment cost per km for type x_i.

(b) $C_{exp\text{-}sub}$

Due to the expansion of the interconnected grid, some existing substations may require expansion such that the operational limits are not violated. So, an expansion cost ($C_{exp\text{-}sub}$) is incurred as follows

$$C_{exp\text{-}sub} = \sum_{j \in Lt} C_T(y_j) \quad (9.3)$$

where *Lt* is the set of transformer candidates, y_j is the transformer type of the *j*th transformer candidate (various typical transformers available according to the utility practices) and $C_T(y_j)$ is the investment cost for type y_j.

(c) $C_{chn\text{-}sub}$

As already discussed in Chap. 7, the voltage of a new substation was assumed to be known say, 230 or 400 kV; its supply cost was approximately considered in terms of its closest distance to a nearby line. In NEP, we have to calculate this cost accurately. It may happen that, based on the objective function terms and the constraints, such a new substation may have to be upgraded to a higher voltage. If *Nc* represents the set of such new upgraded substations, the upgrading cost has to be observed in our modeling. This cost (C_s) is a function of its carrying loading P_{Dk}. As a result

$$C_{chn\text{-}sub} = \sum_{k \in Nc} C_s(P_{Dk}) \tag{9.4}$$

(d) $C_{up\text{-}sub}$

As already noted, the voltage level of an existing substation may be upgraded to a higher level if technically and economically justified. The cost of upgrading ($C_{up\text{-}sub}$) is defined as

$$C_{up\text{-}sub} = \sum_{l \in Ns} C_u(TP_l) \tag{9.5}$$

where Ns is the set of multi-voltage substations (As in NEP, a substation voltage may be changed from one level (say, 230 kV) to two levels (say, 230 and 400 kV), Ns consists of these substations, both existing and new, so that NEP should determine their respective new transformers costs).

TP_l is the power transmitted through substation l and $C_u(TP_l)$ is the upgrading cost for the substation carrying power TP_l.

(e) $C_{sw\text{-}sub}$

The algorithm developed in Chap. 7 was looking for finding the load carrying substations. However, as already noted, sometimes switching substations (nominated by the user) may be justified by which no local load is supplied. It may be either single voltage or dual voltage. In the former, the costs do not involve those due to the transformers while for the latter, they have to be considered. As a result

$$C_{sw\text{-}sub} = \sum_{n \in Nw} \left(C^f_{swn} + C^t_{sw}(TP_n)\right) \tag{9.6}$$

where Nw represents the set of switching substations, selected from available candidates (nominated by the user). C^f_{swn} is the cost of substation n, irrespective of voltage transformation[4] and $C^t_{sw}(TP_n)$ is the cost of transformers required, dependent on the carrying loading (TP_n) on the substation.

(f) $C_{sp\text{-}line}$

As discussed earlier, one way to supply a substation is to split a nearby line as input/output to that substation. The cost of such a procedure is as follows

$$C_{sp\text{-}line} = \sum_{m \in Nsp} C_{spm} \tag{9.7}$$

where Nsp is the set of splitting options selected from available candidates. C_{spm} is the cost of such splitting (m).

(g) C_{loss}

The total active power losses (C_{loss}) are determined as

$$C_{loss} = CP_{loss}(A + B + C) \tag{9.8}$$

[4] It represents the costs of land, protection systems, etc. which obviously depend on the voltage involved.

where

$$A = \left(\sum_{j \in Lt} R_t(y_j)\left(\frac{P_j}{\cos\phi}\right)^2\right) \quad \text{Losses of new transformers}$$

$$B = \left(\sum_{i \in Lc} R_l(x_i)L_i\left(\frac{P_i}{\cos\phi}\right)^2\right) \quad \text{Losses of new lines}$$

$$C = \left(\sum_{k \in Le} R_k\left(\frac{P_k}{\cos\phi}\right)^2\right) \quad \text{Losses of existing transformers and lines}$$

$R_t(y_j)$ is the resistance of transformer type y in position j, $R_l(x_i)$ is the per unit length resistance of line type x in position i, R_k is the resistance of existing transformer and/or line k, Le is the set of existing lines and transformers, CP_{loss} is the cost of per unit losses, P_j is the active power flow of a new transformer, j, P_i is the active power flow of a new line, i, P_k is the active power flow of an existing transformer or line k and $cos\phi$ is an average power factor.

Various constraints should be met. Some of the constraints are already described in Chap. 8. However, they are repeated here for convenience. Others are specific to this chapter.

9.3.3 Constraints

Various constraints should be met during the solution process, as detailed in this subsection.

9.3.3.1 Load Flow Equations

For large-scale power systems, it is of normal practice to use DC load flow equations; otherwise the solution time may be exceptionally high. Moreover, the planner avoids any anxiety about voltage problems and possible convergence difficulties. It is obvious that in the final stage, AC load flow should be performed to have an acceptable voltage profile during normal as well as contingency conditions.

The DC load flow equations are in the form of (9.9).

$$\begin{aligned} &\sum_{j=1}^{N} B_{ij}(\theta_i - \theta_j) = P_{Gi} - P_{Di} \quad \forall i \subset n \\ &\sum_{j=1}^{N} B_{ij}^{m}(\theta_i^{m} - \theta_j^{m}) = P_{Gi}^{m} - P_{Di} \quad \forall i \subset n \cap m \subset C \end{aligned} \tag{9.9}$$

where θ_i, θ_j are the voltage phase angles of buses i and j, respectively, B_{ij} is the imaginary part of the element ij of the admittance matrix; P_{Gi} is the power generation at bus i, P_{Di} is the power demand at bus i, and n is the set of system buses. The index m shows the contingency parameters and variables. C is the rest of contingencies. N is the system number of buses.

9.3.3.2 Transmission Limits

For each of the transmission elements (lines and transformers), power transfer should not violate its rating during both normal and contingency (N − 1, in this book) conditions, so

$$\begin{aligned} & b_k(\theta_i - \theta_j) \leq \overline{P}_k^{No} \quad \forall k \in (Lc + Lt + Le) \\ & b_k^m(\theta_i^m - \theta_j^m) \leq \overline{P}_k^{Co} \quad \forall k \in (Lc + Lt + Le) \cap m \end{aligned} \tag{9.10}$$

where $\overline{P}_k^{No}$, $\overline{P}_k^{Co}$ are the element k ratings during normal and contingency conditions, respectively; θ_i, θ_j are the voltage phase angles of line k during normal conditions; θ_i^m, θ_j^m are the voltage phase angles of line k following contingency m; C is the set of contingencies, and Lc, Lt, Le are as defined earlier.

9.3.3.3 Substation Limitations

A new as well as an already existing substation may have some limitations in terms of the number of possible connections (input or output lines/feeders). Hence

$$\sum_{i \in Lc} M_i^j \leq \overline{M}^j \qquad \forall j \in n \tag{9.11}$$

where $\overline{M}^j$ is the maximum limit of the number of connecting lines to bus j; M_i^j is a counter set = 1 if line i is connected to bus j, otherwise zero, and n is as defined earlier.

9.3.3.4 Islanding Conditions

The systems should be so planned that no island appears during normal and contingency conditions. So

$$N_{island} = 0 \tag{9.12}$$

As a line contingency (outage) is modeled in the algorithm by choosing a very high value for the line reactance, an islanding is detected by checking the phase angle difference across a line to be a large number. This happens due to the fact that the far end of the line terminates at a load bus.

9.4 Solution Methodology

The solution methodologies may be the ones described in Sect. 8.5. Here we focus, only, on the hybrid approach, outlined in Sect. 8.5.2.4.[5] However, as described below, since for a practical system, various transmission alternatives (for instance, in terms of capacity and voltage) are available between any two substations, we may note the following observations.

Assume that between substations A and B, the alternative links may be of the following types

- 230 kV (single or double-circuit), if both A and B are existing substations, with 230 kV primary voltages.
- 400 kV (single or double-circuit), if both A and B are existing substations, with 230 kV primary voltages (The secondary may be say 63 kV); however, both of them are upgraded to 400 kV (So that the upgraded substation would be 400 to 230 to 63 kV).
- 230 kV (single or double-circuit), if both A and B are new substations and 230 kV is chosen (by the algorithm) as the favorable choice for the primary. The secondary is at sub-transmission level.
- 400 kV (single or double-circuit), if both A and B are new substations and 400 kV is chosen (by the algorithm) as the favorable choice.
- 230 kV (single or double-circuit), if A (or B) is an existing substation with 230 kV primary voltage and B (or A) is a new substation; its voltage is chosen to be 230 kV (by the algorithm) as the favorable choice. The secondary is at sub-transmission level.
- The same as above with 400 kV, instead of 230 kV.

Between any of these two substations, initially all alternatives are simply defined. These alternatives are ranked according to their capacities. For instance, if between substations A and B, both 230 and 400 kV lines are assumed as candidates, the highest capacities (e.g. double-circuit, 2-bundle for 230 kV and double-circuit, 4-bundle for 400 kV) are chosen as initial candidates (Later on, in the *decrease* stage, lower capacities are tried). Using the higher ranked large capacity alternatives normally results in the least number of right-of-way requirements for the network (practically favorable). The solution procedure is the same as the one discussed in Sect. 8.5.2.4. However, this time, in (8.6), (8.2) should be replaced by (9.1).

9.5 Candidate Selection

For any NEP problem, we should first select a number of candidate paths. Even by using the three stage algorithm, the solution time may be too high, if the number of

---

[5] The reader may follow other alternatives. However, they may only be applied for small scale systems.

candidates is large. To reduce the solution time, three mechanisms are employed to reduce the number of candidates

- *All Possible Candidates (APC)*
  In this stage, all possible candidates between any of two substations (either existing or new) are generated.
- *All Feasible Candidates (AFC)*
  The non-feasible solutions (due to environmental limitations, constraints violations, and so on) are then removed. AFC consists of feasible paths, by which all constraints are met during normal as well as contingency conditions.
- *All Good Candidates (AGC)*
  At this stage, the aim is to select the most attractive candidates. In fact, adding or removing a candidate may have the following three effects

  – Connects one or more buses to the system.
  – Reduces overloads on other elements.
  – Improves power transfer profile of the network.

The candidates of option 1 are considered to be the most attractive as they remove islanding conditions. If a candidate removes any overload on other elements, it is considered as the next attractive choice. The next options (candidates) improve the power transfer profile of the network (the least attractive).

For illustration purposes, suppose there are 1000 candidates. Initially all of them are added to the network and checked to find out removing which one of them results in an islanding. There may be, say, 50 of these lines (call it list *L1*). Then with list *L1* added to the network, add each of the remaining candidates (950) to the network and calculate the following index in each case

$$\text{Candidate Evaluation Function } 1(CEF1) = \sum_{i \in L} OC_i \tag{9.13}$$

where L is the set of candidates; OC_i is the overloaded capacity $= LNC_i - ACL_i$ of element i if $ACL_i \leq LNC_i$ (LNC is the loading during normal conditions and ACL is the available capacity limit) and $OC_i = 0$ if $ACL_i > LNC_i$.

Then select a number of the most attractive candidates (i.e. candidates with the lowest *CEF1*), in terms of reducing overloading conditions (say another 100 candidates); call it list *L2*. It should be mentioned that the number of the candidates of list *L2* depends on the computer facilities available. The lager it is chosen by the user, the better results will be achieved.

With lists *L1* and *L2* added to the network (150 candidates), add each of the remaining candidates (850) to the network and calculate the following index in each case

$$\text{Candidate Evaluation Function } 2(CEF2) = \sum_{i \in L} FC_i \tag{9.14}$$

where free capacity $FC_i = ACL_i - LNC_i$ if $ACL_i \geq LNC_i$ and $FC_i = 0$ if $ACL_i < LNC_i$. ACL_i, LNC_i and L are as defined earlier.

Then select a number of the most attractive candidates (i.e. the candidates with the highest *CEF2*), in terms of improving power transfer profile of the network (say another 125 candidates) (call it list *L3*). Again, this number depends on the computer facilities available.

Finally, by combining the high-ranked candidates *L1*, *L2* and *L3*, the final list of the candidates is formed.

9.6 Numerical Results

To assess the capability of the proposed hybrid approach and the detailed modeling, a 77-bus test system as depicted in Fig. 9.5 is employed. It should be mentioned, however, that in our test example, we have not considered some of the objective function terms ($C_{chn\text{-}sub}$, $C_{sw\text{-}sub}$, $C_{sp\text{-}line}$ and C_{loss}).

This is a dual voltage network with the detailed information as outlined in Appendix J. The general data is provided in Table 9.6.

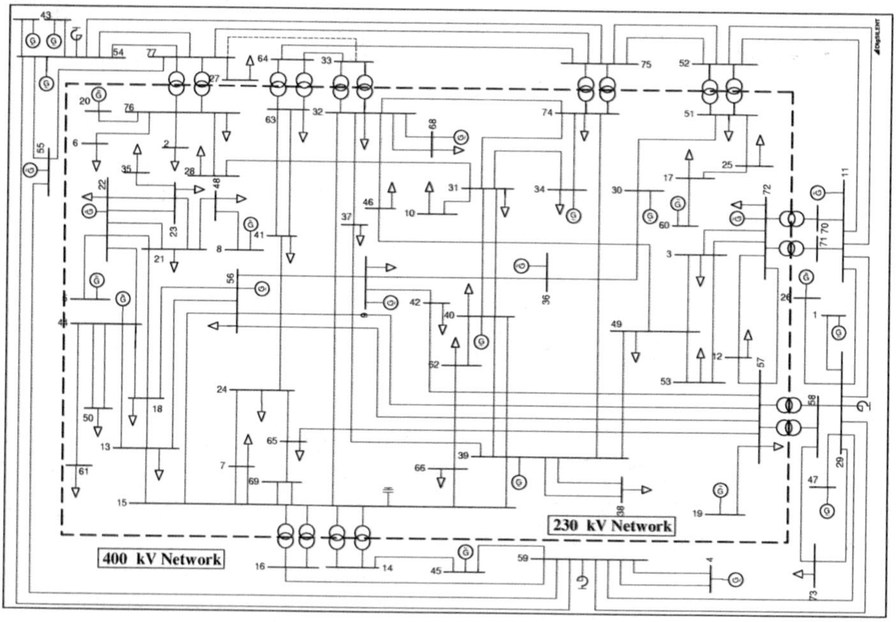

Fig. 9.5 77-bus test system

Table 9.6 General data of 77-bus network

No. of buses	77	Total load (MW)	8,210
No. of generation buses	26	No. of lines	137

Table 9.7 A summary of the violations with no expansion

The number of contingencies resulting in islands	11
Sum of lines overloads in normal conditions (per unit)	3.93
Sum of lines overloads in generation contingencies (10% reductions) (per unit)	0.00
Sum of 230 kV lines overloads (for all contingencies) in contingency conditions (per unit)	27.54
Sum of 400 kV lines overloads (for all contingencies) in contingency conditions (per unit)	5.75
Sum of transformers overloads (for all contingencies) in contingency conditions (per unit)	13.69

Table 9.8 Parameters of the candidate elements

	Parameters			
	R (p.u.)	X (p.u.)	S (p.u.)	Cost
Single 230 kV line	0.00025	0.001	1.1	R 150,000/km
Double 230 kV line	0.000125	0.0005	2.2	R 250,000/km
Single 400 kV line	0.000015	0.00025	3.3	R 250,000/km
Double 400 kV line	0.0000075	0.000125	6.6	R 400,000/km
400 kV:230 kV transformer	0.013	0.257	2.75	R 12500,000

DCLF results in observing some violations in either normal or contingency conditions, a summary of which is provided in Table 9.7. The contingencies are assumed to be tripping out of any single transmission line or 10% reduction in the generation level of any generation bus. As shown, the system is confronted with some violations for which NEP has to provide some types of solutions.

In this test case, we have not considered any switching substations. Moreover, splitting of the lines and the losses are also not considered.

In order to perform NEP, some candidate lines should be initially selected. To do so, in our test example, we have assumed that any route between any two of the buses with a straight length of less than 200 km may be a choice. The geographical information of the buses is provided in Appendix J.

With the technical data provided in Table 9.8 for the candidate elements, an overall 673 candidate paths (1346 candidate lines; each path with a 400 kV line and a 230 kV line) and 11 transformer candidates are selected for further process.

The hybrid algorithm is then applied. The steps are summarized in Table 9.9. The details are shown in Appendix K. As shown, following the backward stage, from 1346 of 230 kV and 400 kV candidate double circuit lines and also 11 candidate transformers, eleven 400 kV paths (22 lines) and twenty nine 230 kV paths (58 lines) are justified to remove all violations for the normal conditions. The network is made robust in response to all contingencies, provided extra 400 kV (10 paths; 20 lines) and 230 kV (6 paths; 12 lines) lines are added. These are justified using the forward stage. In none of the above stages, a transformer candidate is justified. The decrease stage is then tried. At this stage, four 400 kV and twelve 230 kV double circuit lines are reduced to single circuit lines.

Table 9.9 A summary of the results

Algorithm stage	Number of elements, justified					Total length, justified				$C_{new\text{-}line}$ (10^3 R)	$C_{exp\text{-}sub}$ (10^3 R)	$C_{up\text{-}sub}$ (10^3 R)
	Single 230 kV	Double 230 kV	Single 400 kV	Double 400 kV	Transformer 400 kV:230 kV	Single 230 kV	Double 230 kV	Single 400 kV	Double 400 kV			
Backward	0	29	0	11	0	0	264.8	0	75.5	96431	0.0	287500
Forward	0	35	0	21	0	0	658.4	0	591.0	401016	0.0	362500
Decrease	12	21	4	15	0	312.8	309.8	216.8	357.3	321461	0.0	362500

It should be mentioned that at each stage, based on the justified 400 and 230 kV lines, some single voltage substations may be required to be upgraded to 400 kV:230 kV type. The capacity can be determined. The costs of the lines ($C_{new\text{-}line}$) are also shown. The total cost of the lines would be R 321.461 $\times$ 10^6.

Problems

1. From available resources in your area of living, prepare a table similar to Table 9.8 on various parameters (Resistance, reactance, susceptance, thermal capacity, construction cost, etc.) of existing HV, EHV and UHV transmission lines.
2. For the Garver test system with the modified load (with the details given in Fig. 9.2 and Table 9.2), with due attention to (9.13) and (9.14), determine APC and AGC. For simplicity, assume the system to be single voltage [#DCLF.m; Appendix L: (L.5)].
3. For problem 2 [#DCLF.m; Appendix L: (L.5)]
   (a) Suggest a third scenario.
   (b) Comparing the scenario suggested above with those scenarios within the chapter, find out the transmission lines construction costs.
   (c) Repeat (b), provided the substations costs (both expanding and new) are also observed.
   (d) Repeat (b), if the cost of the losses is also considered.
   (e) Comparing (a) through (d), select the optimum plan.
4. Using [#Hybridsearch.m; Appendix L: (L.4)] and for problem 2
   (a) Find out an optimal plan based on minimization of the construction cost.
   (b) Compare the optimal plan, in terms of the transmission lines construction costs, substation costs and the losses with those scenarios of problem 3.

References

The references of this chapter are same as the references of Chap. 8. Reference [1] is the typical reference of most transmission planning studies. Reference [2] reviews the research up to 2003. Test models are covered in [3]. Some mathematical based approaches are given in [4], [5–8]. Non-mathematical based algorithms are quite a few. Some are introduced in [9–12]. Some of these are compared in [13]. If the transmission system comprises of several voltages; the substation configuration is to be determined in combination with transmission network; GEP and TEP are to be analyzed together or a multi-year approach is to carried out, the problem becomes more complex. Some of these issues are covered in [14–17].

1. Villasana R, Garver LL, Salon SJ (1985) Transmission network planning using linear programming. IEEE Trans Power Apparatus Syst PAS-104(2):349–356
2. Latorre G, Cruz RD, Areiza JM, Villegas A (2003) Classification of publications and models on transmission expansion planning. IEEE Trans Power Syst 18(2):938–946

3. Romero R, Monticelli A, Garcia A, Haffner S (2002) Test systems and mathematical models for transmission network expansion planning. IEE Proc Gener Transm Distrib 149(1):27–36
4. Rider MJ, Garcia AV, Romero R (2008) Transmission system expansion planning by a branch-and-bound algorithm. IET Gener Transm Distrib 2(1):90–99
5. Haffner S, Monticelli A, Garcia A, Mantovani J, Romero R (2000) Branch and bound algorithm for transmission system expansion planning using a transportation model. IEE Proc Gener Transm Distrib 147(3):149–156
6. Alguacil N, Motto AL, Conejo AJ (2003) Transmission expansion planning: a mixed-integer LP approach. IEEE Trans Power Syst 18(3):1070–1077
7. Romero R, Monticelli A (1994) A hierarchical decomposition approach for transmission network expansion planning. IEEE Trans Power Syst 9(1):373–380
8. Romero R, Rocha C, Mantovani M, Mantovani JRS (2003) Analysis of heuristic algorithms for the transportation model in static and multistage planning in network expansion systems. IEE Proc Gener Transm Distrib 150(5):521–526
9. Oliveira GC, Costa APC, Binato S (1995) Large scale transmission network planning using optimization and heuristic techniques. IEEE Trans Power Syst 10(4):1828–1834
10. Sum-Im T, Taylor GA, Irving MR, Song YH (2009) Differential evolution algorithm for static and multistage transmission expansion planning. IET Gener Transm Distrib 3(4):365–384
11. Romero R, Rider MJ, IdeJ Silva (2007) Metaheuristic to solve the transmission expansion planning. IEEE Trans Power Syst 22(4):2289–2291
12. Galiana FD, McGillis DT, Marin MA (1992) Expert systems in transmission planning. Proc IEEE 80(5):712–726
13. Gallego RA, Monticelli A, Romero R (1998) Comparative studies on nonconvex optimization methods for transmission network expansion planning. IEEE Trans Power Syst 13(3):822–828
14. Seifi H, Sepasian MS, Haghighat H, Foroud AA, Yousefi GR, Rae S (2007) Multi-voltage approach to long-term network expansion planning. IET Gener Transm Distrib 1(5):826–835
15. Li W, Lu J (2005) Risk evaluation of combinative transmission network and substation configurations and its application in substation planning. IEEE Trans Power Syst 20(2):1144–1150
16. Sepasian MS, Seifi H, Foroud AA, Hatami AR (2009) A multiyear security constrained hybrid generation-transmission expansion planning algorithm including fuel supply costs. IEEE Trans Power Syst 24(3):1609–1618
17. Xie M, Zhong J, Wu FF (2007) Multiyear transmission expansion planning using ordinal optimization. IEEE Trans Power Syst 22(4):1420–1428

Chapter 10
Reactive Power Planning

10.1 Introduction

We have, so far, covered GEP, SEP and NEP so that the system planned is capable of meeting the loads in both normal and N−1 conditions. However, as detailed in Chaps. 8 and 9, DCLF was used as the basic governing equations, due to the reasons cited there. Obviously, in a practical case, the assumptions on which DCLF equations are based, are not strictly valid. For instance, the flat voltage assumption and reactive power ignorance may lead to some results, a bit far from the actual conditions. As a result, we have to follow a detailed ACLF analysis to make sure that the system performance is acceptable from those senses, too. It is apparent that looking at the voltage problem of a system is not an easy task, at all, as several related aspects of the problem, such as, voltage profile and stability should also be covered. Moreover, the allocation and sizing of reactive power resources, as the main control devices affecting the voltage conditions, should also be investigated. In a power system context, these aspects are studied in a so called *Reactive Power Planning* (RPP) problem. We assume here that the reader is already familiar with the studies carried out in a basic *power system analysis* course. Instead, we will focus on some practical aspects of ACLF.[1] Moreover, we study, in some details, how a RPP problem may be formulated as an optimization problem by which, reactive power resources may be allocated, while voltage performance conditions are optimized. Initially, we will discuss the voltage performance issues in Sect. 10.2. Following that, we briefly review some aspects of the problem in Sect. 10.3. The optimization problem formulation and numerical results are then provided in Sects. 10.4 and 10.5, respectively.

[1] The interested reader may refer to the list of the references at the end of the chapter.

H. Seifi and M. S. Sepasian, *Electric Power System Planning*, Power Systems,
DOI: 10.1007/978-3-642-17989-1_10,

10.2 Voltage Performance of a System

Over the years, voltage performance of a power system has received attention from both analysis and improvement points of view. Although voltage magnitudes are, normally, of main concern, during the last, perhaps, two decades, voltage stability is also received attention in literature. In this section, we try to briefly, differentiate between these two aspects of voltage performance of a system. Following that, we review some of the indices which may be used for each case.

10.2.1 Voltage Profile

A good word for acceptable voltage magnitudes of a system buses is *voltage profile*. Normally a voltage magnitude of 1.0 p.u. is considered to be favorable. For PQ (load) buses, in practical conditions, the voltages may not be kept strictly at this value. A range of say 0.95–1.05 p.u. may be considered acceptable. A generation bus (PV bus), is considered to be a voltage controlled bus and its voltage is set by the operators. The reactive power of a generating unit is controlled by changing its reference set point.

An index is constructed to show an acceptable performance for voltage profile. The following index, P_{prof}, is considered in this chapter

$$
\begin{aligned}
P_{prof} &= \sum_{i=1}^{N} (V_i - V_i^{set})^2 \\
V_{i,set} &= \begin{cases} 1.0 & i \in \text{PQ buses} \\ V_{set\,point} & i \in \text{PV buses} \end{cases}
\end{aligned}
\tag{10.1}
$$

where V_i is the voltage magnitude of bus i, V_i^{set} is the reference voltage of bus i and N is the system number of buses.

The sum may be calculated for all PV and PQ buses. In other words, if all PQ bus voltage magnitudes are 1.0 p.u. and PV bus voltages are held at their respective set points, the index would be zero. Further the voltage magnitudes are from their set points, the higher the index would be. As a result, a lower P_{prof} is considered favorable. This index can be easily calculated if an AC load flow is performed.

10.2.2 Voltage Stability

Let us assume that for a typical power system, the voltage magnitude of a specific bus is 1.0 p.u., once the bus apparent power is 2.0 + j1.0 p.u. Now assume that the load (both P and Q) is increased by 20% to 2.4 + j1.2 p.u. If an ACLF is

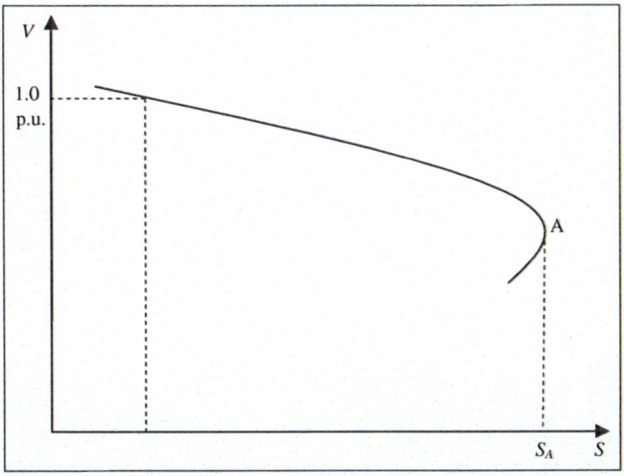

Fig. 10.1 S–V curve

performed, the bus voltage may be reduced to 0.95 p.u. What happens if we continue increasing the load? A simple trajectory for the case is shown in Fig. 10.1. After a certain point (point A), no solution may be found by running an ACLF. The system is considered to be voltage unstable for any load higher than S_A. This curve is commonly referred to S–V curve in power system terminology.

Instead of *S*, either, *P* or *Q* may be increased; and Q–V or P–V curves generated. Even if S is increased, the voltage performance may be drawn in terms of *Q* or *P*. Moreover, instead of increasing the load of a specific bus, the load of the entire system may be increased. These aspects are normally studied in the so called *static* voltage stability analysis (as opposed to *dynamic* type[2]). The static term is used as the approach followed in V-curves generations is based on *algebraic* load flow equations. The dynamic type is based on detailed differential equations, beyond the scope of this book.

Lets us come back to a typical Q–V curve for two cases (Fig. 10.2). In both cases, the voltage of operating point is 1.0 p.u. In other word, the voltage profiles of both cases are considered to be the same and acceptable. However, in case I, the distance to the nose point (the so called *critical point*[3]) is lower. In other words, the voltage stability performance of case II is better. This distance may be considered as the relative merit of voltage stability performance, denoted by P_{stab}. To find P_{stab} for the whole system, the reactive power loads of all buses are proportionally increased until the nose point is reached for the weakest bus. The total reactive power increase is considered as P_{stab}.[4]

[2] See the list of the references at the end of the chapter.

[3] Also called *collapse* point.

[4] Other indices may also be used. For further details, see the list of the references at the end of this chapter.

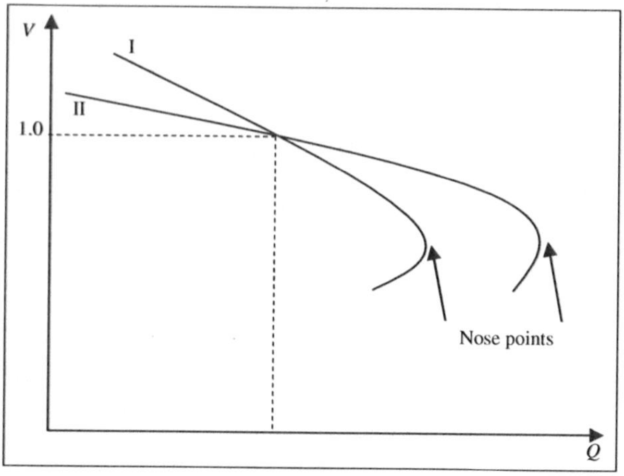

Fig. 10.2 Q–V curve

10.2.3 Voltage Performance Control Parameters

The following actions may affect the voltage performance (both profile and stability) of a system

- Changing the taps of tap changing transformers,
- Changing the voltage set points of voltage controlled buses (PV buses),
- Switching *in* or *out* of capacitors and/or reactors, or any reactive power resource.

These options may be employed by the system operator to improve the voltage performance in various operating conditions. As power system planning is the main concern of this book, we will focus mainly on the third option, and leave the first two unchanged for operational performances. Later on, we develop an optimization problem in which reactive power resources may be allocated and sized. The reactive power resources are introduced in Sect. 10.2.4. Through some numerical examples, the problem description is provided in Sect. 10.3.

10.2.4 Static Versus Dynamic Reactive Power Resources

The reader is familiar with the basic elements of reactive power resources, namely, capacitors and reactors. A capacitor may generate reactive power while a reactor, absorbs reactive power. The reactive power generated/absorbed by a capacitor/reactor, is equal[5] to V^2/X. Its generation is fixed and proportional to X, but the voltage of its connecting bus cannot be directly controlled. In other words, a bus

[5] $X = \frac{1}{\omega C}$ for a capacitor and $X = \omega L$ for a reactor.

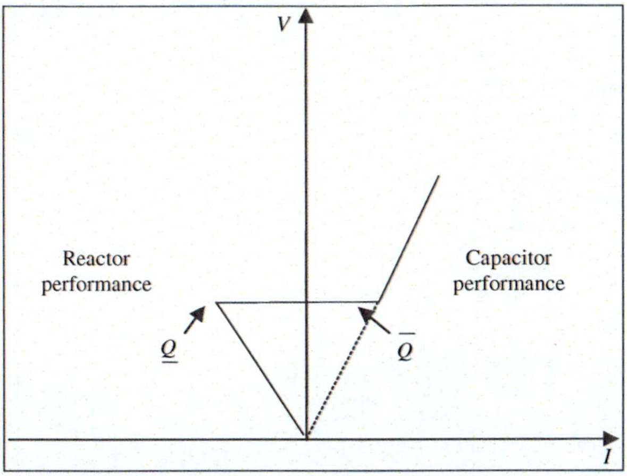

Fig. 10.3 V–I characteristics of an SVC

with a connected capacitor/reactor is, in fact, a PQ bus. Besides being fixed, its reactive power generation can not be, instantly, changed. These types of reactive power resources are named as *static* resources.

Now consider a PV bus in which its voltage may be kept fixed at a specified value. A simple example is a bus with a connecting generator (P, nonzero) or with a synchronous condenser.[6] From a load flow analysis, the reactive power generation, Q, is determined. It is generated/absorbed by the generator or the condenser. Provided it is within the reactive power capability of the resource, it may be generated/absorbed, instantly, while its value is dictated by the system conditions and not fixed. These types of reactive power resources are considered as *dynamic* types.

Another type of a dynamic resource is an SVC[7] with an almost instantaneous response and a V–I characteristics such as the one shown in Fig. 10.3. Within its reactive power capability range, its voltage is fixed. Outside the range, it behaves as either a capacitor (more than $\bar{Q}$) or a reactor (lower than $\underline{Q}$).

Why do we have to use a dynamic resource, while it is a more expensive element in comparison with a static type?

Suppose that the voltage profile of a system is acceptable for normal conditions. Now if a contingency (such as a line outage) happens, the voltages on some specific buses may drop to unacceptable values, even though the reactive power generations of some PV buses are increased. One way to overcome the problem is to switch in a capacitor, if available, at the problematic buses. The difficulty,

---

[6] See the list of the references at the end of this chapter.

[7] Static Var Compensator (For further details, see the list of the references at the end of this chapter). Note that although *static* is used in its name, an SVC is considered to be a dynamic resource due to the explanations cited above. The term static is used here to show that an SVC does not have any moving element.

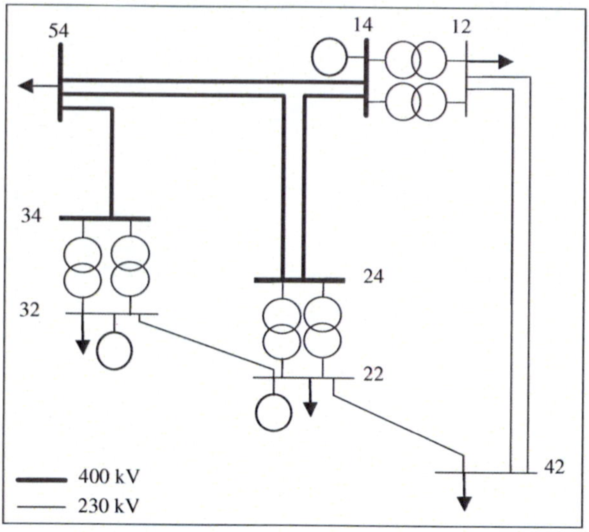

Fig. 10.4 A typical case

however, is the fact that switching is not instantaneous. We may not able to tolerate such a condition even for some milliseconds.

Now consider a situation even worse. Suppose that after the aforementioned contingency, the load flow does not converge at all. In other words, following the contingency in question, the system is voltage unstable. The only way to prevent such an undesirable condition is to provide an *instant* control action which may be able to solve the problem. Obviously, an SVC may provide a solution, but a capacitor/reactor may not.

Briefly speaking, both static and dynamic resources affect voltage profile as well as voltage stability of a system. However, for the studies referred to in this chapter, static types are employed for acceptable voltage profile and stability during normal conditions, while dynamic types (such as an SVC[8]) are employed for acceptable performances in response to contingencies (N$-$1 in our examples).

10.3 Problem Description

Let us consider the same network of Fig. 9.2 of Chap. 9 (repeated here as Fig. 10.4). The NEP problem was performed based on DCLF formulation. With the additional data as detailed in Table 10.1, ACLF is performed with the results

[8] Hereon, in terms of a dynamic type, a so called Reactive Power Compensator (RPC) terminology is used to demonstrate a compensator with instant control action. A PV bus with $P = 0$ and specified $\overline{Q}$ and $\underline{Q}$ is used to model its response.

Table 10.1 Additional data

Bus	Reactor (p.u.)	Generation		
		$\overline{Q}$ (p.u.)	$\underline{Q}$ (p.u.)	V_{set} (p.u.)
12	1.0	–	–	–
14	0.5	5.0	−5.0	1.0
22	0.5	0.4	−0.4	1.0
24	1.0	–	–	–
32	–	0.45	−0.45	1.0
34	–	–	–	–
42	1.0	–	–	–
54	0.5	–	–	–

Table 10.2 ACLF results

Bus	Voltage		Generation	
	V (p.u.)	Angle (rad.)	P_G (p.u.)	Q_G (p.u.)
12	0.995	−0.025	–	–
14	1.000	0.000	3.868	1.376
22	1.000	−0.101	0.500	0.187
24	1.014	−0.079	–	–
32	1.000	−0.041	0.650	0.393
34	1.005	−0.043	–	–
42	0.934	−0.251	–	–
54	1.011	−0.066	–	–

Table 10.3 Comparison of ACLF and DCLF

Element	AC load flow			DC load flow	Difference
	P (p.u.)	Q[a] (p.u.)	S (p.u.)	P (p.u.)	ΔP/P (%)
12–42	0.722	0.103	0.730	0.700	−3.14
14–54	1.636	−0.351	1.673	1.610	−1.62
22–32	−0.283	0.080	0.294	−0.287	1.39
22–42	0.373	0.098	0.386	0.339	−10.03
34–54	0.102	−0.056	0.116	0.103	0.97
14–12	1.243	0.188	1.257	1.22	−1.88
32–34	0.102	−0.256	0.276	0.104	1.92
22–24	−1.150	−0.647	1.319	−1.112	−3.42
14–24	0.989	−0.213	1.012	0.959	−3.13
24–54	−0.169	0.052	0.177	−0.154	9.74

[a] It should be mentioned that the values of *Q* reported in this chapter are the values given after the line charging is accounted for

shown in Table 10.2 [#ACLF.m; Appendix L: (L.6)]. A comparison between ACLF and DCLF results (Table 10.3) shows that the differences are quite small in terms of the active power flows through the elements.

Table 10.4 Results for the base case

Contingency on element	Voltage profile index	Voltage stability index (%)
12–42	No convergence	No convergence
14–54	0.0062	15.00
22–32	0.0044	30.05
22–42	No convergence	No convergence
34–54	0.0044	41.50
14–12	0.0049	14.00
34–32	0.0044	45.50
24–22	0.0045	33.00
14–24	No convergence	No convergence
24–54	0.0084	5.50

Table 10.5 Results with a ±0.5 p.u. reactive power compensator addition

Contingency on element	Voltage profile index	Voltage stability index (%)
12–42	No convergence	No convergence
14–54	0.000	24.00
22–32	0.000	33.00
22–42	No convergence	No convergence
34–54	0.000	34.50
14–12	0.000	34.50
34–32	0.000	35.50
24–22	0.000	32.00
14–24	0.000	12.00
24–54	0.000	17.50

Now we calculate P_{prof} and P_{stab}, based on the procedure discussed in Sect. 10.2. With $\underline{V} = 0.95$ p.u. and $\overline{V} = 1.05$ p.u., $P_{prof} = 0.004$. To calculate P_{stab}, we increase the active powers of all load buses gradually until the load flow diverges. In doing so, we increase the reactive loads of the buses in such a way that power factors remain unchanged. Moreover, we assume that the active power increases are compensated by the slack bus (bus 14). In this way, P_{stab} is found to be 1.45 which shows that if the total load of the system increases by 45%, the system encounters difficulty, in terms of, voltage stability.

Let us repeat the same tests, but this time with a single contingency in each case. A summary of the results is provided in Table 10.4 [#ACLF.m; Appendix L: (L.6)]. For three contingencies, the ACLF does not converge at all. For the last contingency, the profile index is not good. In calculating the voltage profile index, it is assumed that the acceptable voltage range is 0.95–1.05 p.u. for both normal and contingency conditions. In (10.1), only the buses (either PQ or PV) with voltages out of the above range are considered. The voltage stability performance is the worst for the contingency on bus 24-bus 54 with the least $P_{stab} = 1.055$.

Let us add a dynamic reactive power resource (RPC) at bus 42, rated ±0.5 p.u., as a synchronous condenser (which is in fact a synchronous generator with P = 0.0. It is modeled as a PV bus; its voltage (V) should be specified), with

Table 10.6 Results with a ±1.0 p.u. reactive power compensator addition

Contingency on element	Voltage profile index	Voltage stability index (%)
12–42	0.000	18.50
14–54	0.000	30.50
22–32	0.000	58.50
22–42	0.000	25.00
34–54	0.000	70.00
14–12	0.000	69.00
34–32	0.000	70.50
24–22	0.000	61.50
14–24	0.000	15.50
24–54	0.000	23.00

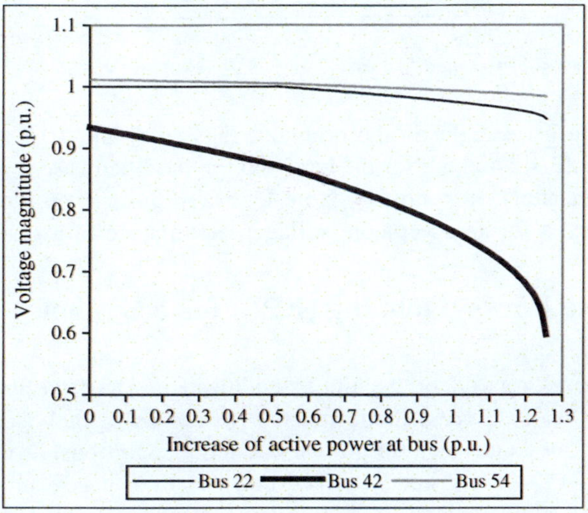

Fig. 10.5 P–V curves for the base case

$V = 1.0$ p.u. The results shown in Table 10.5 demonstrate that two divergences still happen this time [#ACLF.m; Appendix L: (L.6)]. The voltage profile is however improved in comparison with that shown in Table 10.4. P_{stab} is overall improved; although for some specific contingencies (such as the one on 34–32), it may be degraded.[9] If we repeat the tests with a ±1.0 p.u. reactive power compensator addition at bus 42 (instead of ±0.5 p.u.), we notice from Table 10.6 that the difficulties are resolved[10] [#ACLF.m; Appendix L: (L.6)].

[9] Comment whether this is due to numerical problems of the algorithm employed or it may happen in practice.

[10] Is this solution optimal? In other words, can we still find a better solution? We will come to this point, later on, in this chapter.

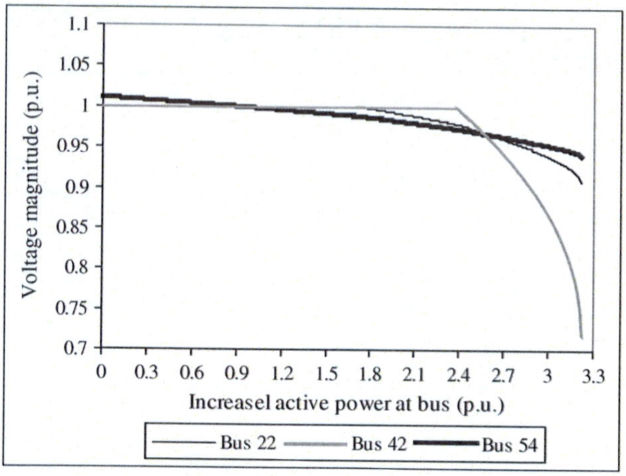

Fig. 10.6 P–V curves for the modified case

To have a further insight on the problem, Figs. 10.5 and 10.6 show the P–V curves of buses 22, 42 and 54, for the base and the modified cases (with ±1.0 p.u. dynamic compensator), respectively. Note that both voltage profile and stability should be observed for an acceptable voltage performance of a system.

10.4 Reactive Power Planning (RPP) for a System

In a RPP problem for a system, the aim is to allocate and to determine the sizes of the reactive power resources. Static reactive resources, namely, capacitors and reactors are allocated and sized for normal operating conditions. Dynamic reactive compensators (RPCs) are properly placed and sized so that secure operation of transmission grid is guaranteed following any single contingency, namely, transformers, transmission lines and power plant units.

To properly allocate and size the aforementioned static resources, a multi-objective optimization problem is solved while various constraints are checked to be met. The optimization problem is further discussed in Sect. 10.4.1. For proper placement and sizing of RPCs, a special procedure is discussed in Sect. 10.4.2.

10.4.1 Static Reactive Resource Allocation and Sizing

Static reactive resources affect both voltage profile (i.e. the voltage magnitudes) and voltage stability (i.e. the distance of current operating voltage to voltage collapse point) while at the same time, they affect system losses. A four-objective optimization problem (namely, voltage profile, voltage stability, system losses and reactive power resources cost) is considered as described below (see (a)–(d)). A solution procedure is outlined in part (e).

(a) *Voltage profile*
The voltage profile performance is evaluated based on the index defined earlier in Sect. 10.2.1 (P_{prof}). However, as in the optimization problem, the voltages are forced to be within the limits due to the constraints (see sect. 10.4.2.1), P_{prof} is calculated based on (10.1), irrespective of their magnitudes.

(b) *Voltage stability*
The voltage stability performance is evaluated based on the index defined earlier is Sect. 10.2.2 (P_{stab}).

(c) *System losses*
Minimizing active losses may be considered as another objective function. This index is described as

$$P_{loss} = \sum_{m=1}^{Nb} g_m[(V_m^s)^2 + (V_m^r)^2 - 2\, V_m^s V_m^r \cos\theta_m] \tag{10.2}$$

where V_m^s and V_m^r are the sending and the receiving end voltage magnitudes of line m, g_m is the line m conductance, θ_m is the phase angle difference of line m and Nb is the number of lines.

(d) *Reactive power resources cost*
The cost incurred due to the installation of reactive power resources should be minimized. This index can be described as

$$P_{cost} = \sum_{i=1}^{Nc} (C_{fi} + C_{vi}Q_i) \tag{10.3}$$

where C_{fi} is the fixed installation cost of reactive power resource at bus i, C_{vi} is variable cost (per kVAr) of reactive power resource at bus i (the investment cost), Q_i is the capacity of reactive power resource at bus i and the Nc is the total number of allocation points of these resources.

(e) *Overall evaluation function*
The resulting multi objective optimization problem described as

- Min. P_{prof}
- Max. P_{stab}
- Min. P_{loss}
- Min. P_{cost}

subject to $\boldsymbol{H} = 0$ (load flow equations) and $\boldsymbol{G} \leq 0$ (inequality constraints such as limits on voltage magnitudes, active (reactive) power generations of power plants, etc.) may be solved by an optimization method. As the objective function terms are not of the same units, a normalization procedure is used and a fitness function as described by (10.4) is employed.

$$F_e = -\alpha_1 \frac{P_{prof,e}}{\underline{P}_{prof}} + \alpha_2 \frac{P_{stab,e}}{\overline{P}_{stab}} - \alpha_3 \frac{P_{loss,e}}{\underline{P}_{loss}} - \alpha_4 \frac{P_{cost,e}}{\overline{P}_{cost}} \tag{10.4}$$

The sign notations used are due to the fact that while stability index is going to be maximized, others have to be minimized. Moreover, they are so normalized near 1 that they may be added together. Note that P_{cost} is normalized based on its maximum value as its minimum value is zero. Any other normalization procedure may be employed and the way represented by (10.4) is not unique. α_1 through α_4 are introduced so that for each objective function, different weighting (based on relative importance) may be assigned (all assumed to be equal to 1.0 in this chapter). $P_{obj,e}$, obj $\in \{prof, stab, loss, cost\}$ is the value of each objective function. – on top or below letter P denotes the maximum value or the minimum value, respectively.

10.4.2 Dynamic Reactive Resource Allocation and Sizing

10.4.2.1 Basics

As already noted, a dynamic reactive compensator (RPC) is employed to enhance voltage security of the system in response to any single contingency of the transmission elements. The system is considered secure if load flow converges and besides satisfying power flow limits, all voltages are within, say, 0.95–1.05 p.u. In response to a single contingency, the following conditions may occur. All static reactive resources already allocated in Sect. 10.4.1, are assumed to be in service

(a) Load flow converges and the system shows an acceptable condition in terms of voltage magnitudes. No further action is required.
(b) Load flow does not converge due to an islanding condition following the contingency. No RPC may solve the problem.
(c) Load flow does not converge, but not due to an islanding condition. Further action is required to solve the problem.
(d) Load flow converges but some of the voltages are out of range $0.95 \text{ p.u.} \leq V_i \leq 1.05 \text{ p.u.}$ Further action is required.

In the studies conducted, the optimum sizes and locations of RPCs are found to solve (c) and (d) above. For any single contingency, a single RPC is checked to solve the problem. Maximum and minimum reactive power capacities of a RPC are considered equal.[11] Multi-RPC application to solve all contingency cases is considered in the following subsection.

10.4.2.2 Determination of the Maximum Number, Allocations and Sizing of RPCs

In order to find the optimum allocations and sizes of RPCs, a preliminary list of buses together with maximum permissible RPCs capacities should be initially

[11] Unequal limits may also be considered.

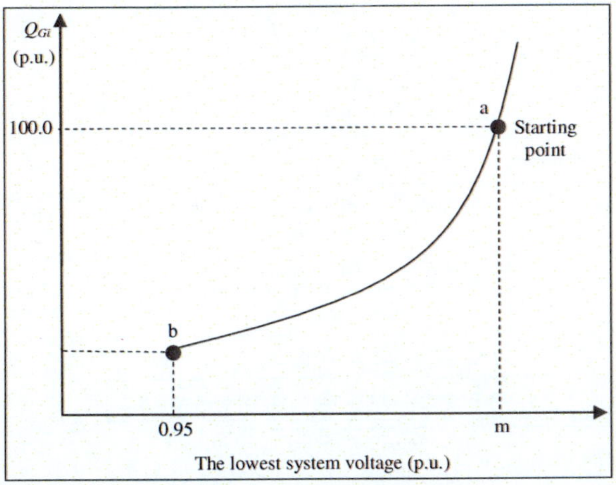

Fig. 10.7 Step by step procedure for RPC capacity determination

generated. To do so, an iterative procedure is followed in which for any contingency mentioned in Sect. 10.4.2.1 (parts (c) and (d)), a single high capacity (say 100 p.u.) RPC is applied in each PQ bus while its reference voltage is set at 1.0 p.u. (At that time, the lowest system voltage may be m p.u., i.e., point a, Fig. 10.7). Thereafter, its capacity is gradually reduced while all grid voltages are monitored to be greater than or equal to 0.95 p.u. (point b, Fig. 10.7). The least capacity option; to keep all voltages above 0.95 p.u., is selected for this particular bus. As already mentioned, the followings are checked

(a) Load flow converges,
(b) All voltage magnitudes are within an acceptable range.

For any single contingency j, the application of RPC at bus i may result in the following situations

(A) The conditions (a) and/or (b) above are not met even with a high capacity RPC at bus i. Bus i will not be a candidate bus for contingency j.
(B) The conditions (a) and (b) are met. The reference voltage is reduced and the lowest reactive resource is found so that both conditions are met. Q_{Gij} found is the minimum RPC capacity at bus i for convergence in response to contingency j.
(C) Repeat the procedure (A) and (B) for all PQ buses and find $\underline{Q}_{GJ} = \min_{\{i\}}\{Q_{Gij}\}$.
(D) Repeat the procedure (A–C) for all contingencies (Sect. 10.4.2.1).

It is evident that the number of applied buses will be less than or equal to the contingencies. However, an optimization procedure may be followed to reduce both the number and the sizes of RPCs.

As a matter of fact, simultaneous application of all RPCs determined in steps (C) and (D), may not be required and may result in over design. All the

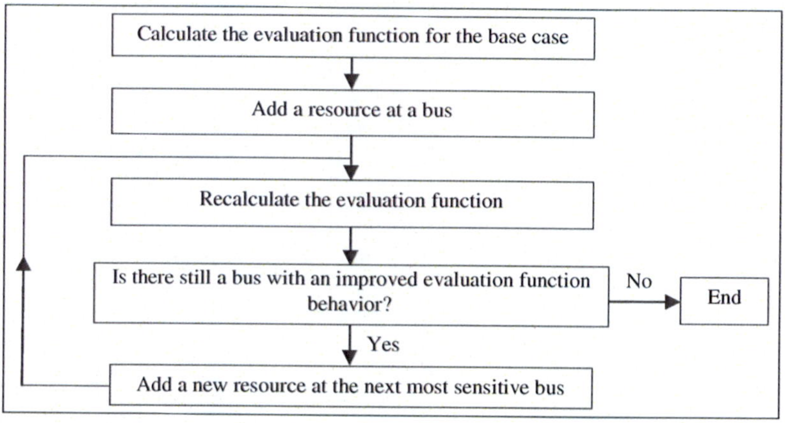

Fig. 10.8 Sensitivity approach

aforementioned RPCs may be considered as candidate RPCs and an optimization procedure[12] can be performed to find final optimum allocations and sizing. The objective function is considered as the minimum total RPC application while already mentioned conditions (a) and (b) are still checked to be met.

10.4.3 Solution Procedure

In Sects. 10.4.1 and 10.4.2, we developed two optimization problems. The former was seeking for allocations and sizing of static resources; with the evaluation function defined by (10.4). The latter focused on finding the minimum total RPC applications (i.e. the least cost); with the details given there.

We may use one of existing powerful optimization algorithms[13] to solve the above optimization problems. If the system under study is small and the search space is limited in terms of the resources candidates, we may search the entire space and calculate the evaluation function; in order to find the optimum solution.

If the search space is large, a powerful metaheuristic approach is Genetic Algorithm (GA) by which the solution may be found; quickly and in an efficient manner. We present some numerical examples in Sect. 10.5 on using GA. An alternative, yet simple solution procedure is depicted in Fig. 10.8. This is called the *sensitivity* approach in which the evaluation function is initially calculated for the base case. Following that, a small reactive resource[14] is applied at each bus,

[12] For some details, see Sect. 10.4.3.

[13] See Chap. 2 for details.

[14] Either static or dynamic resource may be applied; although the application for static resources is more straightforward and used hereon. For dynamic resources, initially the maximum resources should be applied and then gradually reduced.

one-by-one and the evaluation function, recalculated.[15] Based on the resulting calculations, the most sensitive buses are determined. Thereafter, a small reactive resource (say 0.1 p.u. of capacitor) is applied at the most sensitive bus and the whole procedure is repeated. For instance, in the second run, the first bus may be still the most sensitive and a second 0.1 p.u. resource may be added to that bus. The procedure is repeated until no further bus may be found which results in improving the evaluation function.

The proposed approach may be used for both small and large systems. Some numerical examples are provided in Sect. 10.5.

10.5 Numerical Results

Two test systems are considered here. The first one is one shown in Fig. 10.4. The second one is a large test system as already used in Chap. 9 (Fig. 9.5).[16]

10.5.1 Small Test System

As the system shown in Fig. 10.4 is a small test system, the following two cases are considered

- *Case I.* To find the global optimum, the entire space is searched to allocate and size the static reactive resources. All buses are considered as candidates. Moreover, it is assumed that at each bus a maximum five blocks of 0.1 p.u. capacitor banks may be applied. C_{fi} is considered to be five times $C_{vi.}$[17]
- *Case II.* The same as above, but this time using the sensitivity approach proposed in Fig. 10.8.

The results for Case I are shown in Table 10.7. Four conditions are tabulated. The first 3 focus on optimizing a single objective function term. The fourth condition considers a multi-objective optimization case. For each of the conditions above, the resulting objective function terms as well as the justified buses and their respective capacities are also shown.

The results for case II are the same as above except that in minimizing P_{loss}, the result is 7.558 (instead of 7.536).

---

[15] The constraint terms are added to the evaluation function with large penalty coefficients, so that the final solution will end up with the optimum objective function while all constraints are met.

[16] It should be mentioned that some of the results shown in this section may not be readily regenerated by the Matlab codes attached to this book; as they are generated by a software with slightly different algorithms. For details, see [13] at the end of this chapter.

[17] C_{vi} = R 1.0/p.u.

Table 10.7 Results for Case I

Conditions	Bus number : (p.u. justified)	P_{prof}	P_{stab} (p.u.)	P_{loss} (MW)	P_{cost} (R)
Minimize P_{prof}	12: (0), 22: (0), 32: (0) 42: (0.3), 54: (0)	0.000	1.458	7.778	5.3
Maximize P_{stab}	12: (0.5), 22: (0.5), 32: (0) 42: (0.4), 54: (0.3)	0.005	1.689	8.292	21.7
Minimize P_{loss}	12: (0.5), 22: (0.5), 32: (0.2) 42: (0.1), 54: (0)	0.003	1.486	7.536	21.3
Optimize F_i (see (10.4))	12: (0), 22: (0.4), 32: (0) 42: (0.3), 54: (0)	0.000	1.550	7.778	10.7

Table 10.8 ACLF results for buses

Bus	Voltage		Generation	
	V (p.u.)	Angle (rad.)	PG (p.u.)	QG (p.u.)
12	1.000	−0.025	–	–
14	1.000	0.000	3.868	−1.593
22	1.000	−0.101	0.500	−0.385
24	1.014	−0.079	–	–
32	1.000	−0.040	0.650	−0.393
34	1.005	−0.043	–	–
42	1.005	−0.254	–	–
54	1.0113	−0.0656	–	–

Table 10.9 ACLF results for transmission flows

Element	AC load flow		
	P (p.u.)	Q (p.u.)	S (p.u.)
12–42	0.731	−0.112	0.739
14–54	1.632	−0.351	1.669
22–32	−0.282	0.080	0.293
22–42	0.364	−0.073	0.371
34–54	0.103	−0.056	0.117
14–12	1.252	−0.029	1.252
32–34	0.103	−0.257	0.277
22–24	−1.142	−0.648	1.313
14–24	0.983	−0.214	1.006
24–54	−0.166	0.052	0.174

With those capacitors justified in the case of optimizing F_i (see (10.4)), the ACLF results for the normal (no contingency) conditions are tabulated in Tables 10.8 and 10.9 [#ACLF.m; Appendix L: (L.6)]. With these static resources added, the contingency results are shown in Table 10.10 [#ACLF.m; Appendix L: (L.6)]. As shown, for contingency on element 22–42, the load flow diverges. Based on the approach already presented, we apply a high capacity RPC at all buses,

Table 10.10 Contingency analysis for the network with the added capacitors

Contingency on element	Voltage profile index	Voltage stability index (%)
12–42	0.061	8.50
14–54	0.000	30.50
22–32	0.000	51.00
22–42	No convergence	No convergence
34–54	0.000	59.00
14–12	0.000	55.00
32–34	0.000	59.00
22–34	0.000	53.50
14–24	0.000	25.50
24–54	0.000	30.00

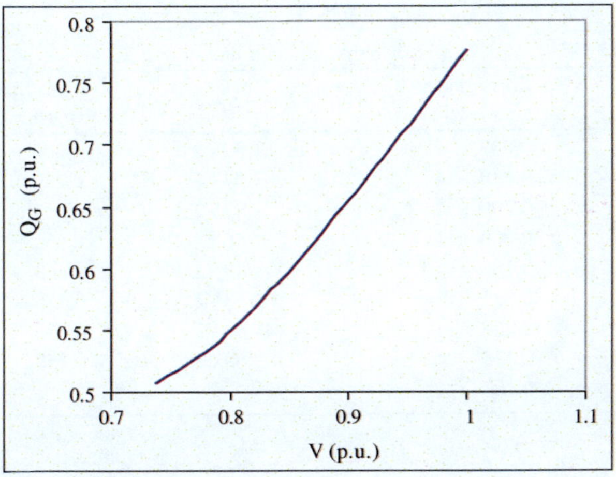

Fig. 10.9 The process of finding the minimum RPC capacity for bus 42

one-by-one, to find the most effective bus. Bus 42 is chosen as a result. Its minimum capacity is then found so that the constraints (voltage magnitudes) are met. As shown in Fig. 10.9, the minimum RPC capacity required is 0.72 p.u. Note that for this specific simple test case, as a single RPC is found to solve the problem, no multi-RPC optimization procedure as outlined in Sect. 10.4.2 is required.

10.5.2 Large Test System

The 77-bus test system described in Chap. 9[18] is used here for assessing the proposed RPP procedures. As shown in Table 10.11, the ACLF for the normal

[18] For details, see Appendix J.

Table 10.11 Out of range voltages for the 77-bus network

Bus	Voltage (p.u.)	Bus	Voltage (p.u.)
B 2V2	0.9324	B 32V2	0.9469
B 3V2	0.9430	B 34V2	0.9477
B 6V2	0.9280	B 37V2	0.9422
B 10V2	0.9259	B 46V2	0.9377
B 12V2	0.9471	B 49V2	0.9400
B 17V2	0.9038	B 53V2	0.9405
B 25V2	0.9273	B 60V2	0.9035
B 28V2	0.9403	B 68V2	0.9439
B 31V2	0.9430		

Table 10.12 Maximum capacitor banks at each bus

Bus	Capacitance (p.u.)	Bus	Capacitance (p.u.)
B 2V2	0.087 × 4	B 32V2	0.300 × 4
B 3V2	0.338 × 4	B 34V2	0.250 × 4
B 6V2	0.200 × 4	B 37V2	0.313 × 4
B 10V2	0.200 × 4	B 46V2	0.550 × 4
B 12V2	0.188 × 4	B 49V2	0.463 × 4
B 17V2	0.075 × 4	B 53V2	0.350 × 4
B 25V2	0.063 × 4	B 60V2	0.075 × 4
B 28V2	0.175 × 4	B 68V2	0.350 × 4
B 31V2	0.400 × 4		

conditions results in 17 bus voltage magnitudes to be out of the range (0.95–1.05 p.u.).[19]

Initially, we apply static reactive compensator of capacitor type at each individual bus with a voltage of less than 0.95 p.u., to make it 0.95 p.u. The results are shown in Table 10.12. For instance, a reactive power of 1.20 p.u. is required to make the voltage of bus B 32V2 equal to 0.95 p.u.

Thereafter, it is assumed that the values shown in Table 10.12 are the maximum capacitor banks which may be applied at each bus. However, for having extra flexibility, an extra stage is considered for each bus. For instance, for bus B 32V2, a 5-stage 0.3 p.u. bank (i.e. maximum 1.5 p.u. installation) is considered as the maximum bank applicable.

We now come to the sensitivity algorithm, already described. Four conditions are considered as the evaluation function (see Fig. 10.8). They are the same as

[19] This range is also considered for the contingency conditions.

Table 10.13 Results for 77-bus network

Conditions	Bus number[a]: (p.u. justified)	P_{prof}	P_{stab} (p.u.)	P_{loss} (MW)	P_{cost} (R)
Minimize P_{prof}	2: (0.435), 3: (1.69), 6: (1.0) 10: (1.0), 12: (0.94), 17: (0.375) 25: (0.315), 28: (0.875), 31: (2.0) 32: (1.5), 34: (1.25), 37: (1.56) 46: (2.75), 49: (2.31), 53: (1.75) 60: (0.075), 68: (1.75)	0.047	22.055	77.668	106.585
Maximize P_{stab}	2 : (0.435), 3: (0.338), 6: (1.0) 10: (1.0), 12: (0.94), 17: (0.375) 25: (0.315), 28: (0.875), 31: (2.0) 32: (1.5), 34: (1.25), 37: (1.56) 46: (1.1), 49: (2.31), 53: (0.7) 60: (0.075), 68: (1.75)	0.051	22.079	76.157	102.533
Minimize P_{loss}	2: (0.435), 3: (1.69), 6: (1.0) 10: (1.0), 12: (0.94), 17: (0.375) 25: (0.315), 28: (0.875), 31: (2.0) 32: (1.5), 34: (1.25), 37: (1.56) 46: (0.55), 49: (0.92), 53: (1.05) 60: (0.075), 68: (1.75)	0.051	21.935	75.915	102.296
Optimize F_i (see (10.4))	2: (0.435), 3: (1.69), 6: (1.0) 10: (1.0), 12: (0.94), 17: (0.0) 25: (0.063), 28: (0.875), 31: (2.0) 32: (1.5), 34: (1.25), 37: (1.56) 46: (2.75), 49: (2.31), 53: (1.75) 60: (0.3), 68: (1.75)	0.028	22.055	77.397	101.183

[a] For simplicity, only the bus number is shown in this table

those shown in Table 10.7. With C_{fi} = R 5.0 and C_{vi} = R 1.0/p.u. in (10.3), the results are shown in Table 10.13.[20]

[20] As you see, P_{prof} is 0.047 (where a single objective is considered) in comparison with 0.028, where multi-objectives are involved. This is a typical difficulty that may happen with simple (such as sensitivity) algorithms.

Table 10.14 ACLF results for buses; after compensation

Bus	Voltage (p.u.)	Bus	Voltage (p.u.)
2	0.9888	32	0.9862
3	0.9894	34	0.9763
6	1.0046	37	0.9815
10	1.0079	46	0.9929
12	0.9784	49	0.9943
17	0.9643	53	0.9953
25	0.9736	60	1.0150
28	0.9840	68	0.9867
31	0.9766		

Table 10.15 Results based on GA

Conditions	Bus number: (p.u. justified)	P_{prof}	P_{stab} (p.u.)	P_{loss} (MW)	P_{cost} (R)
Optimize F_i	2: (0.435), 3: (1.69), 6: (0.8) 10: (1.0), 12: (0.94), 17: (0.075) 25: (0.315), 28: (0.875), 31: (2.0) 32: (1.5), 34: (1.25), 37: (1.56) 46: (2.75), 49: (2.31), 53: (1.75) 60: (0.0), 68: (1.75)	0.028	22.055	77.294	101.01

Table 10.16 Results for contingency conditions

Contingencies with violated voltage	Violated voltage buses	Compensated buses	Required capacity (p.u.)
B 12V2 B 72V2	B 12V2	B 12V2	0.5
B 15V2 B 66V2	B 66V2 and B 62V2	B 62V2	1
B 62V2 B 66V2	B 62V2		

It should be noted that the candidate buses as well as the maximum permissible capacitor bank of each bus are the ones shown in Table 10.12. Upon compensation, the voltages are improved (Table 10.14).

For the condition where the evaluation function is considered to be a combination of all terms, reactive power compensation is repeated using GA (Table 10.15). As expected, the results are improved in comparison with the sensitivity approach.

The results of Table 10.14 are for the normal conditions. To check for RPC requirements, N−1 conditions are tested on each individual element. 11 single contingencies result in islanding for which RPC can not provide a solution.[21] None of the others results in load flow divergence. Therefore no RPC is required. However, for three contingencies, some voltages are violated. Based on a trial and

[21] Transmission enhancement may be tried.

error approach,[22] the application of some level of reactive compensation can solve the problem. The results are shown in Table 10.16.

Problems[23]

1. Investigate in your area of living what the reactive power resources are available and how they are managed and controlled in keeping voltages.
2. In the test system, as shown in Fig. 10.4.

   (a) Analyze the relationship between the loads power factors and inaccuracies involved in using DCLF (in comparison with ACLF).
   (b) Analyze the inaccuracies involved in using DCLF whenever the voltage reference set points of PV buses are either 0.95, 1.00 or 1.05 p.u. (similar for all buses).

3. The thermal capacity of a transmission line is defined in terms of MVA while in DCLF, the line flow of a transmission line is calculated based on MW. In problem 2 (a), find the difference between the apparent power flowing through a line and its active power flow from DCLF. Can a relationship be defined?

4. (a) For a RPP problem, introduce some other indices for both voltage profile and stability.
   (b) For the modified Garver system, find the voltage profile performance using the new defined indices [#ACLF.m; Appendix L: (L.6)].
   (c) For the same system, do the same for voltage stability performance [#ACLF.m; Appendix L: (L.6)].

5. In the modified Garver system [#ACLF.m; Appendix L: (L.6)]

   (a) Assuming maximum 0.5 p.u. capacitor to be installed in buses 12, 14 and 54 which are identically switched in or out (for all buses) in steps of 0.1 p.u., analyze their switching on voltage profile and stability. Consider the capacitor banks to be equal for all buses in each case.
   (b) Assuming the transformers taps to be identical, analyze the effect of changing taps on voltage profile and stability (from 0.95 to 1.05, in steps of 0.01).
   (c) Assuming the voltage reference set points to be identical, analyze the effect of changing set points on voltage profile and stability (from 0.95 to 1.05 p.u., in steps of 0.01).

---

[22] Optimization based approaches may also be checked.

[23] In problems 2, 5, 7 and 8, the system is, in fact, the one shown in Fig. 10.4 with the additional details given in Table 10.1.

6. Repeat problem 5 for analyzing the effect of control parameters on system losses. Use $\sum_{i=1}^{Nl} R_i(P_{Li}^2 + Q_{Li}^2)$, as an approximate formula in calculating the losses [#ACLF.m; Appendix L: (L.6)].
7. Propose a heuristic capacitor allocation procedure to allocate capacitor for the modified Garver system, once the losses are to be minimized [#ACLF.m; Appendix L: (L.6)].
8. Perform ACLF for the modified Garver system for a minimum load (60% of the peak values. By reducing the load, assume the generation is compensated by the generation in bus 14). Analyze and discuss voltage profile and stability performance.

References

Reference [1] is a basic power system analysis book. Reference [2] is an early book devoted to RPP. A review of the problem is given in [3]. A comparative study and an overview of some new techniques are provided in [4–6]. The problem of the allocation and sizing both the static and dynamic reactive resources are covered in many references. While the work reported in this chapter regarding dynamic resource allocation is based on [7], some of these research are covered in [8–16].

1. Saadat H (1999) Power systems analysis. McGraw-Hill, New York
2. Miller THE (ed) (1982) Reactive power control in electric systems. Wiley, New York
3. Zhang W, Li Tolbert LM (2007) Review of reactive power planning: objectives, constraints, and algorithms. IEEE Trans Power Syst 22(4):2177–2186
4. Lee KY, Yang FF (1998) Optimal reactive power planning using evolutionary algorithms: a comparative study for evolutionary programming, evolutionary strategy, genetic algorithm and linear programming. IEEE Trans Power Syst 13(1):101–108
5. Eghbal M, El-Araby EE, Yorino N, Zoka Y (2007) Application of metaheuristic methods to reactive power planning: a comparative study for GA, PSO and EPSO. In: IEEE international conference on Systems, man and cybernetics—ISIC
6. Fernao PD, Henggeler AC, Gomes MA (2007) Multi-objective evolutionary approaches for reactive power planning in electrical networks—an overview. In: International conference on Power engineering, energy and electrical drives—Power eng
7. http://www.wecc.biz
8. Osborn DL (1989) Factors for planning a static VAR system. Electric Power Syst Res 17(1):5–12
9. Ebrahimi S, Farsangi MM, Nezamabadi-Pour H, Lee KY (2006) Optimal allocation of static Var compensators using modal analysis, simulated annealing and tabu search. In: Proceeding of IFAC symposium on Power plants and power systems, 5(1):377–382
10. Tiwari A, Ajjarapu V (2008) Optimal allocation of dynamic VAR for enhancing stability and power quality. In: IEEE power and energy society general meeting—Conversion and delivery of electrical energy in the 21st century
11. Alabduljabbar AA, Milanovic JV (2006) Genetic algorithm based optimization for allocation of static VAr compensators. In: Proceeding of 8th IEE International Conference on AC and DC Power Transmission–ACDC, 115–120
12. Venkataramana A, Carr J, Ramshaw RS (1987) Optimal reactive power allocation. IEEE Trans Power Syst 2(1):138–144

13. Yousefi GR, Seifi H, Sepasian MS, Haghighat H, Riahi R, Hosseini H, Kazerouni AK, Mohseni E, Akhavan A (2004) A new reactive power planning procedure for Iranian power grid. Electric Power Syst Res 72(3):225–234
14. EL-Dib AA, Youssef HKM, EL-Metwally MM, Osman Z (2007) Optimum VAR sizing and allocation using particle swarm optimization. Electric Power Syst Res 77(8):965–972
15. Fernandes RA, Lange F, Burchett RC, Happ HH, Wirgau KA (1983) Large scale reactive power planning. IEEE Trans Power Appar Syst PAS-102(5):1083–1088
16. Ajjarapu V, Ping LL, Battula S (1994) An optimal reactive power planning strategy against voltage collapse. IEEE Trans Power Syst 9(2):906–917

Chapter 11
Power System Planning in the Presence of Uncertainties

11.1 Introduction

We have so far covered various power system planning issues, namely, load forecasting, GEP, SEP, NEP and RPP. We assumed, implicitly, that all decisions are made by a single entity. Moreover, we assumed that the information used lacks any uncertainty. None of the above is strictly true. In terms of the former, due to power system de-regulating, GEP, from one side, is unbundled from the others (SEP, NEP and RPP). Some new market participants act as major players for investing on new generation facilities. These generation companies try to make the most profit from their investments. They should, somehow predict the rivals behaviors. They should, have their own input information (such as the system load forecasting) for proper decision makings. From this viewpoint, GEP is a completely different story in comparison with the traditional environment. We will see, however, that a modified traditional GEP may also be used in the de-regulated environment; now, from other entities viewpoints. If GEP is decided by some entities based on their own judgements, how can a different or some different entities proceed towards the other steps (SEP, NEP and RPP) if they cannot make sure what the GEP players do in actual life. Still, there are more uncertainties involved for their various decision makings.[1] So, briefly speaking, uncertainties play major roles in power system planning issues of the new environment.

The above points have received much attention in literature over the last one or two decades. We will briefly review the topics in this chapter so that the reader can follow up the relevant issues in literature. Initially, we will briefly review power system de-regulating in Sect. 11.2. Section 11.3 is devoted to uncertainties involved in power system planning issues. In Sect. 11.4, we discuss more some practical considerations of observing uncertainties and/or planning issues in a

[1] The uncertainties hold for the traditional environment, too. However, they are more pronounced in a de-regulated environment.

H. Seifi and M. S. Sepasian, *Electric Power System Planning*, Power Systems,
DOI: 10.1007/978-3-642-17989-1_11, © Springer-Verlag Berlin Heidelberg 2011

deregulated power system. How to deal with uncertainties in power system planning is covered in Sect. 11.5.

11.2 Power System De-regulating

Over the last two decades or even more, power system industry has experienced a drastic change in terms of economical observations. In the traditional or the so *called regulated* industry, power system structure consists of generation, transmission and distribution owned by a single entity; or if owned by different owners, controlled or *regulated* by a single entity. In other words, the single entity decides on where and how to allocate generation and/or transmission facilities. The investment and the operational costs as well as an appropriate level of profit for the owners are compensated by regulated tariffs imposed on the customers.

For long, economists criticized this approach as being inefficient as it may impose large burden on the customers, due to the fact that there is no incentive of reducing the costs by the owners.[2]

In the so called new *de-regulated* environment, the tariffs are not regulated anymore. The electricity is provided by some suppliers (known as *GenCos*[3]) as a commodity. The customers may wish to buy this commodity either from an specific supplier or from the wholesale market (known as *power pool*); directly or indirectly (through the so called *DisCos*,[4] *aggregators* or *retailers*). The electricity price is determined based on the bids provided by the suppliers from one side and those asked by the customers. Nearly all types of economic rules are applicable to an electricity market, too.

Once the winners are decided based on the type of the market,[5] the commodity (electricity) should be transferred from the suppliers to the customers through the available transmission facilities (transmission lines and/or cables, etc.). Open access of all market participants to these facilities is vital in having a fair electricity market, so that no participant is given any unjustified priority in using them. On the other hand, the owners of such facilities have to be compensated for their investments as well as operation costs. That is why the transmission system (owned by the so called *TransCo*[6]) is still fully or partially[7] regulated by an entity, assigned directly or indirectly by the government. The costs of *TransCos* are

---

[2] Of course, the aforementioned single entity tries to control the costs by imposing some legal regulations.

[3] Generation Companies.

[4] Distribution Companies.

[5] There are different types of electricity markets such as power pool, bilateral, hybrid, etc.

[6] Transmission Companies.

[7] Through some options such as FTR (Firm Transmission Right).

compensated by transmission service tariffs determined by the aforementioned entity.[8]

We will see in Sect. 11.4 how the power system de-regulating affects power system planning issues. Before that, we discuss uncertainties involved in power system planning, both in *regulated* and *de-regulated* environments, in Sect. 11.3.

11.3 Power System Uncertainties

Two terms of *uncertainty* and *risk* are widely used in power system literature. There are no fixed definitions for these terms. Some believe that they are the same, while some believe one is the result of another. Still, some think of these to be quite independent. We are not going to define these terms precisely in this book. Instead we assume that the uncertainties involved may result in risk. For instance, as GEP, SEP and NEP are based on the forecasted load, any *uncertainty* in the predicted load may result in risk (measured in terms of a predefined index) so that the network planned may be unable to fulfill its functions properly (i.e. to supply all loads). From here on, we focus on uncertainties.

One of the difficult tasks of observing uncertainties in our decisions is the fact that uncertainly should be, somehow, modeled; while there are various types of such uncertainties such as *economic* or *technical*; *controllable* or *uncontrollable*; *non-stochastic* or *stochastic*; and *measurable* or *unmeasurable*. Whatever the type is, the uncertainties may be modeled by some approaches such as those based on scenarios. A more detailed discussion is given in Sect. 11.5.

On the other hand, the uncertainties affect all short-term and long-term decisions. From here on, we focus on long-term decisions, namely, power system planning issues. However, we differentiate between the *regulated* and the *de-regulated* environments in the following subsections.

11.3.1 Uncertainties in a Regulated Environment

As discussed so far in this book, power system planning is based on load forecasting (LF) and consists of GEP, SEP, NEP and RPP, each with its own input parameters. The studies are to be carried out for some years in the future; so the input parameters should be, accordingly, predicted. However, these parameters are in turn, dependent upon some other parameters. As a result, the input parameters to

[8] The entity may vary from one market to another. Some typical ones are Market Operator (MO), Independent System Operator (ISO), etc.

power system planning modules may face uncertainties which obviously affect our decisions. Some of these parameters are

- Economic growth (LF)
- Economic parameters, such as inflation, depreciation and interest rates (LF, SEP, NEP, RPP and GEP)
- Fuel cost (directly on GEP and indirectly on SEP, NEP and RPP due to its effect on cost of the losses)
- Technological developments (LF, GEP, SEP, NEP and RPP)
- Electricity price (LF)
- Environmental limitations (directly or indirectly on GEP, SEP and NEP)
- Investment costs (GEP, SEP, NEP and RPP)
- Regulatory and legal acts (LF, GEP, SEP, NEP and RPP)
- Demand side management programs (LF)
- Operation and maintenance costs (GEP, SEP, NEP and RPP)
- Resource (such as fuel and water,) availability (GEP)
- Social factors (such as population growth rate) (LF)

It is obvious that the uncertainties involved in above or similar parameters are case dependent for each electric power industry.

11.3.2 Uncertainties in a De-regulated Environment

We discussed earlier in this chapter that power system de-regulating has resulted in appearing new independent entities such as GenCos, TransCos, DisCos, etc.; each aiming at making, perhaps, the maximum profit (revenues minus costs) from its properties. A system operator tries to coordinate the behaviors of market players in such a way that the system is operated reliably and in an efficient manner.

Each entity now should make its own decisions. Obviously it should, somehow, take the behaviors of the other players into consideration. In this new situation, the electricity price is determined based on the supply–demand rule. Now there is no guarantee of investment costs recoveries.

On the other hand, in most parts of the world, the de-regulating is still going on. New rules and legal acts are continuously appearing. Moreover, any national or even international economic decision and/or crisis influences the electric power industry; directly or indirectly. The single-player environment has replaced by a multi-player game, with its risks and uncertainties involved.

The power system planning in a de-regulated environment is a challenging area which has received much attention in literature. In Sect. 11.4, we discuss some of its basic issues. In the following subsections, we differentiate between the uncertainties involved in GEP from one side and SEP, NEP and RPP (as Transmission Expansion Planning, TEP) from the other side.

11.3.2.1 Uncertainties in GEP

Besides those uncertainties introduced for the regulated environment, as an owner should now make its own decision in investing on a power plant, it faces new uncertainties. The *electricity price* is the most important example. The investor may invest in a location with an anticipated high electric price. However, the behaviors of the other players should also be predicted and taken into account. This prediction is not an easy task at all and is uncertain.

On the other hand, while a generation investor tries to invest in a location with the maximum possible profit, any separate investment on transmission system (TEP) may have positive or negative effects on the suppliers profits. So, a power plant investor should take this uncertain TEP, also into account.

The reader should note that some types of uncertainties already present in a regulated industry may have quite more dominant effects in the de-regulated case. For instance, the costs of primary resources (such as gas, oil, etc.) may make an owner to defer its investment in a place or changes its decision and invests in another place. In a regulated environment, although these costs are still effective, the investor may still invest at the same place and time; as the money is guaranteed to be back by some appropriate tariffs.

11.3.2.2 Uncertainties in TEP

Besides those uncertainties in a regulated environment, the most important uncertain factor which influences TEP (SEP, NEP and RPP) is the uncertain GEP output. How TEP may be properly performed if GEP is decided upon by the other market players? As the costs of TEP should be recovered from the market participants (both the suppliers and the customers),[9] an overdesign may result in players dissatisfactions. Underdesign can result in similar effects as the suppliers may be unable to sell and the loads of the customers may not be fulfilled.

11.4 Practical Issues of Power System Planning in a De-regulated Environment

Having discussed, so far, various aspects of regulated and de-regulated environments and the uncertainties involved, let us now review some practical issues of power system planning problem in a de-regulated environment.

The load still, has to be predicted (Chap. 4). The important consideration is the fact that, now, the driving factors may be different or some driving parameters may

[9] In fact the recovery is only from the customers, as the suppliers, somehow, increase the prices, if they have to pay something.

have more pronounced effects. For instance, *electricity price* may exhibit more fluctuations. Due to *elasticity* of power system loads, the demands may have more variations in comparison with the regulated environment in which controlled tariffs apply. Moreover, economic factors, such as GDP, has normally stronger effects. The basic algorithms, however, remain the same as in Chap. 4.

Another factor that influences the forecasted load is the so called *Demand Side Management* (DSM) or *Demand Response* (DR). DSM or DR is an issue of concern in both regulated and de-regulated environments. It is a process of controlling the electric demand (reducing, shifting, etc.). In a regulated environment, there has been less incentive for a customer to change its demand. Due to various penalties and rewards set for cooperations in DSM and DR programs, in a de-regulated case, it may have stronger effects on the forecasted load. Although the DSM or DR past performance is achievable, its future performance prediction is not an easy task at all and depends on various parameters and conditions.

Beside DSM or DR, as already noted, electricity price should be forecasted, too, in order the load to be forecastable. In a de-regulated environment, the long-term electricity price forecasting is an important issue of concern which affects load forecasting (as detailed above) as well as TEP (as we will discuss later). So, briefly speaking, in a de-regulated environment and from the load forecasting viewpoint

- The basic algorithms are essentially the same.
- The driving parameters may be different.
- Long-term price forecasting requires extensive, sometimes complicated, algorithms.
- DSM requires special considerations.

Now let us move towards GEP. Based on some deterministic input parameters, GEP was considered in Chap. 5 on a single bus basis. GEP in combination with an approximate consideration of transmission system was considered in Chap. 6. What happens in a de-regulated environment?

In this new environment, there are, in fact, *two different* entities, thinking of electricity supply. The *first*, is *an independent entity*, belonging or somehow assigned by the government; directly or indirectly; which should worry about meeting the generation requirements of the system. Still, this entity may perform the same studies, noting the following points

- Some of the input parameters are not deterministic, anymore. For instance, the type of available power plants (see Chap. 5) are not known in advance, as this entity is not now the real investor and the final decision makers for investing on power plants are different.
- The studies carried out in Chaps. 5 and 6 may be used as guidelines for the investors. If sufficient investors are available to invest on all power plants studied by the independent entity (capacity and location), the generation requirements are fulfilled. As the transmission system enhancement cost is also at minimum from the studies of Chap. 6 (although approximate), the investor may make sufficient benefit from its decision, as it has to, normally, pay for the

transmission enhancement needs, too. However, this situation may not happen due to the following two reasons

- The objective functions defined in Chaps. 5 and 6 were primarily cost and/or technically based. The investor may not obey the independent entity suggestions as it tries to invest in a location and with a capacity to make the most profit from the market. For instance, it may invest in a place with the maximum forecasted electric price, yet with minimum fuel cost; provided its generation may be predicted to be sold to either local loads or can be transmitted to remote loads. If the independent entity wishes to perform the studies as in Chaps. 5 and 6 in such a way that the GenCos may obey the results with a higher probability, it should consider new objective function terms (such as those for observing possible GenCos profits).
- Enough investors may not be found to invest on all locations or with the same capacities as suggested by the independent entity. Still, there may be more generations in some places than what suggested by this entity.

The *second group of entities* thinking about the generations, are the GenCos, or the real investors. In order to invest in a place, an investor should perform detailed studies to decide on location and capacity of its generation so that the maximum possible profit is anticipated. In doing so, it should model the behaviors of its rivals. This is a completely new study required in a de-regulated environment.

Briefly speaking, in terms of GEP, in a de-regulated environment

- The basic algorithms as outlined in Chaps. 5 and 6 may be used by the independent entity with some modifications in terms of objective function terms, and modeling.
- The input parameters may not be deterministic. Moreover, some new input parameters, such as the predicted long-term electricity price, may also be needed.
- GEP from the view points of GenCos should be developed.

Now, we move towards SEP, NEP and RPP. These problems were addressed in Chaps. 7–10. The objective functions were primarily cost based, while various technical constraints had to be met during the planning process. When we come to a de-regulated environment, we come across the following points

- Although the transmission system is still regulated (or somehow provides fair and open access to all participants; in other words, its access is not competence based), its design should not only be cost based. New objective functions regarding market behaviors should be also observed so that electric power transactions are facilitated in a fairly and indiscriminate manner. Moreover, some other objective functions or constraints such as *reliability* indices may have more pronounced effects and need special considerations in this environment. Some of new objective functions or constraints may need new input parameters, such as the predicted long-term electricity price.
- The major difficulty in TEP problems is the fact that the GEP results to be used as the input decisions are not deterministic anymore.

Briefly speaking, TEP undergoes little variations in comparison with GEP. Later on, we will talk about how to deal with the nondeterministic nature of GEP on TEP.

When we come to RPP as the final stage of the planning process (Chap. 10), we note that this step requires the least modifications in a de-regulated environment. Any reactive power resource which is primarily intended for improved voltage performance of the system may, however have some effects on electric power market performance. This is due to the fact that acceptable voltage performance may facilitate market transactions. Moreover, reactive power is also transacted in a market as an *ancillary service*. As a result, sometimes new objective functions or constraints may be added to those already considered in Chap. 10 to make the situation more appropriate for a de-regulated environment.

11.5 How to Deal with Uncertainties in Power System Planning

So far, we have introduced power system planning in both regulated and de-regulated environments. We discussed that new objective functions and constraints may be required to be added in the latter case. We also talked about the uncertainties which are normally more pronounced in the latter case. One of the current approaches in dealing with the uncertainties is scenario technique. A *scenario* is one of the possible conditions that may happen in the future.[10] A *plan* is a combination of *options* (such as lines, cables, transformers, etc.) employed for the problem solution. An *attribute* or a *criterion* (such as total cost, LOLE, etc.) may be used to evaluate a *plan* performance. If a_{ij} denotes the *attribute* of *plan i* in *scenario j* and $a_{opt\ j}$ denotes the *optimum* plan for that *scenario*, r_{ij} is defined as the *regret index* as follows

$$r_{ij} = a_{ij} - a_{opt\, j} \tag{11.1}$$

A *robust plan* is a plan for which its *regret index* is zero for all scenarios.[11]

Let us move onward with a simple example. Suppose there are three different scenarios A, B and C for each, three plans 1, 2 and 3 are the optimum ones, respectively. For instance, plan 1 results in the least cost (*attribute*) of R 120 for scenario A. The probability of scenario A occurrence is assumed to be 0.25. It is assumed that 120 is obtained through the approaches detailed in this book. The details are shown in Table 11.1.

---

[10] Say, for different load forecasts.

[11] For instance, if attribute is defined as LOLE to be less than a prespecified value, a robust plan is the one for which LOLE is less than that value if any of the possible scenarios happens.

Table 11.1 Plan–scenario matrix

Plan/scenario	A	B	C
1	120[a]	–	–
2	–	140	–
3	–	–	110
Probability	0.25	0.50	0.25

[a] R

Table 11.2 Plan–scenario costs

Plan/scenario	A	B	C
1	120	120 + 30	120 + 8
2	140 + 0	140	140 + 16
3	110 + 15	110 + 35	110
Probability	0.25	0.5	0.25

Table 11.3 Plan–scenario summary

Plan/scenario	A	B	C
1	120	150	128
2	140	140	156
3	125	145	110
Probability	0.25	0.5	0.25

Now assume that to make plan 1 robust for scenario B, too, we have to invest more extra R 30. This figure may be 8 to make it robust for scenario C. The values are shown in Table 11.2 and summarized in Table 11.3.

Looking at the results, we note that if plan 3 is selected as the primary choice, an extra cost of 35 and a separate extra cost of 15 for scenarios B and A, respectively, make it robust for these scenarios, too. However, we should check for the possible overlaps and interactions of the solutions provided for scenarios B and A. It may happen that instead of 35 + 15 = 50 extra cost, an extra cost of 35 (110 + 35 = 145) may make a robust solution for all scenarios. We should make the same tests on other rows for the final decision.

Although this approach may be applied in principle, it is a costly solution, as it totally ignores the probability of scenarios occurrences. There are some systematic approaches to deal with it, as detailed below.

11.5.1 Expected Cost Criterion

According to this criterion, the sum of the costs times their respective probabilities are calculated. The plan with the least expected cost is the final choice. As shown in Table 11.4, plan 3 with the expected cost of 131.25 is selected.

Table 11.4 Expected cost results

Plan/scenario	A	B	C	Expected cost
1	120	150	128	137
2	140	140	156	144
3	125	145	110	131.25
Probability	0.25	0.50	0.25	–

11.5.2 Min-max Regret Criterion

Three steps are employed here.

(a) For each scenario, the plan–scenario regret matrix is formed using the following relationship and as shown in Table 11.5.

$$r_{ij} = a_{ij} - \min\{a_{ij}, i = 1, \ldots, \text{Number of plans}\} \tag{11.2}$$

where

a_{ij} The attribute of plan i in scenario j

r_{ij} The i–j th element of plan–scenario regret matrix

(b) For each plan i, its maximum regret is calculated for various scenarios as shown in Table 11.6, i.e.

$$r_i = \max\{r_{ij}, j = 1, \ldots, \text{Number of scenarios}\} \tag{11.3}$$

(c) The plan with the minimum r_i is selected, i.e.

$$\text{Final plan} = \min\{r_i\}, \quad i = 1, \ldots, \text{Number of plans} \tag{11.4}$$

In other words, plan 1 is selected as the final choice.

Table 11.5 Plan–scenario regret matrix

Plan/scenario	A	B	C
1	0	10	18
2	20	0	46
3	25	5	0

Table 11.6 The maximum regrets results

Plan	Maximum regret
1	18
2	46
3	25

11.5.3 Laplace Criterion

According to this criterion, the plan with the minimum total cost is selected as the final plan, as shown in Table 11.7.

11.5.4 The Van Neuman–Morgenstern (VNM) Criterion

According to this criterion, either the most pessimistic or the most optimistic cases are generated as shown in Tables 11.8 and 11.9. In the most pessimistic case, it is assumed that this case would happen in practice and the plan attribute (VNM) is calculated for that most pessimistic scenario. The final choice would be the one with the lowest VNM (i.e. plan 3 in Table 11.8).

In the most optimistic case, it is assumed that this case would happen. The rest is as above. So, plan 3 with a VNM equal to 110 would be the final choice.

11.5.5 Hurwicz Criterion

Based on this criterion, a compromise is made between the optimistic and pessimistic scenarios as follows

1. Assign a value to α in the range [0, 1] so that zero implies a pessimistic decision while 1 shows an optimistic choice.
2. Calculate the attribute for the most pessimistic scenario (A).
3. Calculate the attribute for the most optimistic scenario (B).
4. Calculate the attribute of each plan from

$$\text{Attribute} = \alpha B + (1 - \alpha)A$$

5. Select the plan with the best attribute.

For instance, for the above case and if $\alpha = 0.8$, we have

$$\text{Attribute}_1 = 0.8(120) + 0.2(150) = 126$$
$$\text{Attribute}_2 = 0.8(140) + 0.2(150) = 143.2$$
$$\text{Attribute}_3 = 0.8(110) + 0.2(145) = 115$$

So, plan 3 would be the final choice.

Table 11.7 Laplace criterion results

Plan/scenario	A	B	C	Laplace criterion
1	120	150	128	398
2	140	140	156	436
3	125	145	110	380

Table 11.8 VNM pessimistic results

Plan/scenario	A	B	C	VNM
1	120	150	128	150
2	140	140	156	156
3	125	145	110	145

Table 11.9 VNM optimistic results

Plan/scenario	A	B	C	VNM
1	120	150	128	120
2	140	140	156	140
3	125	145	110	110

11.5.6 Discussion

Other criteria may also be used. The problem with all the aforementioned criteria is the fact that the selected plan is optimum only for a single scenario and may be unable to fulfill all constraint requirements for other scenarios. If we are going to select a plan, robust for all scenarios, we may make some modifications in objective functions. For instance in NEP, we may add the costs over various scenarios, so that the final plan is robust for all scenarios. Obviously, the solution will be much more complicated and heuristic based algorithms, such as Genetic Algorithm (GA) may be used to solve the problem.

As the research trends are discussed in Chap. 12, no reference is cited here.

Chapter 12
Research Trends in Power System Planning

12.1 Introduction

Various aspects of power system planning were covered in Chaps. 4–10. Planning in the presence of uncertainties was addressed in Chap. 11. We discussed there some basic concepts appearing in power system planning literature. In this chapter, we are going to cite some references; addressing research trends in power system planning.

12.2 General Observations

We mentioned some references at the end of each chapter which were specific to the materials covered there. As we have often mentioned so far, the models and the solution algorithms are not unique and various versions may be developed. Moreover, once we come to a de-regulated environment, the distinctions are more pronounced. In two earlier papers [1, 2 of Sect. 12.3.1], the models and the publications are classified to date. To save space and to avoid repeating, some references from 2003 onward are cited here; except otherwise specified. We should emphasize that the list of the references are, by no means, complete and the interested reader may consult the vast literature available on the subjects through following some cited references. An important point worth mentioning is that although some references deal with either the traditional or the de-regulated environment, the concepts developed may apply to both with some modifications. Moreover, the techniques and the solution algorithms developed for a specific case (for instance GEP) may be, somehow, used to solve another case (for instance TEP). So, while we classify the publications according to the basic planning issues (LF, GEP, TEP, etc.), some other classifications may also be tried; for instance, based on models, solution algorithms, etc. We do not go into the details of the references, as the book is intended to be a text

H. Seifi and M. S. Sepasian, *Electric Power System Planning*, Power Systems,
DOI: 10.1007/978-3-642-17989-1_12,

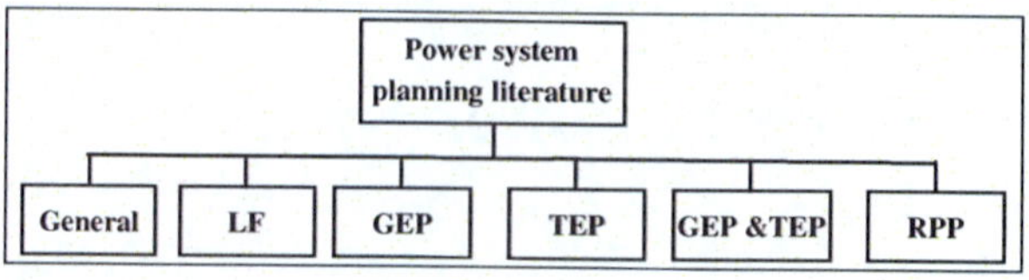

Fig. 12.1 A basic classification of power system planning literature

book. However, we encourage the students and the experts to analyze the references, more thoroughly and classify them as they wish.

As a simple case, consider Fig. 12.1 in which the power system planning literature is categorized as shown.

We discussed earlier in Chap. 11 how in each case, the modeling, the objective functions, the constraints, the solution algorithms, the uncertainties involved, etc. may be also affected by moving toward the de-regulated environment. We have not addressed these details in the references cited in Sect. 12.3. However, as Distributed Generations (DGs) are more and more appearing in practice, some references are mentioned under GEP heading. Moreover, both SEP and NEP research trends are mentioned under the heading of TEP. However, due to large number of papers under TEP, this heading is categorized as *traditional* and *de-regulated* environments. To save space, we have stressed only on journal papers except some limited number of conference papers.

12.3 References

12.3.1 General

[1] Hirst E, Kirby B. Key transmission planning issues. The Electricity Journal 2001; 14 (8): 59–70.

[2] Romero R, Monticelli A, Garcia A, Haffner S. Test systems and mathematical models for transmission network expansion planning. IEE Proceeding of Generation, Transmission and Distribution, 2002; 149 (1): 27–36.

[3] Latorre G, Cruz RD, Areiza JM, Villegas A. Classification of publications and models on transmission expansion planning. IEEE Trans. Power Syst 2003; 18 (2): 938–46.

[4] CIGRE WG 37.30. Network panning in a de-regulated environment. Paris: CIGRE Publication; 2003.

[5] Nadira R, Austria RR, Dortolina CA. Transmission planning today: A challenging undertaking. The Electricity Journal 2004; 17 (4): 24–32.

[6] Yousefi GR, Seifi H. An application of GA and heuristic algorithms in Power Systems planning. In: Proceeding of 6th Annual Int. Conference on Statistics, Mathematics and Related Fields 2007. Honolulu, Hawaii.

[7] CIGRE WG C1.7. Managing the complexity and uncertainties of load, generation and markets in system development planning. Paris: CIGRE Publication; 2008.

12.3.2 LF (2000 Onward)

[1] Da, X, Jiangyan Y, Jilai Y. The physical series algorithm of mid-long term load forecasting of power systems. Electric Power Systems Research 2000; 53 (1): 31–37.

[2] Chen GJ, Li KK, Chung TS, Sun HB, Tang GQ. Application of an innovative combined forecasting method in power system load forecasting. Electric Power Systems Research 2001; 59 (2): 131–7.

[3] Jia NX, Yokoyama R, Zhou YC, Gao ZY. A flexible long-term load forecasting approach based on new dynamic simulation theory-GSIM. International Journal of Electrical Power & Energy Systems 2001; 23 (7): 549–56

[4] Kermanshahi B, Iwamiya H. Up to year 2020 load forecasting using neural nets. International Journal of Electrical Power & Energy Systems 2002; 24 (9): 789–97.

[5] Metaxiotis K, Kagiannas A, Askounis D, Psarras J. Artificial intelligence in short term electric load forecasting: a state-of-the-art survey for the researcher. Energy Conversion and Management 2003; 44 (9): 1525–34.

[6] Hsu CC, Chen CY. Regional load forecasting in Taiwan-applications of artificial neural networks. Energy Conversion and Management 2003; 44 (12): 1941–9.

[7] Pai PF, Hong WC. Forecasting regional electricity load based on recurrent support vector machines with genetic algorithms. Electric Power Systems Research 2005; 74 (3): 417–25.

[8] Wang J, Ma Z, Li L. Detection, mining and forecasting of impact load in power load forecasting. Applied Mathematics and Computation 2005; 168 (1): 29–39.

[9] Soares LJ, Souza LR. Forecasting electricity demand using generalized long memory. International Journal of Forecasting 2006; 22 (1): 17–28.

[10] Pai PF. Hybrid ellipsoidal fuzzy systems in forecasting regional electricity loads. Energy Conversion and Management 2006; 47 (15–16): 2283–9.

[11] Carpinteiro OAS, Leme RC, Souza ACZ, Pinheiro CAM, Moreira EM. Long-term load forecasting via a hierarchical neural model with time integrators. Electric Power Systems Research 2007; 77 (3–4): 371–8.

[12] Hong WC. Chaotic particle swarm optimization algorithm in a support vector regression electric load forecasting model. Energy Conversion and Management 2009; 50 (1): 105–17.

[13] Hong WC. Electric load forecasting by support vector model. Applied Mathematical Modelling 2009; 33 (5): 2444–54.

[14] Niu D, Li J, Liu D. Middle-long power load forecasting based on particle swarm optimization. Computers & Mathematics with Applications 2009; 57 (11–12): 1883–9.

[15] Hahn H, Meyer-Nieberg S, Pickl S. Electric load forecasting methods: Tools for decision making. European Journal of Operational Research 2009; 199 (3): 902–7.

[16] Pedregal DJ, Trapero JR. Mid-term hourly electricity forecasting based on a multi-rate approach. Energy Conversion and Management 2010; 51 (1): 105–11.
[17] AlRashidi MR, EL-Naggar KM. Long term electric load forecasting based on particle swarm optimization. Applied Energy 2010; 87 (1): 320–6.
[18] Wu Q. A hybrid-forecasting model based on Gaussian support vector machine and chaotic particle swarm optimization. Expert Systems with Applications 2010; 37 (3): 2388–94.
[19] Wang J, Zhu S, Zhang W, Lu H. Combined modeling for electric load forecasting with adaptive particle swarm optimization. Energy 2010; 35 (4): 1671–8.

12.3.3 GEP

[1] Kannan S, Slochanal SMR, Subbaraj P, Padhy NP. Application of particle swarm optimization technique and its variants to generation expansion planning problem. Electric Power Systems Research 2004; 70 (3): 203–10.
[2] Chung TS, Li YZ, Wang ZY. Optimal generation expansion planning via improved genetic algorithm approach. International Journal of Electrical Power & Energy Systems 2004; 26 (8): 655–9.
[3] Kagiannas AG, Askounis DT, Psarras J. Power generation planning: a survey from monopoly to competition. International Journal of Electrical Power & Energy Systems 2004; 26 (6): 413–21.
[4] Chen H, Wang X, Zhao X. Generation planning using Lagrangian relaxation and probabilistic production simulation. International Journal of Electrical Power & Energy Systems 2004; 26 (8): 597–605.
[5] Kannan S, Slochanal SMR, Padhy NP. Application and comparison of metaheuristic techniques to generation expansion planning problem. IEEE Trans. Power Syst 2005; 20 (1): 466–75.
[6] Agalgaonkar AP, Kulkarni SV, Khaparde SA. Evaluation of configuration plans for DGs in developing countries using advanced planning techniques. IEEE Trans. Power Syst 2006; 21 (2): 973–81.
[7] Dragoon K, Dvortsov V. Z-method for power system resource adequacy applications. IEEE Trans. Power Syst 2006; 21 (2): 982–8.
[8] Voropai NI, Ivanova EY. Shapley game for expansion planning of generating companies at many non-coincident criteria. IEEE Trans. Power Syst 2006; 21 (4): 1630–7.
[9] Chen SL, Zhan TS, Tsay MT. Generation expansion planning of the utility with refined immune algorithm. Electric Power Systems Research 2006; 76 (4): 251–8.
[10] Kannan S, Slochanal SMR, Baskar S, Murugan P. Application and comparison of metaheuristic techniques to generation expansion planning in the

partially de-regulated environment. IET Generation, Transmission & Distribution 2007; 1 (1): 111–8.

[11] Meza JLC, Yildirim MB, Masud ASM. A model for the multiperiod multiobjective power generation expansion problem. IEEE Trans. Power Syst 2007; 22 (2): 871–8.

[12] Sirikum J, Techanitisawad A, Kachitvichyanukul V. A new efficient GA-Benders' decomposition method: For power generation expansion planning with emission controls. IEEE Trans. Power Syst 2007; 22 (3): 1092–100.

[13] Kannan S, Baskar S, McCalley JD, Murugan P. Application of NSGA-II algorithm to generation expansion planning. IEEE Trans. Power Syst 2009; 24 (1): 454–61.

[14] Wang J, Shahidehpour M, Li Z, Botterud A. Strategic generation capacity expansion planning with incomplete information. IEEE Trans. Power Syst 2009; 24 (2): 1002–10.

[15] Piccolo A, Siano P. Evaluating the impact of network investment deferral on distributed generation expansion. IEEE Trans. Power Syst 2009; 24 (3): 1559–67.

[16] Meza JLC, Yildirim MB, Masud ASM. A multiobjective evolutionary programming algorithm and its applications to power generation expansion planning. IEEE Trans. Systems, Man and Cybernetics, Part A: Systems and Humans 2009; 39 (5): 1086–96.

[17] Shayanfar HA, Lahiji AS, Aghaei J, Rabiee A. Generation expansion planning in pool market: A hybrid modified game theory and improved genetic algorithm. Energy Conversion and Management 2009; 50 (5): 1149–56.

[18] Murugan P, Kannan S, Baskar S. NSGA-II algorithm for multi-objective generation expansion planning problem. Electric Power Systems Research 2009; 79 (4): 622–8.

[19] Chen Q, Kang C, Xia Q, Zhong J. Power generation expansion planning model towards low-carbon economy and its application in China. IEEE Trans. Power Syst 2010; 25 (2): 1117–25.

[20] Pereira AJC, Saraiva JT. A decision support system for generation expansion planning in competitive electricity markets. Electric Power Systems Research 2010; 80 (7): 778–87.

[21] Kaymaz P, Valenzuela J, Park CS. Transmission congestion and competition on power generation expansion. IEEE Trans. Power Systems 2007; 22 (1): 156–63.

[22] Murugan P, Kannan S, Baskar S. Application of NSGA-II algorithm to single-objective transmission constrained generation expansion planning. IEEE Trans. Power Syst 2009; 24 (4): 1790–7.

[23] El-Khattam W, Hegazy YG, Salama MMA. An integrated distributed generation optimization model for distribution system planning. IEEE Trans. Power Syst 2005; 20 (2): 1158–65.

[24] Celli G, Ghiani E, Mocci S, Pilo F. A multiobjective evolutionary algorithm for the sizing and siting of distributed generation. IEEE Trans. Power Syst 2005; 20 (2): 750–7.

12.3.4 TEP

12.3.4.1 TEP in Traditional Environment

[1] Yu H, Chung CY, Wong KP, Zhang JH. A chance constrained transmission network expansion planning method with consideration of load and wind farm uncertainties. IEEE Trans. Power Syst 2009; 24 (3): 1568–76.
[2] Romero R, Rocha C, Mantovani M, Mantovani JRS. Analysis of heuristic algorithms for the transportation model in static and multistage planning in network expansion systems. IEE Proceedings-Generation, Transmission and Distribution 2003; 150 (5): 521–6.
[3] Hashimoto SHM, Romero R, Mantovani JRS. Efficient linear programming algorithm for the transmission network expansion planning problem. IEE Proceedings—Generation, Transmission and Distribution 2003; 150 (5): 536–42.
[4] Chung TS, Li KK, Chen GJ, Xie JD, Tang GQ. Multi-objective transmission network planning by a hybrid GA approach with fuzzy decision analysis. International Journal of Electrical Power & Energy Systems 2003; 25 (3): 187–92.
[5] Al-Hamouz ZM, Al-Faraj AS. Transmission-expansion planning based on a non-linear programming algorithm. Applied Energy 2003; 76 (1–3): 169–77.
[6] Escobar AH, Gallego RA, Romero R. Multistage and coordinated planning of the expansion of transmission systems. IEEE Trans. Power Syst 2004; 19 (2): 735–44.
[7] Manso LAF, Leite da Silva AM. Probabilistic criteria for power system expansion planning. Electric Power Systems Research 2004; 69 (1): 51–8.
[8] Romero R, Rocha C, Mantovani JRS, Sanchez IG. Constructive heuristic algorithm for the DC model in network transmission expansion planning. IEE Proceedings—Generation, Transmission and Distribution 2005; 152 (2): 277–82.
[9] Sanchez IG, Romero R, Mantovani JRS, Rider MJ. Transmission expansion planning using the DC model and nonlinear-programming technique. IEE Proceedings—Generation, Transmission and Distribution 2005; 152 (6): 763–9.
[10] de J Silva I, Rider MJ, Romero R, Garcia AV, Murari CA. Transmission network expansion planning with security constraints. IEE Proceedings—Generation, Transmission and Distribution 2005; 152 (6): 828–36.
[11] Sepasian MS, Seifi H, Foroud AA, Hosseini SH, Kabir EM. A new approach for substation expansion planning. IEEE Trans. Power Syst 2006; 21 (2): 997–1004.
[12] Silva IdJ, Rider MJ, Romero R, Murari CAF. Transmission network expansion planning considering uncertainty in demand. IEEE Trans. Power Syst 2006; 21 (4): 1565–73.

[13] Choi J, Mount TD, Thomas RJ, Billinton R. Probabilistic reliability criterion for planning transmission system expansions. IEE Proceedings—Generation, Transmission and Distribution 2006; 153 (6): 719–27.

[14] Romero R, Asada EN, Carreno E, Rocha C. Constructive heuristic algorithm in branch-and-bound structure applied to transmission network expansion planning. IET Generation, Transmission & Distribution 2007; 1 (2): 318–23.

[15] Xie M, Zhong J, Wu FF. Multiyear transmission expansion planning using ordinal optimization. IEEE Trans. Power Syst 2007; 22 (4): 1420–8.

[16] Oliveira GC, Binato S, Pereira MVF. Value-based transmission expansion planning of hydrothermal systems under uncertainty. IEEE Trans. Power Syst 2007; 22 (4): 1429–35.

[17] Romero R, Rider MJ, Silva I de J. A metaheuristic to solve the transmission expansion planning. IEEE Trans. Power Syst 2007; 22 (4): 2289–91.

[18] Rider MJ, Garcia AV, Romero R. Power system transmission network expansion planning using AC model. IET Generation, Transmission & Distribution 2007; 1 (5): 731–42.

[19] Seifi H, Sepasian MS, Haghighat H, Foroud AA, Yousefi GR, Rae S. Multi-voltage approach to long-term network expansion planning. IET Generation, Transmission & Distribution 2007; 1 (5): 826–35.

[20] Jin YX, Cheng HZ, Yan JY, Zhang L. New discrete method for particle swarm optimization and its application in transmission network expansion planning. Electric Power Systems Research 2007; 77 (3–4): 227–33.

[21] Barros JRP, Melo ACG, Leite da Silva AM. An approach to the explicit consideration of unreliability costs in transmission expansion planning. European Trans. Electrical Power 2007; 17: 401–12.

[22] Rider MJ, Garcia AV, Romero R. Transmission system expansion planning by a branch-and-bound algorithm. IET Generation, Transmission & Distribution 2008; 2 (1): 90–9.

[23] Shayeghi H, Jalilzadeh S, Mahdavi M, Hadadian H. Studying influence of two effective parameters on network losses in transmission expansion planning using DCGA. Energy Conversion and Management 2008; 49 (11): 3017–24.

[24] Fan H, Cheng H. Transmission network expansion planning with security constraints based onbi-level linear programming. European Trans. Electrical Power 2009; 19: 388–99.

[25] Sum-Im T, Taylor GA, Irving MR, Song YH. Differential evolution algorithm for static and multistage transmission expansion planning. IET Generation, Transmission & Distribution 2009; 3 (4): 365–84.

[26] Bustamante-Cedeño E, Arora S. Multi-step simultaneous changes constructive heuristic algorithm for transmission network expansion planning. Electric Power Systems Research 2009; 79 (4): 586–94.

[27] Mahdavi M, Shayeghi H, Kazemi A. DCGA based evaluating role of bundle lines in TNEP considering expansion of substations from voltage level point of view. Energy Conversion and Management 2009; 50 (8): 2067–73.

[28] Shayeghi H, Mahdavi M, Bagheri A. Discrete PSO algorithm based optimization of transmission lines loading in TNEP problem. Energy Conversion and Management 2010; 51 (1): 112–21.

12.3.4.2 TEP in De-regulated Environment

[1] Fang R, Hill DJ. A new strategy for transmission expansion in competitive electricity markets. IEEE Trans. Power Syst 2003; 18 (1): 374–80.

[2] Buygi, MO, Balzer G, Shanechi HM, Shahidehpour M. Market-based transmission expansion planning. IEEE Trans. Power Syst 2004; 19 (4): 2060–7.

[3] Choi J, El-Keib AA, Tran T. A fuzzy branch and bound-based transmission system expansion planning for the highest satisfaction level of the decision maker. IEEE Trans. Power Syst 2005; 20 (1): 476–84.

[4] Gribik PR, Shirmohammadi D, Graves JS, Kritikson JG. Transmission rights and transmission expansions. IEEE Trans. Power Syst 2005; 20 (4): 1728–37.

[5] Buygi MO, Shanechi HM, Balzer G, Shahidehpour M, Pariz N. Network planning in unbundled power systems. IEEE Trans. Power Syst 2006; 21 (3): 1379–87.

[6] Kristiansen T, Rosellón J. A merchant mechanism for electricity transmission expansion problem. Journal of Regulatory Economics 2006; 29 (2): 167–93.

[7] Sauma EE, Oren SS. Economic criteria for planning transmission investment in restructured electricity markets. IEEE Trans. Power Syst 2007; 22 (4): 1394–1405.

[8] Shrestha GB, Fonseka PAJ. Optimal transmission expansion under different market structures. IET Generation, Transmission & Distribution 2007; 1 (5): 697–706.

[9] Tanaka M. Extended price cap mechanism for efficient transmission expansion under nodal pricing. Networks and Spatial Economics 2007; 7 (3): 257–75.

[10] Lu W, Bompard E, Napoli R, Jiang X. Heuristic procedures for transmission planning in competitive electricity markets. Electric Power Systems Research 2007; 77 (10): 1337–48.

[11] Kristiansen T. Allocation of long-term financial transmission rights for transmission expansion. European Journal of Operational Research 2008; 184 (3): 1122–39.

[12] de la Torre S, Conejo AJ, Contreras J. Transmission Expansion Planning in Electricity Markets. Trans. Power Systems 2008; 23 (1): 238–48.

[13] Hua-zheng J, Hao-zhong C. A New Method of Transmission Network Flexible Planning Using the Connection Number. J. Shanghai Jiaotong Univ. (Sci.) 2008; 13 (4): 436–42.

[14] Littlechild SC, Skerk CJ. Transmission expansion in Argentina 1: The origins of policy. Energy Economics 2008; 30 (4): 1367–84.

[15] Littlechild SC, Skerk CJ. Transmission expansion in Argentina 2: The Fourth Line revisited. Energy Economics 2008; 30 (4): 1385–419.

[16] Littlechild SC, Skerk CJ. Transmission expansion in Argentina 3: The evolution of policy. Energy Economics 2008; 30 (4): 1420–61.

[17] Littlechild SC, Skerk CJ. Transmission expansion in Argentina 4: A review of performance. Energy Economics 2008; 30 (4): 1462–90.

[18] Littlechild SC, Skerk CJ. Transmission expansion in Argentina 5: A The regional electricity forum of Buenos Aires province. Energy Economics 2008; 30 (4): 1491–526.

[19] Littlechild SC, Skerk CJ. Transmission expansion in Argentina 6: Distribution companies, regulation and the Public Contest method. Energy Economics 2008; 30 (4): 1527–35.

[20] Léautier TO, Thelen V. Optimal expansion of the power transmission grid: why not?. Journal of Regulatory Economics. 2009; 36 (2): 127–53.

[21] Zhao JH, Dong ZY, Lindsay P, Wong KP. Flexible transmission expansion planning with uncertainties in an electricity market. IEEE Trans. Power Syst 2009; 24 (1): 479–88.

[22] Maghouli P, Hosseini SH, Buygi MO, Shahidehpour M. A multi-objective framework for transmission expansion planning in deregulated environments. IEEE Trans. Power Syst 2009; 24 (2): 1051–61.

[23] Garces LP, Conejo AJ, Garcia-Bertrand R, Romero R. A bilevel approach to transmission expansion planning within a market environment. IEEE Trans. Power Syst 2009; 24 (3): 1513–22.

[24] Eliassi M, Seifi H, Haghifam MR. Multi-objective value-based reliability transmission planning using expected interruption cost due to transmission constraint. In: Proceeding of International Conference on Electric Power and Energy Conversion Systems–EPECS 2009. Sharjah, United Arab Emirates.

[25] Georgilakis PS. Market-based transmission expansion planning by improved differential evolution. International Journal of Electrical Power & Energy Systems 2010; 32 (5): 450–6.

[26] Alguacil N, Motto AL, Conejo AJ. Transmission expansion planning: a mixed-integer LP approach. IEEE Trans. Power Syst 2003; 18 (3): 1070–7.

[27] Yang N, Wen F. A chance constrained programming approach to transmission system expansion planning. Electric Power Systems Research 2005; 75 (2–3): 171–7.

[28] Choi J, Tran T, El-Keib AA, Thomas R, Oh H, Billinton R. A Method for transmission system expansion planning considering probabilistic reliability criteria. IEEE Trans. Power Syst 2005; 20 (3): 1606–15.

[29] Choi J, Mount TD, Thomas RJ. Transmission expansion planning using contingency criteria. IEEE Trans. Power Syst 2007; 22 (4): 2249–61.

[30] Wang Y, Cheng H, Wang C, Hu Z, Yao L, Ma Z, Zhu Z. Pareto optimality-based multi-objective transmission planning considering transmission congestion. Electric Power Systems Research 2008; 78 (9): 1619–26.
[31] Kazerooni AK, Mutale J. Transmission network planning under security and environmental constraints. IEEE Trans. Power Syst 2010; 25 (2): 1169–78.

12.3.5 GEP and TEP

[1] Sepasian MS, Seifi H, Akbari A, Rae S. A multi-area fuel based generation-transmission planning procedure. In: Proceeding of International Conference on Power and Energy Systems–EuroPES 2007. Palma deMallorca, Spain.
[2] Roh JH, Shahidehpour M, Fu Y. Market-based coordination of transmission and generation capacity planning. IEEE Trans. Power Syst 2007; 22 (4): 1406–19.
[3] Tor OB, Guven AN, Shahidehpour M. Congestion-driven transmission planning considering the impact of generator expansion. IEEE Trans. Power Syst 2008; 23 (2): 781–9.
[4] Roh JH, Shahidehpour M, Wu L. Market-based generation and transmission planning with uncertainties. IEEE Trans. Power Syst 2009; 24 (3): 1587–98.
[5] Sepasian MS, Seifi H, Foroud AA, Hatami AR. A multiyear security constrained hybrid generation-transmission expansion planning algorithm including fuel supply costs. IEEE Trans. Power Syst 2009; 24 (3): 1609–18.
[6] Liu G, Sasaki H, Yorino N. Application of network topology to long range composite expansion planning of generation and transmission lines. Electric Power Systems Research 2001; 57 (3): 157–62.

12.3.6 RPP (2000 Onward)

[1] Yorino N, El-Araby EE, Sasaki H, Harada S. A new formulation for FACTS allocation for security enhancement against voltage collapse. IEEE Trans. Power Syst 2003; 18 (1): 3–10.
[2] Yousefi GR, Seifi H, Sepasion MS, Haghighat H, Riahi R, Hosseini H, Akhavan A. A new reactive power planning procedure for Iranian power grid. Electric Power System Research 2004; 72 (3): 225–34.
[3] Chen YL, Ke YL. Multi-objective VAr planning for large-scale power systems using projection-based two-layer simulated annealing algorithms. IEE Proceedings—Generation, Transmission and Distribution 2004; 151 (4): 555–60.
[4] Zhang W, Li F, Tolbert LM. Review of reactive power planning: objectives, constraints, and algorithms. IEEE Trans. Power Syst 2007; 22 (4): 2177–86.

[5] Eghbal M, Yorino N, El-Araby EE, Zoka Y. Multi-load level reactive power planning considering slow and fast VAR devices by means of particle swarm optimisation. IET Generation, Transmission & Distribution 2008; 2 (5): 743–51.
[6] Liu H, Jin L, McCalley JD, Kumar R, Ajjarapu V, Elia N. Planning reconfigurable reactive control for voltage stability limited power systems. IEEE Trans. Power Syst 2009; 24 (2): 1029–38.

12.3.7 Miscellaneous

[1] Abedi M, Taher SA, Sedigh AK, Seifi H. Controller design using μ-synthesis for static VAR compensator to enhance the voltage profile for remote induction motor loads. Electric Power Systems Research 1998; 46 (1): 35–44.
[2] Haghighat H, Seifi H, Yazdian A. An instantaneous power theory based control scheme for unified power flow controller in transient and steady state conditions. Electric Power Systems Research 2003; 64 (3): 175–80.
[3] Keypour R, Seifi H, Yazdin A. Genetic based algorithm for active power filter allocation and sizing. Electric Power System Research 2004; 71 (1): 41–9.
[4] Ghazizadeh MS, Sheikh K, Seifi H. Iran electricity restructuring, a compromise between competition and stability. IEEE Power and Energy Magazine 2007; 5 (2): 16–20.
[5] Yousefi GR, Seifi H, Shirmohammadi D. A new algorithm for reactive power management and pricing in an open access environment. European Transaction on Electrical Power 2008; 18 (2): 109–26.
[6] Seifi S, Imhof K. Voltage stability analysis in an energy management system, a practical implementation. In: Proceeding of International Conference on Power System Computation Conference (PSCC) 1996. Dresden, Germany.
[7] Kazerooni A, Seifi H. A new approach for phase shifter transformer allocation. In: Proceeding of fifth IEE Conference on AC and DC Power Transmission-ACDC 2006. London, UK.
[8] Akbari-foroud A, Seifi H, Asaie K. An alogorithm for voltage uprating in a sub-transmission network. In: Proceeding of 31th IAEE International Conference on Bridging Energy Supply and Demand 2008. Istanbul, Turkey.

12.4 Exercise 1

Let's focus on TEP literature and try to classify, in more details, the papers according to what is proposed in Fig. 12.2 for TEP in traditional environment and Fig. 12.3 for TEP in de-regulated environment. The following points should be observed

- Try to assign each reference of Sect. 12.3.4 to each category mentioned in Figs. 12.2 and 12.3.
- Add and modify these two figures, if you find a reference for which there is no category.
- Try a more detailed search (in terms of the time of the publication, the conference papers, the databases available, etc.) on TEP.

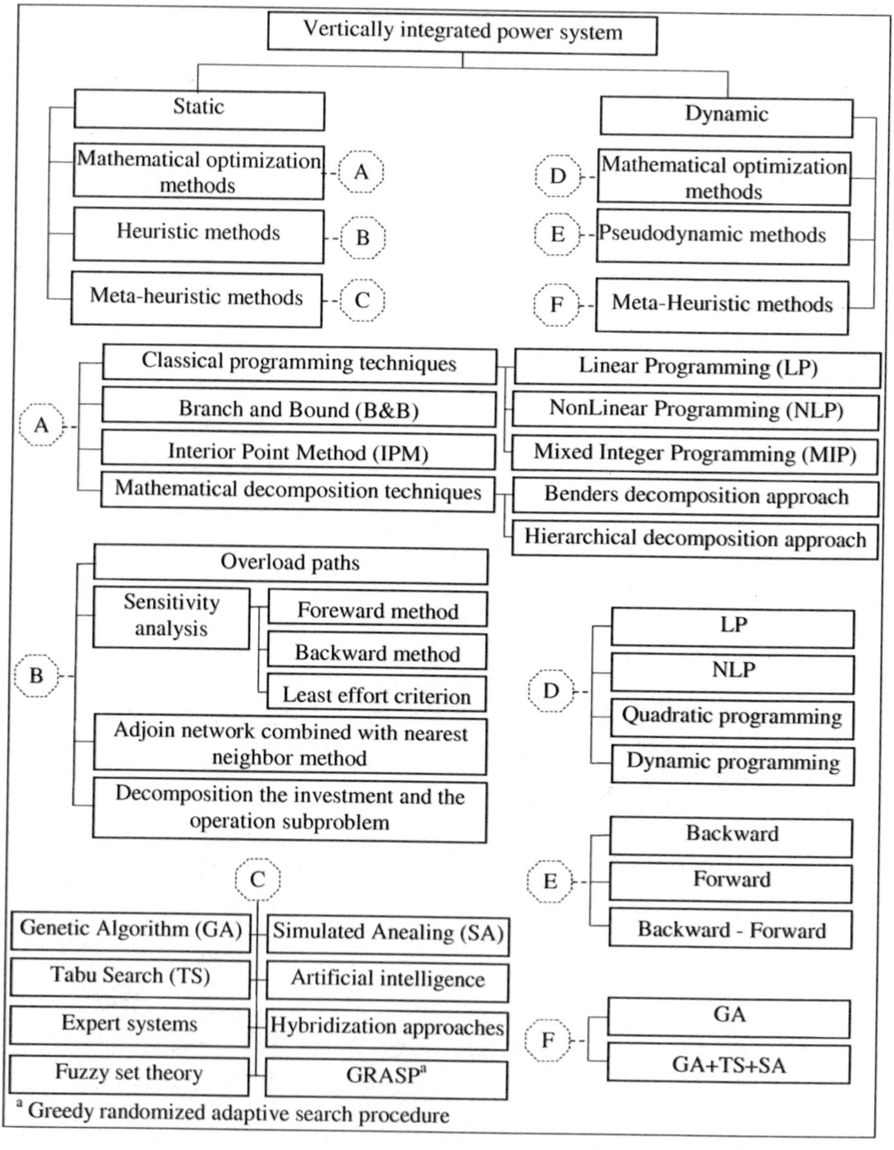

Fig. 12.2 Classification of TEP in traditional environment-research trend

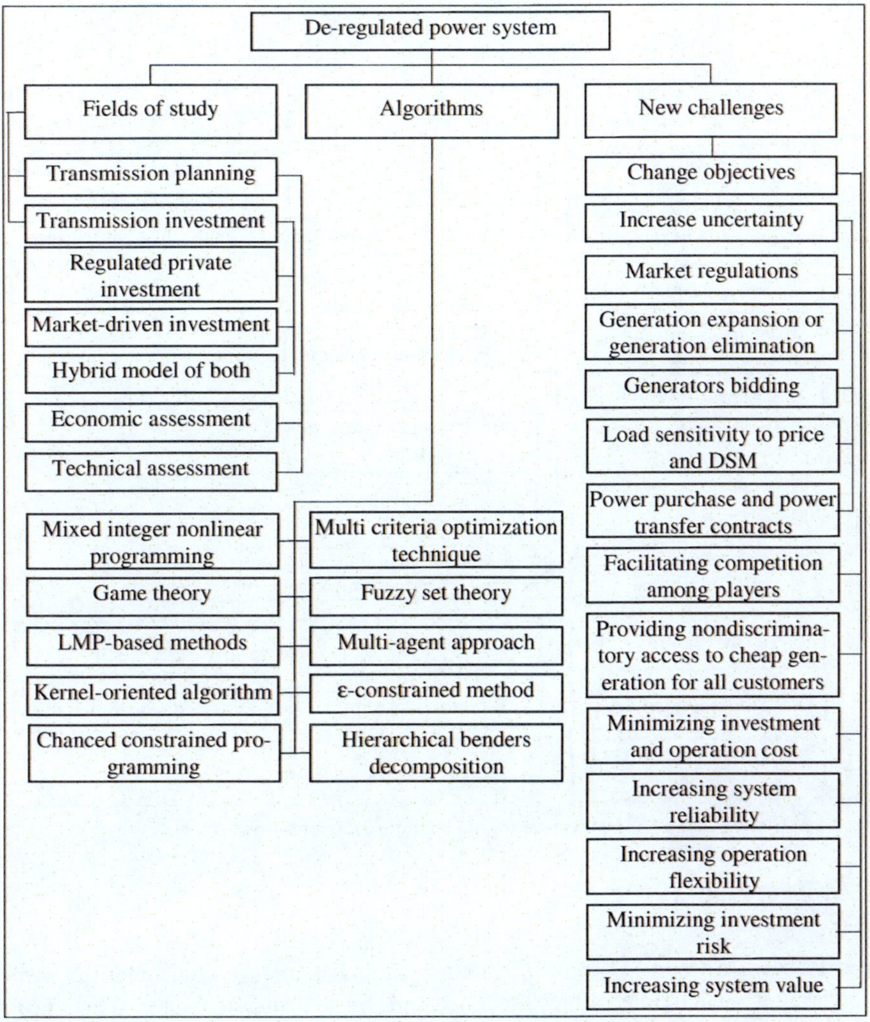

Fig. 12.3 Classification of TEP in de-regulated environment-research trend

- Augment Figs. 12.2 and 12.3; based on the materials found from the previous step. The points raised in Chap. 11 can be quite useful in this regard.
- Try to propose a similar approach for other cases (LF, GEP, RPP, etc.).

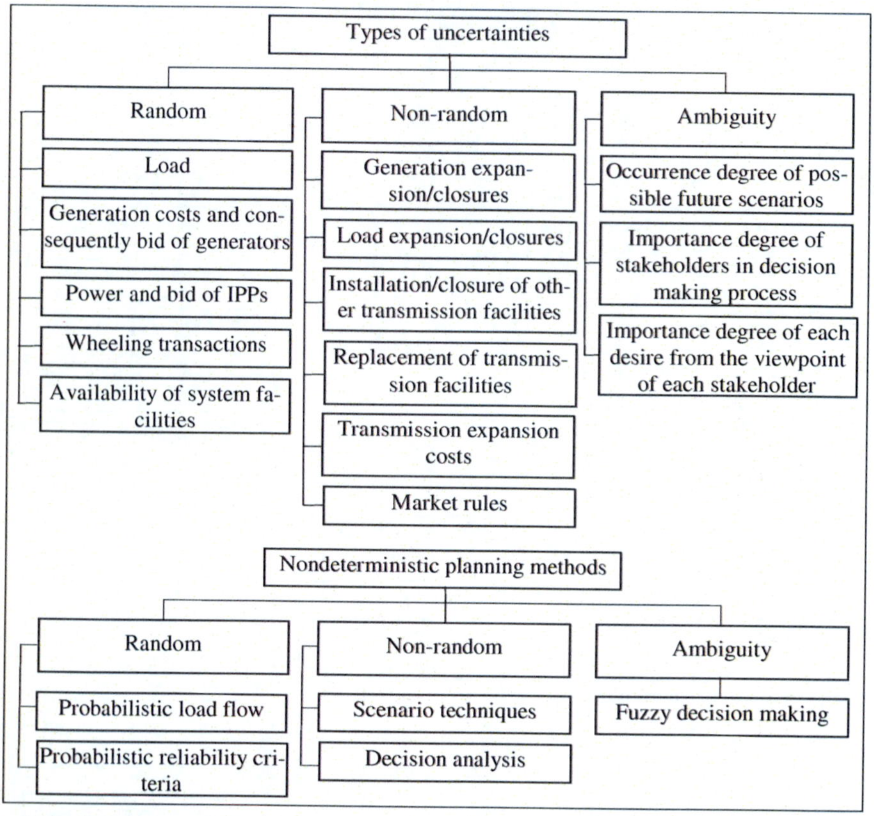

Fig. 12.4 Some types of classification on uncertainties-research trend

12.5 Exercise 2

As discussed in Chap. 11, power system planning is always confronted by some uncertainties which may affect the decisions by a decision maker. The uncertainties may be

- Random
- Non-random

Let's propose a classification of the subjects as shown in Fig. 12.4. Follow the procedure as outlined in Sect. 12.4 for exercise 1. This time, try initially a detailed search on the subject. Modify the classification of Fig. 12.4 to stress the uncertainties involved only in power system planning issues.

Chapter 13
A Comprehensive Example

13.1 Introduction

In the earlier chapters, we discussed the basic issues of power system planning, namely, LF (Chap. 4), GEP (Chaps. 5 and 6), SEP (Chap. 7), NEP (Chaps. 8 and 9) and RPP (Chap. 10). In this chapter, we intend to cover an example with more details so that the reader can, more readily, follow up the steps involved.

We assume the system to be the same as the one for which load forecasting was done in Chap. 4. Initially, SEP is carried out in Sects. 13.2 and 13.3 so that sub-transmission (in Sect. 13.2) and transmission (in Sect. 13.3) substation requirements are determined. Assuming GEP to be known using the approaches as discussed in Chaps. 5 and 6, the NEP results are then shown in Sect. 13.4. The results for RPP are demonstrated in Sect. 13.5.

13.2 SEP Problem for Sub-transmission Level

13.2.1 Basics

The aim is to allocate and size the sub-transmission substation[1] requirements of a system. Current year is assumed to be 2010 and the aim is to solve the problem for 2011 and 2015, separately. The base information (2010) is assumed to be known. Initially the problem is solved for 2015 by which the new substation requirements are determined for that time. Thereafter, those justified for 2015 are assumed as candidates for 2011 and SEP is solved to find those substations which have to be

[1] In this section, by *substation* we mean *sub-transmission substation*.

H. Seifi and M. S. Sepasian, *Electric Power System Planning*, Power Systems,
DOI: 10.1007/978-3-642-17989-1_13, © Springer-Verlag Berlin Heidelberg 2011

constructed in 2011.[2] GA is used to find the solutions. The reader may try any other solution algorithm, however, with observing the modeling requirements discussed in Chap. 7.

13.2.2 System Under Study

The system under study is depicted in Fig. 13.1. It shows an eight-area system. These areas are, typically geographic based. However, if we want to analyze SEP, we may treat some areas together provided they are operated and/or planned by a single entity. For this system, it is assumed that areas E, F and G may be analyzed together. We discuss in details the results for these areas. From here on, we call it area EFG. At the end, we will present the final results for the other areas so that the overall results for the system may be concluded.

13.2.3 Input Data

Area EFG consists of 63 load nodes with the details given in Table 13.1. The load of each node is determined using the approaches presented in Chap. 4.[3]

Moreover, it is assumed that in 2010, there are 7 existing substations[4] with the details provided in Table 13.2.

It is also assumed that in 2011 and 2015, we have to provide some loads, specifically, by substations E1, E2 and E4. In other words, they should be removed from the optimization process; however, have to be observed in substation loadings. These are shown in Table 13.3.

13.2.4 Solution Information

Besides the information as outlined above, some other information are required and considered as follows

- The acceptable voltage drop in a downward grid feeder is considered to be 5%.
- The land cost is considered to be identical in all places.

[2] In practice, the studies should be performed well in advance, as constructing a substation requires enough time. Therefore, the base year is normally not the current year, but, it is the year in which the data are known and already decided upon. For instance, current year may be 2010, base year may be 2013 and study period may be 2014–2020.

[3] For details, see Chap. 4.

[4] Existing substations are denoted by "E".

Fig. 13.1 Eight-area system

- To observe the point that a downward feeder is capable of feeding a maximum of 5 MW load, each load is considered to be dividable to four parts; but considered as the same geographical point.
- The reserve of each substation is considered to be, at least, 30%.
- The loads power factors are considered to be 0.9.
- The cost of the losses is considered to be R1500/kW.
- The distance of each candidate substation to the upward grid is considered to be 5 km; typical costs of upward grid lines/cables are given afterwards.
- It is assumed that some capacity of an existing substation may be reduced (i.e. the substation is derated) or even the substation can be totally removed; provided technically and economically justified. If a capacity is removed, the recovery cost is assumed to be 75% of the substation remaining life. The life of a substation is considered to be 40 years. If for instance in the horizon year of 2015, the life of an existing substation is 20 years and 10 MVA of the capacity is removed (this capacity is determined by the algorithm), $0.75 \times 0.5 = 0.375$ of a 10 MVA substation cost is considered to be recovered (negative cost).
- Various costs of upward and downward feeders are as shown in Table 13.4.

The cost of a substation consists, typically, of the costs of its components, namely, transformers, inward and outward feeders, construction, etc. Table 13.5

Table 13.1 The details of load nodes of area EFG

No.	X	Y	Load in 2011 (MW)	Load in 2015 (MW)
1	54.0636	31.3053	2.26	2.73
2	54.1460	31.6011	2.77	3.34
3	53.6489	31.3161	2.76	3.33
4	53.8627	31.4757	2.44	2.95
5	54.1899	31.7595	3.44	4.15
6	53.8284	31.6996	3.45	4.17
7	53.9779	31.6869	2.46	2.97
8	54.2037	31.7438	2.96	3.57
9	54.1526	31.4586	3.35	4.05
10	54.2589	31.7868	3.46	4.18
11	54.2300	31.7784	3.51	4.23
12	54.1680	31.7510	3.91	4.72
13	53.6597	31.2552	1.00	1.00
14	53.6939	31.3701	0.50	0.50
15	53.7262	31.4917	2.00	2.00
16	53.7246	31.4376	1.00	1.00
17	53.7477	31.5660	1.00	1.00
18	53.7353	31.4608	2.00	2.00
19	54.4700	31.5846	2.52	3.22
20	54.2342	31.5850	2.57	3.29
21	54.9170	31.3268	2.53	3.23
22	54.2946	31.4244	2.61	3.33
23	54.4460	31.6961	1.49	1.91
24	54.4921	31.6176	2.25	2.88
25	54.5295	31.5555	2.25	2.88
26	54.8490	31.4498	1.69	2.16
27	54.4437	31.5791	2.68	3.43
28	54.4267	31.5466	3.15	4.03
29	54.2435	31.2994	2.69	3.45
30	54.5118	31.6341	2.39	3.05
31	54.4653	31.6138	2.66	3.40
32	54.4276	31.5890	2.96	3.79
33	53.0130	31.0467	3.64	4.23
34	53.1931	31.0891	3.39	3.94
35	53.3096	30.9926	3.18	3.69
36	53.5504	30.8559	3.41	3.96
37	53.3012	31.0709	3.97	4.61
38	53.3469	31.1703	3.87	4.49
39	53.3023	31.1219	4.14	4.81
40	53.3665	31.0236	3.52	4.08
41	53.2010	31.1668	3.85	4.47
42	53.2727	31.1560	3.87	4.50
43	53.3962	30.9477	3.29	3.82
44	53.2792	31.1299	3.90	4.53
45	53.2474	31.1244	4.06	4.71

(continued)

Table 13.1 (continued)

No.	X	Y	Load in 2011 (MW)	Load in 2015 (MW)
46	53.0867	31.0543	4.08	4.74
47	53.0399	31.4564	2.00	2.00
48	53.1621	31.1025	11.00	11.00
49	53.4532	30.9277	3.00	3.00
50	54.4913	31.6198	2.00	2.00
51	54.4587	31.6881	3.00	3.00
52	54.5128	31.6233	7.00	7.00
53	54.5127	31.6233	2.00	2.00
54	54.4704	31.6263	4.00	4.00
55	54.4960	31.5715	1.00	1.00
56	54.4937	31.5693	1.00	1.00
57	54.5056	31.5773	1.25	1.25
58	54.5085	31.5789	1.25	1.25
59	54.5113	31.5780	1.25	1.25
60	54.5142	31.5769	1.25	1.25
61	54.2214	31.7674	1.95	2.00
62	53.9886	31.7137	1.50	4.00
63	54.4628	31.6023	2.60	2.60

Table 13.2 Details of existing substations for area EFG in 2010

No.	Name	X	Y	Capacity (MVA)
1	E1	54.2214	31.7674	2 × 30
2	E2	53.9886	31.7137	2 × 15
3	E3	54.1196	31.4471	2 × 15
4	E4	54.4628	31.6023	4 × 15
5	E5	53.2268	31.1167	30 + 15
6	E6	53.2610	31.1230	5 + 7.5
7	E7	53.3820	30.9949	2 × 15

shows the costs of 63 kV:20 kV substations. Such costs for 132 kV:20 kV substations are shown in Table 13.6.

13.2.5 Results

The choice of a candidate substation is an important step.[5] The higher the numbers, the longer solution time would be required. For this example, six candidate substations are assumed as shown in Table 13.7.

[5] Although some mathematical approaches may be proposed for candidate selection, the distribution planners are the best in recommending such locations; as they know current deficiencies in supply points.

Table 13.3 Specific loads in area EFG

No.	Name	X	Y	Specific load (MW)	
				2010	2015
1	E1	54.2214	31.7674	1.95	2
2	E2	53.9886	31.7137	1.5	4
3	E4	54.4628	31.6023	2.6	2.6

Table 13.4 Costs of feeders

Type	Voltage (kV)	Grid	Capacity (p.u.)	Cost (R/km)
Line	63	Upward	0.60	550
Line	63	Upward	1.20	700
Line	63	Upward	2.40	1400
Cable	63	Upward	0.47	6000
Cable	63	Upward	0.93	9500
Line	132	Upward	0.63	432
Line	132	Upward	1.26	540
Line	132	Upward	2.51	972
Line	20	Downward	0.14	76.3
Line	20	Downward	0.29	130
Cable	20	Downward	0.15	420
Cable	20	Downward	0.26	829

Table 13.5 The costs of 63 kV:20 kV substations

Type (MVA)	2×50	3×30	2×40	2×30	2×18.25	2×15	1×30	2×12.5	2×9.33	1×15	2×7.5	1×7.5
Cost (R)	36,500	34,700	34,400	28,450	21,750	22,150	17,250	20,950	20,350	15,950	19,250	15,250

Table 13.6 The costs of 132 kV:20 kV substations

Type (MVA)	2 × 50	2 × 30	2 × 15	1 × 15
Cost (R)	34,700	28,600	24,750	19,250

Table 13.7 The details of candidate substations

No.	Name	X	Y	Capacity (MVA)	
				From	To
1	C1	53.1294	31.0902	1 × 15	2 × 30
2	C2	53.4365	31.2028	1 × 15	2 × 30
3	C3	53.5504	30.8559	1 × 15	2 × 30
4	C4	54.8893	31.3769	1 × 15	2 × 30
5	C5	53.7485	31.4473	1 × 15	2 × 30
6	C6	53.0399	31.4564	1 × 15	2 × 30

Table 13.8 Summary of the results for area EFG in 2015

No.	Name	X	Y	Existing capacity (MVA)	New capacity (MVA)	Loading (MVA)
1	E1	54.2214	31.7674	2 × 30	2 × 30	42
2	E2	53.9886	31.7137	2 × 15	2 × 15	17.8
3	E3	54.1196	31.4471	2 × 15	2 × 15	20.8
4	E4	54.4628	31.6023	4 × 15	4 × 15	42
5	E5	53.2268	31.1167	2 × 30	30 + 15	41.9
6	E6	53.2610	31.1230	5 + 7.5	5 + 7.5	8.6
7	E7	53.3820	30.9949	30 × 2	2 × 15	20.6
8	N1	53.1295	31.0903	0	1 × 20	14
9	N2	54.8893	31.3769	0	1 × 15	7.6
10	N3	53.7486	31.4473	0	1 × 20	13.6

Table 13.9 Summary of the results for area EFG in 2011

No.	Name	X	Y	Existing capacity (MVA)	New capacity (MVA)	Loading (MVA)
1	E1	54.2214	31.7674	2 × 30	2 × 30	34.9
2	E2	53.9886	31.7137	2 × 15	2 × 15	9.3
3	E3	54.1196	31.4471	2 × 15	2 × 15	18.1
4	E4	54.4628	31.6023	4 × 15	4 × 15	41.9
5	E5	53.2268	31.1167	30 + 15	30 + 15	38.2
6	E6	53.2610	31.1230	5 + 7.5	5 + 7.5	7.8
7	E7	53.3820	30.9949	2 × 15	2 × 15	18.2
8	N1	53.1295	31.0903	0	1 × 20	11.6
9	N2	54.8893	31.3769	0	1 × 15	4.7
10	N3	53.7486	31.4473	0	1 × 20	13

The existing substations are assumed to be unexpandable. From the six candidates, C1, C4 and C5 are determined as new required substations with the capacities of 1 × 20 MVA, 1 × 15 MVA and 1 × 20 MVA, respectively in 2015. These are called N1, N2 and N3. The results for area EFG are summarized in Table 13.8. The loadings of all existing as well as new substations are also given.

Now assuming N1, N2 and N3 as candidates for 2011, the process is repeated for the area. The results are summarized in Table 13.9. As seen, all three candidate substations are required for that year, too.

If the algorithm is repeated for the remaining areas, the overall results for the system (Fig. 13.1) are as shown in Table 13.10. As seen, some existing or new substations are of 132 kV type.

13.3 SEP Problem for Transmission Level

After the sub-transmission substation requirements, as detailed in the earlier section, are finalized, the procedure should be repeated, now, for transmission substations.

Table 13.10 Summary of the results for all areas

No.	Name	X	Y	Voltage (kV:kV)	Capacity (MVA)		Load (MW)	
					2011	2015	2011	2015
1	E1	54.2214	31.7674	63:20	2 × 30	2 × 30	31.41	37.80
2	E2	53.9886	31.7137	63:20	2 × 15	2 × 15	8.37	16.02
3	E3	54.1196	31.4471	63:20	2 × 15	2 × 15	16.29	18.72
4	E4	54.4628	31.6023	63:20	4 × 15	4 × 15	37.71	37.80
5	E5	53.2268	31.1167	63:20	30 + 15	30 + 15	34.38	37.71
6	E6	53.2610	31.1230	63:20	5 + 7.5	5 + 7.5	7.02	7.74
7	E7	53.3820	30.9949	63:20	2 × 15	2 × 15	16.38	18.54
8	E8	54.3255	31.8730	63:20	2 × 40	2 × 40	40.41	50.22
9	E9	54.3393	31.8351	63:20	2 × 30	2 × 30	35.01	37.71
10	E10	54.2763	31.9684	63:20	2 × 30	2 × 30	37.62	37.71
11	E11	54.3309	31.9059	63:20	2 × 30	2 × 30	40.32	39.60
12	E12	54.3011	31.8512	63:20	2 × 22.5	2 × 22.5	25.29	27.99
13	E13	54.3868	31.8084	63:20	2 × 30	2 × 30	37.08	37.08
14	E14	54.3950	31.8493	63:20	2 × 40	2 × 40	49.59	53.91
15	E15	54.3638	31.8741	63:20	2 × 30	2 × 30	32.31	37.53
16	E16	54.1978	32.0483	63:20	2 × 40	2 × 40	50.04	50.31
17	E17	54.3490	31.8939	63:20	2 × 40	2 × 40	43.83	49.86
18	E18	54.2467	31.8964	63:20	3 × 30	3 × 30	60.30	60.66
19	E19	54.3845	31.9193	63:20	2 × 22.5 + 30	2 × 22.5 + 30	41.40	47.16
20	E20	54.3173	31.9312	63:20	3 × 15	3 × 15	28.17	28.35
21	E21	54.5799	31.7745	132:20	2 × 15	2 × 15	18.36	17.19
22	E22	54.0231	32.2923	63:20	2 × 30	2 × 30	37.53	32.49
23	E23	53.9954	32.2146	63:20	2 × 40	2 × 40	48.96	50.22
24	E24	53.9256	32.3371	63:20	2 × 30	2 × 30	37.62	34.38
25	E25	55.4422	31.6286	132:20	2 × 30	2 × 30	31.14	36.81
26	E26	55.9801	31.8949	132:20	1 × 15	1 × 15	9.36	8.64
27	E27	54.3251	29.8251	132:20	30	30	14.31	16.74
28	E28	56.9272	34.3403	132:20	15	15	8.82	10.53
29	E29	54.3926	30.0507	132:20	2 × 15	2 × 15	15.21	19.50
30	E30	55.7466	31.7407	132:20	2 × 30	2 × 30	9.36	13.14
31	E31	54.2114	30.4647	132:20	30	30	17.37	18.27
32	E32	56.9664	33.6088	132:20	2 × 30	2 × 30	28.17	33.66
33	E33	54.1004	32.1385	63:20	2 × 30 + 30	2 × 30 + 30	56.43	56.43
34	N1	53.1295	31.0903	63:20	1 × 20	1 × 20	10.44	12.60
35	N2	54.8893	31.3769	63:20	1 × 15	1 × 15	4.23	6.84
36	N3	53.7486	31.4473	63:20	1 × 20	1 × 20	11.70	12.24
37	N4	54.3641	31.8379	63:20	0	2 × 40	0	50.04
38	N5	54.2405	31.9342	63:20	2 × 30	2 × 40	37.71	50.40
39	N6	53.5572	32.0192	63:20	0	1 × 15	0	2.43
40	N7	54.3448	31.9460	63:20	1 × 15	2 × 40	9.36	49.86
41	N8	54.0567	32.1836	63:20	2 × 30	2 × 30	9.00	29.70
42	N9	57.5374	33.1930	132:20	1 × 20	1 × 20	12.06	12.24
43	N10	54.6507	32.3400	63:20	1 × 15	1 × 15	4.50	5.22
44	N11	53.7777	32.3769	63:20	2 × 30	2 × 30	12.51	26.46

Table 13.11 Extra load nodes for transmission SEP

No.	Name	X	Y	Load (MW)		No.	Name	X	Y	Load (MW)	
				2011	2015					2011	2015
1	P1	53.8719	32.1218	10	20	13	P13	53.7472	32.3808	20	20
2	P2	54.6277	31.7413	35	35	14	P14	54.0368	31.9306	8	10
3	P3	54.7858	31.7475	20	20	15	P15	54.0427	31.9270	5	8
4	P4	54.7410	31.6232	22.5	22.5	16	P16	56.8258	33.0100	12	12
5	P5	54.5676	32.2977	15	20	17	P17	53.1241	31.1153	6.8	6.8
6	P6	54.5676	32.2977	15	20	18	P18	53.8600	32.3515	7	7
7	P7	54.0538	31.9487	15	25	19	P19	56.7927	33.8431	4	14
8	P8	54.0152	31.9500	10	18	20	P20	53.7527	32.3908	15	15
9	P9	54.0584	31.9544	12	15	21	P21	53.8152	31.6647	15	17
10	P10	53.3994	31.1941	12	15	22	P22	53.9213	32.3333	12	12
11	P11	54.3346	32.1902	5	5	23	P23	54.0183	31.9505	6	6
12	P12	54.2851	31.8508	55	55	24	P24	54.0196	31.9497	10	12

Table 13.12 The costs of transmission substations

Type (MVA)	2 × 500	2 × 315	2 × 250	2 × 200	2 × 160	2 × 125
Cost (R)	24,8000	21,8000	18,7000	14,9000	13,3000	11,7000

The load nodes, in this case, are in fact the sub-transmission substations loadings, as already given in Table 13.10. In practice, there may be some extra sub-transmission substations requirements due to a large residential complex, an industrial sector, etc., not already observed in sub-transmission SEP. The reason is that these types of consumers may have to be directly supplied through a sub-transmission voltage and may require separate substations. These extra substations are denoted by "P" and shown in Table 13.11.

The costs of the downward feeders are, in fact, the costs of upward feeders of Table 13.4. The costs of transmission substations are, essentially, independent of voltage level (mainly proportional to its MVA) and are shown in Table 13.12.

There are 11 existing transmission substations, denoted by ET, as detailed in Table 13.13.

The assumptions are similar to what outlined in Sect. 13.2, except

- For substation E31, the acceptable voltage drop of the downward feeder is considered to be 6%.
- The reserve requirement of each substation is considered to be, at least, 40%, except for ET10 which is considered to be 50%.
- It is assumed that each load node is totally supplied through a single transmission substation.

Six substation candidates, denoted by TC, are selected as shown in Table 13.14.

Upon running the SEP for 2015, two new substations are justified. The details of the results are shown in Table 13.15.

Table 13.13 Existing transmission substations

No.	Name	X	Y	Voltage (kV:kV)	Capacity (MVA)	
					Existing	Expandable
1	ET1	53.9263	32.3382	230:63	2 × 125	Yes
2	ET2	54.3297	31.9444	230:63	2 × 160	Yes
3	ET3	54.0573	31.9446	230:63	80 + 125	No
4	ET4	55.4581	31.7079	230:132	1 × 80	No
5	ET5	54.1604	32.0781	230:63	2 × 125	Yes
6	ET6	54.3833	31.8087	400:63	2 × 200	No
7	ET7	54.2015	31.88448	400:63	2 × 200	Yes
8	ET8	56.9522	33.54409	400:63	2 × 200	Yes
9	ET9	54.0182	31.95631	230:63	2 × 125	Yes
10	ET10	52.8333	31.0000	230:63	2 × 125	No
11	ET11	55.1500	30.1000	400:132	2 × 200	No

Table 13.14 The details of candidate transmission substations

No.	Name	X	Y	Capacity (MVA)	
				From	To
1	TC1	53.7472	32.3808	2 × 80	2 × 250
2	TC2	54.5767	31.7420	2 × 80	2 × 250
3	TC3	53.2700	31.1300	2 × 80	2 × 250
4	TC4	53.1800	31.1100	2 × 80	2 × 250
5	TC5	53.9600	31.5500	2 × 80	2 × 250
6	TC6	53.7472	32.3808	2 × 80	2 × 250

Table 13.15 Summary of the results for 2015

No.	Name	X	Y	Existing capacity (MVA)	New capacity (MVA)	Loading (MVA)
1	ET1	53.9263	32.3382	2 × 125	2 × 125	172.1
2	ET2	54.3297	31.9444	2 × 160	2 × 160	238.9
3	ET3	54.0573	31.9446	80 + 125	80 + 125	150
4	ET4	55.4581	31.7079	1 × 80	1 × 80	40.9
5	ET5	54.1604	32.0781	2 × 125	2 × 125	186.6
6	ET6	54.3833	31.8087	2 × 200	2 × 200	233.1
7	ET7	54.2015	31.88448	2 × 200	2 × 200	236
8	ET8	56.9522	33.54409	2 × 200	2 × 200	91.6
9	ET9	54.0182	31.95631	2 × 125	2 × 125	163.2
10	ET10	52.8333	31.0000	2 × 125	2 × 125	41.9
11	ET11	55.1500	30.1000	2 × 200	2 × 200	60.6
12	NT1	54.5767	31.7420	0	2 × 180	247.1
13	NT2	53.2700	31.1300	0	2 × 80	80.9

Table 13.16 Summary of the results for 2011

No.	Name	X	Y	Existing capacity (MVA)	New capacity (MVA)	Loading (MVA)
1	ET1	53.9263	32.3382	2 × 125	2 × 125	149
2	ET2	54.3297	31.9444	2 × 160	2 × 160	236.6
3	ET3	54.0573	31.9446	80 + 125	80 + 125	93.8
4	ET4	55.4581	31.7079	1 × 80	1 × 80	45
5	ET5	54.1604	32.0781	2 × 125	2 × 125	180.4
6	ET6	54.3833	31.8087	2 × 200	2 × 200	224.3
7	ET7	54.2015	31.88448	2 × 200	2 × 200	238
8	ET8	56.9522	33.54409	2 × 200	2 × 200	72.3
9	ET9	54.0182	31.95631	2 × 125	2 × 125	67.4
10	ET10	52.8333	31.0000	2 × 125	2 × 125	96.7
11	ET11	55.1500	30.1000	2 × 200	2 × 200	52.1
12	NT1	54.5767	31.7420	0	2 × 125	171.4
13	NT2	53.2700	31.1300	0	0	0

With the substations justified for 2015 as candidates for 2011, the results for 2011 are shown in Table 13.16.

13.4 NEP Problem for Both Sub-transmission and Transmission Levels

Assuming the base year to be 2010, the aim is plan the network for 2011 and 2015. The existing as well as new expansion requirements for 63, 132 and 230 plus 400 kV grids are shown in Figs. 13.2, 13.3 and 13.4, respectively. The details and the procedure are as follows

(a) Construct the base case grid for 2010. Electrical details of existing (in 2010) lines and transformers in sub-transmission level are shown in Table 13.17; whereas those of transmission level are shown in Table 13.18.
(b) Add the new sub-transmission substations with the details given in Table 13.10 for 2011 and perform NEP problem of sub-transmission level for that year. In doing so
    - Ignore any limitation observed on transmission level.[6]
    - Consider the generation details as given in Table 13.19.
(c) Once done for 2011, repeat performing NEP problem for sub-transmission level in 2015.
(d) Having accomplished the above tasks, we have now to repeat the steps for transmission level. Add the new transmission substations with the details

[6] $\overline{Q}$ and $\underline{Q}$ are taken to be 2/3 and −1/3 of P_G and PV setpoints are assumed to be 1.0.

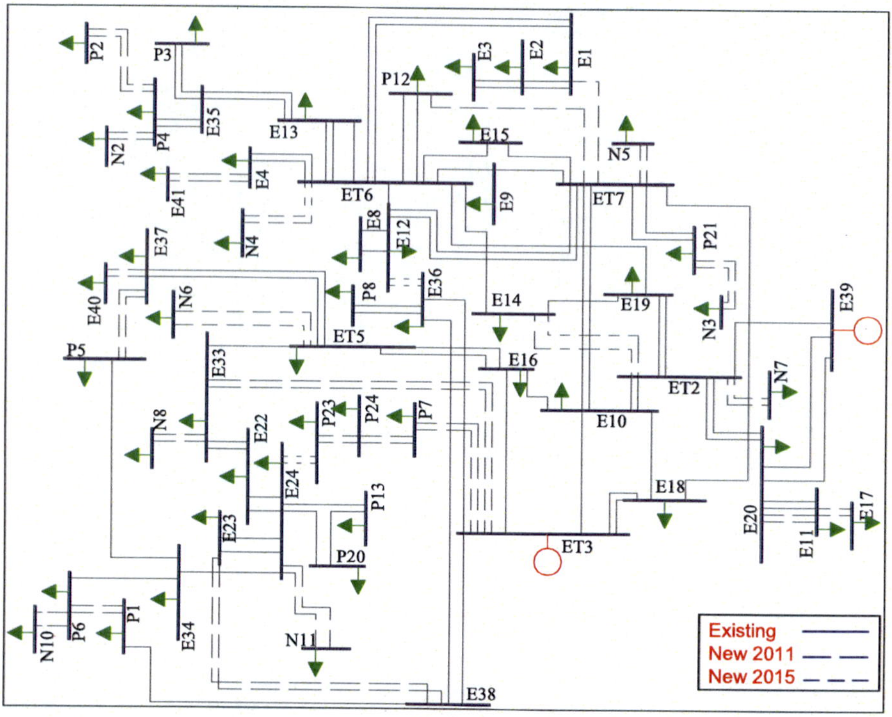

Fig. 13.2 Existing as well as new expansion requirements (63 kV)

given in Table 13.16 for 2011 and perform NEP problem for transmission level for that year. Now observe all limitations on transmission level and consider the generation details as given in Table 13.19.

(e) Once down for 2011, repeat performing NEP problem of transmission level for 2015.

The following points are worth mentioning

- As some of transmission substations are supplied through neighboring grids, they are ignored in Sect. 13.4. These are identified in Table 13.19 as being bold.
- The details of the sub-transmission elements (lines and transformers) costs are shown in Table 13.20; whereas those of transmission level are shown in Table 13.21.
- The results of sub-transmission level for 2011 and 2015 are shown in Tables 13.22 and 13.23, respectively.
- The results of transmission level for 2011 and 2015 are shown in Table 13.24.[7]

[7] No new transmission element is required for 2015.

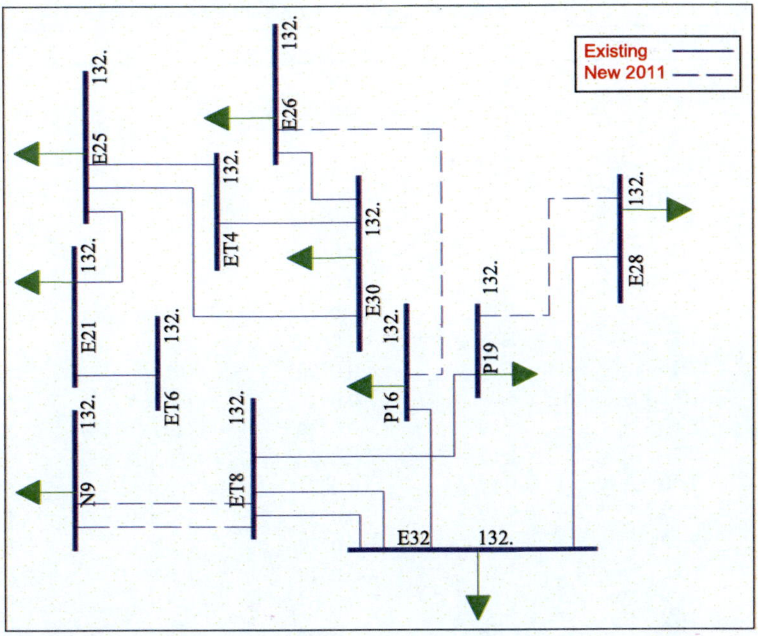

Fig. 13.3 Existing as well as new expansion requirements (132 kV)

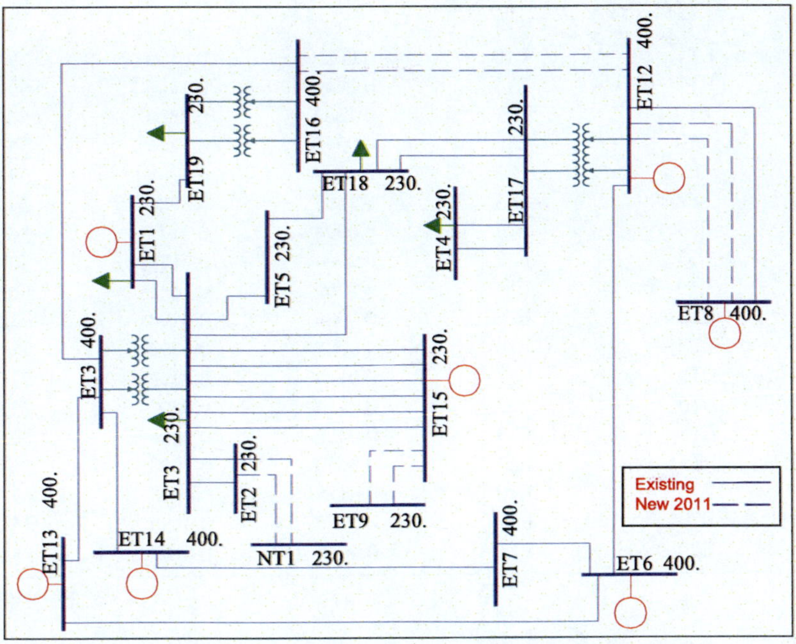

Fig. 13.4 Existing as well as new expansion requirements (230 and 400 kV)

Table 13.17 Existing (in 2010) lines and transformers for sub-transmission level

No.	From bus	To bus	R (p.u.)	X (p.u.)	S (MVA)
1	E25	E30	0.0527	0.1339	125.4
2	E25	ET4	0.0150	0.0458	125.4
3	ET4	E30	0.0546	0.1462	125.4
4	E25	E21	0.1318	0.3113	119.2
5	E30	E26	0.0291	0.1086	153.0
6	E21	ET6	0.0327	0.0945	119.2
7	E32	E28	0.0681	0.2343	153.0
8	P19	ET8	0.0778	0.2678	153.0
9	E32	ET8	0.0078	0.0268	153.0
10	E32	ET8	0.0078	0.0268	153.0
11	P16	E32	0.0506	0.1741	153.0
12	ET2	E19	0.0439	0.0963	56.9
13	ET2	E19	0.0439	0.0963	56.9
14	E20	ET2	0.0092	0.0203	56.9
15	E20	ET2	0.0092	0.0203	56.9
16	ET2	E39	0.0139	0.0304	56.9
17	ET2	E10	0.0279	0.0701	56.9
18	ET2	E10	0.0279	0.0701	56.9
19	E9	ET7	0.0467	0.1165	59.8
20	E12	E8	0.0035	0.0158	74.7
21	E8	E12	0.0035	0.0158	74.7
22	ET7	E10	0.0499	0.1245	59.8
23	ET7	E10	0.0499	0.1245	59.8
24	E12	ET7	0.0120	0.0299	59.8
25	E12	ET7	0.0120	0.0299	59.8
26	ET7	E18	0.0080	0.0199	59.8
27	ET7	E15	0.0440	0.1146	59.8
28	E10	ET3	0.1623	0.4520	56.9
29	E10	E16	0.0555	0.1216	56.9
30	E2	E3	0.1597	0.3988	59.8
31	E2	E3	0.1597	0.3988	59.8
32	E33	E22	0.0994	0.2179	56.9
33	E33	E22	0.0994	0.2179	56.9
34	E19	E14	0.0709	0.1647	56.9
35	ET5	E16	0.0085	0.0294	73.0
36	ET5	E16	0.0085	0.0294	73.0
37	ET5	E37	0.0114	0.0392	73.0
38	ET5	E37	0.0114	0.0392	73.0
39	ET5	E33	0.0450	0.1058	56.9
40	ET5	E33	0.0450	0.1058	56.9
41	E18	E10	0.0375	0.0936	59.8
42	E20	E11	0.0072	0.0198	45.3
43	E20	E11	0.0072	0.0198	45.3

(continued)

Table 13.17 (continued)

No.	From bus	To bus	R (p.u.)	X (p.u.)	S (MVA)
44	E1	E2	0.1109	0.2433	56.9
45	E1	E2	0.1109	0.2433	56.9
46	E15	ET6	0.0277	0.0738	59.8
47	E36	E38	0.0085	0.0294	73.0
48	E38	P1	0.0598	0.2057	73.0
49	E36	ET3	0.0199	0.0686	73.0
50	E14	ET6	0.0362	0.0887	56.9
51	E16	ET3	0.1069	0.3304	56.9
52	E22	E24	0.0439	0.1097	59.8
53	E22	E24	0.0439	0.1097	59.8
54	E23	E24	0.0427	0.1469	73.0
55	E23	E24	0.0427	0.1469	73.0
56	E9	ET6	0.0208	0.0518	59.8
57	E12	ET6	0.0439	0.0963	56.9
58	E12	ET6	0.0439	0.0963	56.9
59	E13	ET6	0.0001	0.0002	218.0
60	ET6	E13	0.0002	0.0003	59.8
61	ET6	E13	0.0002	0.0003	59.8
62	ET6	E4	0.1156	0.2534	56.9
63	E4	ET6	0.1156	0.2534	56.9
64	E19	ET6	0.0994	0.2179	56.9
65	ET6	E1	0.0832	0.1825	56.9
66	E1	ET6	0.0832	0.1825	56.9
67	ET3	E38	0.0114	0.0392	73.0
68	E18	ET3	0.0878	0.2193	59.8
69	ET3	E18	0.0878	0.2193	59.8
70	E39	E20	0.0000	0.0001	218.0
71	E39	E20	0.0000	0.0001	218.0
72	P21	ET7	0.1138	0.3919	73.0
73	P21	ET7	0.1138	0.3919	73.0
74	P12	ET6	0.0797	0.2743	73.0
75	P12	ET6	0.0797	0.2743	73.0
76	P8	E36	0.0427	0.1469	73.0
77	P8	E36	0.0427	0.1469	73.0
78	E35	E13	0.0798	0.1992	59.8
79	E35	E13	0.0798	0.1992	59.8
80	E35	P3	0.0798	0.1992	59.8
81	E35	P3	0.0798	0.1992	59.8
82	E35	P4	0.0798	0.1992	59.8
83	E35	P4	0.0798	0.1992	59.8
84	E24	P20	0.0998	0.2490	59.8
85	E24	P13	0.0998	0.2490	59.8
86	P20	P13	0.0998	0.2490	59.8

(continued)

Table 13.17 (continued)

No.	From bus	To bus	R (p.u.)	X (p.u.)	S (MVA)
87	E34	E24	0.0398	0.1372	73.0
88	E34	P6	0.0285	0.0980	73.0
89	E34	P5	0.0285	0.0980	73.0
90	ET6	ET6	0.0025	0.1100	40.0
91	ET4	ET4	0.0030	0.1500	80.0
92	ET5	ET5	0.0021	0.0993	125.0
93	ET5	ET5	0.0021	0.0993	125.0
94	ET2	ET2	0.0021	0.0993	125.0
95	ET2	ET2	0.0021	0.0993	125.0
96	ET1	E24	0.0021	0.0993	125.0
97	ET1	E24	0.0021	0.0993	125
98	ET3	ET3	0.0049	0.1512	80.0
99	ET3	ET3	0.0021	0.0993	125.0
100	ET7	ET7	0.0030	0.1211	200.0
101	ET7	ET7	0.0030	0.1211	200.0
102	ET6	ET6	0.0030	0.1211	200.0
103	ET6	ET6	0.0030	0.1211	200.0
104	ET8	ET8	0.0030	0.1211	200.0
105	ET8	ET8	0.0030	0.1211	200.0
106	ET9	E36	0.0021	0.0993	125.0
107	ET9	E36	0.0021	0.0993	125.0
108	NT1	P4	0.0021	0.0993	125.0
109	NT1	P4	0.0021	0.0993	125.0

13.5 RPP Problem for Both Sub-transmission and Transmission Levels

Having performed the NEP problem in the earlier stage, RPP has, now, to be performed. In fact, in NEP, any overload is checked to be removed for both normal and contingency conditions. Voltage conditions are checked in RPP stage. Although various operating conditions may be considered, we consider only the peak loading conditions (as normal conditions) and N − 1 contingency conditions (as we did in NEP problem). As noted in Chap. 10, the reactive power requirement is, initially, found for the normal conditions. Once determined, they are assumed to be *in* and new resource requirement is determined for contingency conditions. We start from the earlier period (2011). The results are then used for the next period (2015) and the studies are repeated for that period.

Table 13.18 Existing (in 2010) lines and transformers for transmission level

No.	From bus	To bus	R (p.u.)	X (p.u.)	S (MVA)
1	ET16	ET3	0.0010	0.0177	2598.1
2	ET7	ET6	0.0001	0.0018	2598.1
3	ET7	ET14	0.0003	0.0053	2598.1
4	ET13	ET6	0.0004	0.0071	2598.1
5	ET8	ET12	0.0065	0.0487	1615.0
6	ET6	ET12	0.0027	0.0207	1615.0
7	ET3	ET13	0.0001	0.0009	2598.1
8	ET3	ET14	0.0001	0.0009	2598.1
9	ET17	ET18	0.0098	0.0623	386.4
10	ET17	ET18	0.0098	0.0623	386.4
11	ET17	ET4	0.0004	0.0023	386.4
12	ET17	ET4	0.0004	0.0023	386.4
13	ET15	ET3	0.0001	0.0008	386.4
14	ET15	ET3	0.0001	0.0008	386.4
15	ET15	ET3	0.0001	0.0008	386.4
16	ET15	ET3	0.0001	0.0008	386.4
17	ET15	ET3	0.0001	0.0008	386.4
18	ET15	ET3	0.0001	0.0008	386.4
19	ET5	ET18	0.0168	0.1072	386.4
20	ET3	ET18	0.0202	0.1291	386.4
21	ET3	ET5	0.0034	0.0219	386.4
22	ET2	ET3	0.0059	0.0374	386.4
23	ET2	ET3	0.0059	0.0374	386.4
24	ET3	ET1	0.0057	0.0358	386.4
25	ET3	ET1	0.0057	0.0358	386.4
26	ET19	ET1	0.0037	0.0235	386.4
27	ET12	ET17	0.0031	0.0615	200.0
28	ET12	ET17	0.0031	0.0615	200.0
29	ET3	ET3	0.0031	0.0615	200.0
30	ET3	ET3	0.0031	0.0615	200.0
31	ET16	ET19	0.0031	0.0615	200.0
32	ET6	ET12	0.0031	0.0615	200.0

Table 13.19 Generation data

No.	Bus	P_G (MW)	Voltage (kV)
1	ET15[a]	–	230
2	**ET1**	54.5	230
3	ET13	470	400
4	ET14	470	400
5	**ET12**	17.5	400
6	**ET8**	460	400
7	**ET6**	20	400
8	E39	103	63
9	ET3	124	63

[a] ET15 is considered to be a boundary generation bus so that any generation deficiency (due to increase of load) is assumed to be transferred from this bus

Table 13.20 Sub-transmission elements (lines and transformers) costs

No.	R (p.u./km)	X (p.u./km)	S (MVA)	Voltage	Circuit	Variable cost (R/km)	Fix cost (R)
1	0.0045	0.0098	59.9	63 kV	1	25	310
2	0.0022	0.0050	119.8	63 kV	2	35	620
3	0.0008	0.0025	153.2	132 kV	1	44	360
4	0.0004	0.0012	306.4	132 kV	2	56	720
5	0.0024	0.0960	40.0	132 kV:63 kV	1	–	1057
6	0.0012	0.0480	80.0	132 kV:63 kV	2	–	2114

Table 13.21 Transmission elements (lines and transformers) costs

No.	R (p.u./km)	X (p.u./km)	S (MVA)	Voltage	Circuit	Variable cost (R/km)	Fix cost (R)
1	0.00012	0.000764	397	230 kV	1	42	800
2	0.000067	0.000563	738	230 kV	1	45	800
3	0.00006	0.000382	794	230 kV	2	58	1600
4	0.000034	0.000281	1476	230 kV	2	62	1600
5	0.000018	0.000204	1321	400 kV	1	86.5	1260
6	0.000009	0.000102	2640	400 kV	2	111	2520
7	0.00130	0.050500	200	400 kV:230 kV	1	–	2233
8	0.00065	0.025020	400	400 kV:230 kV	2	–	4466

13.5.1 Results for 2011

The steps followed for 2011 are summarized as shown below

(a) For normal conditions, perform ACLF and determine the out of range (i.e. out of 0.95–1.05 p.u.) voltages. The results are shown in Table 13.25.
(b) For each of the buses as shown in Table 13.25, determine the reactive power resource (capacitor) capacity which makes its voltage equal to 0.95 p.u. This is determined using an ACLF and by applying stepwise capacitor at the mentioned bus. The results are shown in Table 13.26.
(c) As compensation of a bus affects other buses, at this stage an optimization problem should be solved with the values given in Table 13.26 as the candidates. Assuming 0.05 p.u. capacitor banks to be employed with the cost terms given by (10.3) ($C_{fi} = 0.0$ and $C_{vi} =$ R 20000/p.u.), the objective function is assumed to consist of the investment cost as well as the cost of the losses. The cost of the losses is assumed to be R 1500/kW. No normalization is used (10.4) and these two cost terms are directly added together. GA is used to find the solution. The results are shown in Table 13.27.
(d) Once the capacitor banks are determined for the normal conditions, they would be assumed *in* and the contingency conditions are tried to check if any RPC would be required. The results show that for all N − 1 contingencies, the load flow converges and the voltages would be within the acceptable range of 0.9–1.05 p.u. So, no RPC is required.

Table 13.22 The results of sub-transmission level for 2011

No.	From bus	To bus	Length (km)	Voltage (kV)	Circuit	S (MVA)
1	P23	P24	0.15	63	2	119.8
2	P7	ET3	0.56	63	2	119.8
3	N7	ET2	1.43	63	2	119.8
4	E11	E17	2.17	63	2	119.8
5	E11	E20	3.09	63	2	119.8
6	P24	P7	3.23	63	2	119.8
7	N8	E33	6.48	63	2	119.8
8	ET7	N5	6.63	63	2	119.8
9	E4	E41	7.56	63	2	119.8
10	ET7	P12	8.73	63	1	59.9
11	N10	P6	9.11	63	2	119.8
12	E37	E40	9.44	63	2	119.8
13	ET7	E1	13.15	63	1	59.9
14	E24	N11	14.57	63	2	119.8
15	P4	P2	16.94	63	2	119.8
16	E33	ET3	21.92	63	2	119.8
17	E23	E38	23.85	63	2	119.8
18	P21	N3	24.97	63	2	119.8
19	N2	P4	30.76	63	2	119.8
20	E37	P5	38.49	63	2	119.8
21	E28	P19	56.61	132	1	153.2
22	ET8	N9	66.86	132	2	306.4
23	P6	P1	68.26	63	2	119.8
24	E26	P16	147.1	132	1	153.2

Table 13.23 The results of sub-transmission level for 2015

No.	From bus	To bus	Length (km)	Voltage	Circuit	S (MVA)
1	E14	ET2	12.23	63 kV	2	119.8
2	E12	E36	26.14	63 kV	2	119.8
3	E24	P23	43.83	63 kV	2	119.8
4	N6	ET5	18.9	63 kV	2	119.8
5	N4	ET6	38.4	63 kV	2	119.8
6	ET4	ET4	–	230 kV:132 kV	1	80

Table 13.24 The results for transmission level for 2011

No.	From bus	To bus	Length (km)	Voltage (kV)	Circuit	S (MVA)
1	ET15	ET9	3.9	230	2	794
2	ET2	NT1	32.39	230	2	1476
3	ET12	ET16	174.18	400	2	2640
4	ET12	ET8	246.91	400	2	2640

Table 13.25 Out of range voltages for the 2011 network

Bus	Voltage (p.u.)	Bus	Voltage (p.u.)	Bus	Voltage (p.u.)
E3	0.8253	E9	0.9132	E23	0.9303
E2	0.8598	N5	0.9161	ET6	0.9304
E41	0.8623	E8	0.9178	E21	0.9313
E4	0.8642	E34	0.9189	E35	0.9322
N3	0.8674	P5	0.9192	N8	0.9335
P21	0.8827	E12	0.9203	E33	0.9364
P13	0.8917	E22	0.9235	E24	0.9371
E1	0.8932	E13	0.9239	E10	0.9389
E14	0.8953	ET6	0.9239	P1	0.944
P20	0.8959	E19	0.9249	E40	0.9455
P12	0.8982	N11	0.9281	E17	0.9456
N10	0.9093	ET7	0.9287	E16	0.9475
P6	0.9113	E18	0.9295	E37	0.9486
E15	0.912	P2	0.9297	E26	0.9496
P3	0.913				

Table 13.26 Maximum capacitor banks at each bus for the 2011 network

Bus	Capacitance (p.u.)	Bus	Capacitance (p.u.)	Bus	Capacitance (p.u.)
E3	0.32	E9	0.65	E23	0.29
E2	0.46	N5	0.61	ET6	0.18
E41	0.47	E8	0.8	E21	0.12
E4	0.56	E34	0.31	E35	0.25
N3	0.25	P5	0.26	N8	0.22
P21	0.31	E12	0.92	E33	0.3
P13	0.29	E22	0.42	E24	0.33
E1	0.77	E13	0.97	E10	0.36
E14	0.67	ET6	0.97	P1	0.04
P20	0.27	E19	0.45	E40	0.05
P12	0.72	N11	0.21	E17	0.08
N10	0.21	ET7	0.88	E16	0.07
P6	0.25	E18	0.65	E37	0.03
E15	0.59	P2	0.15	E26	0.01
P3	0.22				

13.5.2 Results for 2015

Assuming the values as justified in Table 13.27 to be *in*, the procedure as outlined in (a), (b) and (c) above is repeated with the 2015 network. The results are shown in Tables 13.28, 13.29, 13.30. Step (d) (above) is then checked. It is noted that for contingency from ET6 to N4, the voltage of bus N4 is reduced to 0.8876 p.u. (lower than 0.9 p.u.). It is easily checked that adding a 0.05 p.u. capacitor bank at this bus, solves the problem.

Table 13.27 Optimal capacitor banks at each bus for the 2011 network

Bus	Capacitance (p.u.)	Bus	Capacitance (p.u.)	Bus	Capacitance (p.u.)
E3	0.1	E9	0.15	E23	0.25
E2	0.05	N5	0.2	ET6	0
E41	0.05	E8	0.2	E21	0
E4	0.2	E34	0.05	E35	0
N3	0.05	P5	0.05	N8	0.05
P21	0.1	E12	0.15	E33	0.3
P13	0.1	E22	0.2	E24	0.35
E1	0.15	E13	0.2	E10	0.15
E14	0.25	ET6	0.4	P1	0.05
P20	0.1	E19	0.2	E40	0.05
P12	0.3	N11	0.1	E17	0.15
N10	0	ET7	0.4	E16	0.15
P6	0.1	E18	0.3	E37	0.1
E15	0.15	P2	0.15	E26	0.1
P3	0.1				

Table 13.28 Out of range voltages for the 2015 network

Bus	Voltage (p.u.)	Bus	Voltage (p.u.)	Bus	Voltage (p.u.)
N4	0.8818	E3	0.9188	E2	0.9363

Table 13.29 Maximum capacitor banks at each bus for the 2015 network

Bus	Capacitance (p.u.)	Bus	Capacitance (p.u.)	Bus	Capacitance (p.u.)
N4	0.31	E3	0.08	E2	0.07

Table 13.30 Optimal capacitor banks at each bus for the 2015 network

Bus	Capacitance (p.u.)	Bus	Capacitance (p.u.)	Bus	Capacitance (p.u.)
N4	0.35	E3	0.05	E2	0.05

Appendix A
DC Load Flow

A.1 The Load Flow Problem

Formulation of classic load flow problem requires considering four variables at each bus i of power system. These variables are

1. P_i (Net active power injection)
2. Q_i (Net reactive power injection)
3. V_i (Voltage magnitude)
4. θ_i (Voltage angle)

The active and reactive power injections are calculated as follows

$$P_i = P_{Gi} - P_{Di} \tag{A.1}$$

$$Q_i = Q_{Gi} - Q_{Di} \tag{A.2}$$

in which P_{Gi} and Q_{Gi} are active and reactive power generations at bus i, respectively, whereas P_{Di} and Q_{Di} are active and reactive power demands at this bus, respectively.

Based on the application of Kirchhoff's laws to each bus

$$\mathbf{I} = \mathbf{YV} \tag{A.3}$$

$$I_i = \frac{(P_i - jQ_i)}{|V_i|} e^{j\theta_i} \tag{A.4}$$

where

I_i Net injected current at bus i
V Vector of bus voltages
I Vector of injected currents at the buses
Y Bus admittance matrix of the system

I, **V** and **Y** are complex. $V_i = |V_i|e^{j\theta_i}$ is the ith element of vector **V**. The **Y** matrix is symmetrical. The diagonal element Y_{ii} (self admittance of bus i) contains the sum of admittances of all the branches connected to bus i. The off diagonal element Y_{ij} (mutual admittance) is equal to the negative sum of the admittances between buses i and j. $Y_{ij} = |Y_{ij}|e^{j\delta_{ij}} = G_{ij} + jB_{ij}$ lies in the ith row and the jth column of matrix **Y**. G and B are subsequently called conductance and susceptance, respectively..

Using (A.4) to replace **I** in (A.3) results in (A.5) and (A.6).

$$P_i = \sum_{j=1}^{N} |Y_{ij}||V_i||V_j| \cos(\theta_i - \theta_j - \delta_{ij}) \tag{A.5}$$

$$Q_i = \sum_{j=1}^{N} |Y_{ij}||V_i||V_j| \sin(\theta_i - \theta_j - \delta_{ij}) \tag{A.6}$$

where N is the number of system buses.

To solve full load flow equations, two of four variables must be known in advance at each bus. This formulation results in a non-linear system of equations which requires iterative solution methods. In this formulation, convergence is not guaranteed.

A.2 DC Load Flow Solution

Direct Current Load Flow (DCLF) gives estimations of lines power flows on AC power systems. DCLF looks only at active power flows and neglects reactive power flows. This method is non-iterative and absolutely convergent but less accurate than AC Load Flow (ACLF) solutions. DCLF is used wherever repetitive and fast load flow estimations are required.

In DCLF, nonlinear model of the AC system is simplified to a linear form through these assumptions

- Line resistances (active power losses) are negligible i.e. $R \ll X$.
- Voltage angle differences are assumed to be small i.e. $\sin(\theta) = \theta$ and $\cos(\theta) = 1$.
- Magnitudes of bus voltages are set to 1.0 per unit (flat voltage profile).
- Tap settings are ignored.

Based on the above assumptions, voltage angles and active power injections are the variables of DCLF. Active power injections are known in advance. Therefore for each bus i in the system, (A.5) is converted to

$$P_i = \sum_{j=1}^{N} B_{ij}(\theta_i - \theta_j) \tag{A.7}$$

in which B_{ij} is the reciprocal of the reactance between bus i and bus j. As mentioned earlier, B_{ij} is the imaginary part of Y_{ij}.

As a result, active power flow through transmission line i, between buses s and r, can be calculated from (A.8).

$$P_{Li} = \frac{1}{X_{Li}}(\theta_s - \theta_r) \tag{A.8}$$

where X_{Li} is the reactance of line i.

DC power flow equations in the matrix form and the corresponding matrix relation for flows through branches are represented in (A.9) and (A.10).

$$\theta = [\mathbf{B}]^{-1}\mathbf{P} \tag{A.9}$$

$$\mathbf{P_L} = (\mathbf{b} \times \mathbf{A})\theta \tag{A.10}$$

where

- **P** N × 1 vector of bus active power injections for buses 1, …, N
- **B** N × N admittance matrix with R = 0
- ***θ*** N × 1 vector of bus voltage angles for buses 1, …, N
- $\mathbf{P_L}$ M × 1 vector of branch flows (M is the number of branches)
- **b** M × M matrix (b_{kk} is equal to the susceptance of line k and non-diagonal elements are zero)
- **A** M × N bus-branch incidence matrix

Each diagonal element of **B** (i.e. B_{ii}) is the sum of the reciprocal of the lines reactances connected to bus i. The off-diagonal element (i.e. B_{ij}) is the negative sum of the reciprocal of the lines reactances between bus i and bus j.

A is a connection matrix in which a_{ij} is 1, if a line exists from bus i to bus j; otherwise zero. Moreover, for the starting and the ending buses, the elements are 1 and −1, respectively.

Example A.1 A simple example is used to illustrate the points discussed above. A three-bus system is considered. This system is shown in Fig. A.1, with the details given in Tables A.1 and A.2.

With base apparent power equal to 100 MVA, **B** and **P** are calculated as follows

$$\mathbf{B} = \begin{bmatrix} 23.2435 & -17.3611 & -5.8824 \\ -17.3611 & 28.2307 & -10.8696 \\ -5.8824 & -10.8696 & 16.7519 \end{bmatrix} \quad \mathbf{P} = \begin{bmatrix} \text{Unknown} \\ 0.53 \\ -0.9 \end{bmatrix}$$

As bus 1 is considered as slack,[1] the first row of **P** and the first row and column of **B** are disregarded. θ_2 and θ_3 are then calculated using (A.9) as follows.

[1] With angle = 0.

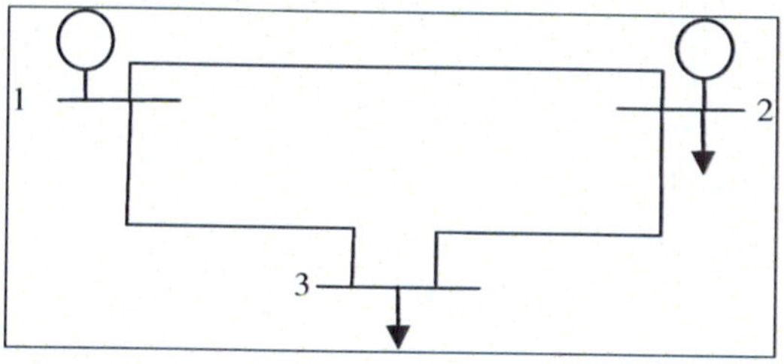

Fig. A.1 Three-bus system

Table A.1 Loads and generations

Bus number	Bus type	P_D (MW)	Q_D (MVAr)	P_G (MW)
1	Slack	0	0	Unknown
2	PV	10	5	63
3	PQ	90	30	0

Table A.2 Branches

Line number	From bus	To bus	X (p.u.)	Rating (MVA)
1	1	2	0.0576	250
2	2	3	0.092	250
3	1	3	0.17	150

$$\begin{bmatrix}\theta_2\\ \theta_3\end{bmatrix}=\begin{bmatrix}28.2307 & -10.8696\\ -10.8696 & 16.7519\end{bmatrix}^{-1}\begin{bmatrix}0.53\\ -0.9\end{bmatrix}=\begin{bmatrix}-0.0025\\ -0.0554\end{bmatrix}\text{Radian}=\begin{bmatrix}-0.1460^\circ\\ -3.1730^\circ\end{bmatrix}$$

A and **b** are then calculated as

$$\mathbf{A}=\begin{bmatrix}1 & -1 & 0\\ 0 & 1 & -1\\ 1 & 0 & -1\end{bmatrix}\quad \mathbf{b}=\begin{bmatrix}17.3611 & 0 & 0\\ 0 & 10.8696 & 0\\ 0 & 0 & 5.8824\end{bmatrix}$$

Therefore, the transmission flows are calculated using (A.10) as follows

$$\begin{bmatrix}P_{L1}\\ P_{L2}\\ P_{L3}\end{bmatrix}=\text{BaseMVA}\times\mathbf{b}\times\mathbf{A}\times\theta$$

$$=100\times\begin{bmatrix}17.3611 & 0 & 0\\ 0 & 10.8696 & 0\\ 0 & 0 & 5.8824\end{bmatrix}\times\begin{bmatrix}1 & -1 & 0\\ 0 & 1 & -1\\ 1 & 0 & -1\end{bmatrix}\times\begin{bmatrix}0\\ -0.0025\\ -0.0554\end{bmatrix}$$

$$=\begin{bmatrix}4.4243\\ 57.4243\\ 32.5757\end{bmatrix}\text{MW}$$

Appendix B
A Simple Optimization Problem

In this appendix, a simple optimization problem is devised and solved by some optimization algorithms.

B.1 Problem Definition

The problem is *Economic Dispatch* (ED), briefly described in Chap. 1 in which the aim is to optimize the total generation cost, F_T, defined as

$$F_T = \sum_{i=1}^{N} F_i(P_i) \qquad i = 1, \ldots, N \tag{B.1}$$

where

P_i	The active power generation of generation unit i
N	The number of generation units
$F_i(P_i)$	Generation cost of unit i

$F_i(P_i)$ is defined as

$$F_i(P_i) = a_i P_i^2 + b_i P_i + c_i \tag{B.2}$$

where a_i, b_i and c_i are known in advance.

Two types of constraints are observed as follows

$$P_{i\min} \leq P_i \leq P_{i\max} \qquad i = 1, \ldots, N \tag{B.3}$$

$$\sum_{i=1}^{N} P_i - P_D = 0 \tag{B.4}$$

where (B.3) refers to satisfying the generation level of each unit to be within its respective minimum and maximum limits and (B.4) refers to the balance of total generation with the total demand (P_D).

B.2 Results

The following five algorithms are applied to solve this problem as bellow

- Interior Point (IP)
- Genetic Algorithm (GA)
- Simulated Annealing (SA)
- Particle Swarm (PS)
- Differential Evolution (DE)

IP is used as an analytical approach while the other four are used as meta-heuristic techniques. The first three are implemented using Matlab Toolbox. Codes are generated for PS and DE. PS is based on the approach detailed in [1], while DE is developed based on [2].

The system under study is *New England test system* with the details given in [3]. The population in GA, PS and DE is taken to be 100. The convergence criterion is taken to be the maximum number of iterations and set to 1,000 (although other criteria may also be employed).

There are three main files developed as

- IP_SA_GA
- DE
- PS

There are seven functions generated with the details given in Table B.1.

Except IP, the other approaches are tried 10 times, using various initial populations. The results are summarized in Table B.2.

Table B.1 Details of the generated functions

Function name	Function description	Called by
call_gendata	Generation units data	IP_SA_GA, DE, PS
costfun	Calculation of total cost	IP and GA in IP_SA_GA
costfunsa	Calculation of the sum of total cost and the penalty function	SA in IP_SA_GA
call_objective	Calculation of the sum of total cost and the penalty function	DE, PS
cut2lim	Applying the generation limits	DE, PS
select_individual	Select individual for mutation	DE
discrete_recombination	Recombination or crossover operator	DE

Table B.2 The results of the different approaches

Method	Best	Average	Worst	Time (second)
SA	39167.16	41400.54	43835.36	11.80
PS	37140.93	38250.73	39790.21	0.84
GA	36931.36	37016.33	37120.11	71.00
DE	36842.22	36842.23	36842.23	0.85
IP	36842.22	–	–	0.90

B.3 Matlab Codes

In the following pages, the Matlab codes are given. It should be mentioned that no specific reason is used in choosing the above methods and based on the type of the problem, alternative algorithms may be tried. The reader is encouraged to try other algorithms for which some details are given in the chapter body.

a) "IP_SA_GA" M-file code

```
clc,clear all,close all
format compact

%      -----------  Solving Method   -------------
% Method -->
% 1:Interior Point
% 2:Genetic Algorithm
% 3:Simulated Annealing
SM=3;

%      ------------  Parameter set up ------------
 demand=6254.23;         % Total load
 pcf=1e4;                % Penalty coefficient
 gen_data=call_gendata; % Returns generator's data
 % Dimension of problem (here;number of units)
 dimnsn=10;
 mni=1000;             % Maximum number of iterations
 npop=100;             % Population size
 lb = gen_data(:,2)'; % Set lower bounds (Pmin in generator)
 ub = gen_data(:,3)'; % Set upper bounds (Pmax in generator)
 Aeq=ones(1,10);      % Equality constraint  P1+P2+...+P10=PD
 beq=demand;
 % Make a starting guess ---> ( Random making )
    x0 = (gen_data(:,3)+ (gen_data(:,2)-gen_data(:,3)).*...
        rand(dimnsn,1))';

% ------------ Switching to solving method  -----------------
if SM==1                     %------  Interior Point  ---------
        options = optimset('Algorithm','interior-point',...
            'Display','iter');
```

```
        [x,fval] = fmincon(@costfun,x0,[],[],Aeq,beq,lb,...
            ub,[],options);
        cost=fval;

elseif SM==2                %------  Genetic Algorithm  ------
      options =
gaoptimset('Generations',mni,'InitialPenalty'...
          ,pcf,'PopulationSize',npop,...
       'TimeLimit',inf,'StallGenLimit',inf,'PlotFcns',...
       @gaplotbestf,'Display','iter');
      [x,fval] = ga(@costfun,dimnsn,[],[],Aeq,beq,lb,...
          ub,[],options);
      cost=fval;

else                        %------  Simulated Annealing  ----
      options = saoptimset('MaxFunEvals',mni,'PlotFcns',...
          @saplotbestf,...
          'StallIterLimit',inf,'TimeLimit',inf,...
          'Display','iter');
      [x,fval] = simulannealbnd(@costfunsa,x0,...
          lb,ub,options);
      cost=fval-pcf*(sum(x)-demand)^2;

end

fprintf('\n'),display('Final solution is:'),Fs=x'
fprintf('\n'),display('Load that not served:'),...
    load_Mismatch=sum(x)-demand
fprintf('\n'),display('Associated cost:'),cost
```

b) "DE" M-file code

```
clc,clear all,close all
format compact

%------------  Parameter set up -------------

npop=100;           % Population size
mni=1000;           % Maximum number of iteration
dimnsn=10;          % Dimension of problem
                    % here;number of units)
demand=6254.23;     % Total load
pcf=1e4;            % Penalty coefficient
F=0.5;              % Mutation factor (scaling factor)
RR=0.9;             % Recombination (crossover) rate

%------------  Generator data   -------------
gen_data=call_gendata;      % Returns generator's data

% Make matrices the same size as population
% from vector Pmin & Pmax
    Pminrep=repmat(gen_data(:,2),1,npop);
    Pmaxrep=repmat(gen_data(:,3),1,npop);
```

```
% Randomly initialize population
    population=Pmaxrep+ (Pminrep-Pmaxrep) .* rand...
        (dimnsn,npop);

% Total objective function includes total cost and
% penalty function
     objective=call_objective(population,gen_data,pcf,demand);

% Determine best solution
 [objmin index_individual]=min(objective);bestsolution=...
     population(:,index_individual);

% -----------  Main loop:'Scheme DE/rand/1/bin'   ---------
for iter=1:mni

   iter
  % Select three different individuals for making
  % each trial vector
    slind=select_individual(npop);

  % Make trial vectors based 'DE/rand/1' : Mutation Operator
    trial_vectors=population(:,slind(1,:))+F*(population...
        (:,slind(2,:))-population(:,slind(3,:)));

  % Implement discrete recombination: Binomial crossover
    unew=discrete_recombination(population,trial_vectors,RR);

  % Limits on decision variables
    unew=cut2lim(unew,Pminrep,Pmaxrep);

  % Total objective function includes total cost
  % and penalty function
    objectivenew=call_objective(unew,gen_data,pcf,demand);

  % Deterministic selection
     replace=objectivenew<objective;
     objective(replace)=objectivenew(replace);population...
         (:,replace)=unew(:,replace);

 % Best solution so far:
   [objmin index_individual]=min(objective);bestsolution=...
       population(:,index_individual);

     evolution(iter)=objmin;

end

fprintf('\n'),display('Final solution is:'),bestsolution
fprintf('\n'),display('Load that not served:'),...
    load_Mismatch=sum(bestsolution)-demand
fprintf('\n'),display('Associated cost:'),cost=objmin-pcf*...
    (sum(bestsolution)-demand)^2

plot(1:mni,evolution)
```

c) "PS" M-file code

```
clc,clear all,close all
format compact

%------------  Parameter set up -------------

sws=100;              % Swarm  size (Population size)
mni=1000;             % Maximum number of iterations
dimnsn=10;            % Dimension of problem
                      % (here; number of units)
demand=6254.23;       % Total load
pcf=1e4;              % Penalty coefficient

%----- Generator data -----
gen_data=call_gendata;      % Returns generator's data

 % Range of decision (control) variables
    rangd=gen_data(:,3)-gen_data(:,2);

 % Make matrices the same size as population
 % from vector Pmin, Pmax & rangd
    Pminrep=repmat(gen_data(:,2),1,sws);
    Pmaxrep=repmat(gen_data(:,3),1,sws);
    rangdrep=repmat(rangd,1,sws);

% Position initialization
     position=Pminrep+ rangdrep .* rand(dimnsn,sws);

% Velocity initialization: it is assumed that maximum
% velocity is limited to 0.1*(Pmax-Pmin)
     velocity=0.1*rangdrep .*(1- 2*rand(dimnsn,sws));

% Total objective function includes total cost
% and penalty function
     objective=call_objective(position,gen_data,pcf,demand);

% Pbest & Gbest (initial assignment)
  [objmin index_particle]=min(objective);
  gbest=position(:,index_particle);gbest_objective=objmin;
  pbest=position;pbest_objective=objective;

% ------------  Main loop   --------------------
for iter=1:mni
    iter
    socialcom=rand(dimnsn,sws).*...
        (repmat(gbest,1,sws)-position); % Social component
    cognitivcom=rand(dimnsn,sws).*...
        (pbest-position);             % Cognitive component

 % Update velocity based on constriction(Clerc's) coefficient
    velocity=0.73*(velocity+2.05*cognitivcom+2.05*socialcom);

 % Limit velocity (step)
```

```
        velocity=cut2lim(velocity,-0.1*rangdrep,0.1*rangdrep);

      % Update position
        position=position+velocity;

      % Limits on decision variables
        position=cut2lim(position,Pminrep,Pmaxrep);

      % Total objective function includes
      % total cost and penalty function
        objective=call_objective(position,gen_data,pcf,demand);

      % Update pbest
        replace=objective<pbest_objective;
        pbest_objective(replace)=objective(replace);
        pbest(:,replace)=position(:,replace);

      % Update gbest
        [objmin index_particle]=min(pbest_objective);
        gbest=pbest(:,index_particle);
        gbest_objective=objmin;

        swarming(iter)=gbest_objective;
    end

    fprintf('\n'),display('Final solution is:'),gbest
    fprintf('\n'),display('Load that not served:'),...
        load_Mismatch=sum(gbest)-demand
    fprintf('\n'),display('Associated cost:'),...
        cost=gbest_objective-pcf*(sum(gbest)-demand)^2
    plot(1:mni,swarming)
```

d) "call_gendata" M-file code

```
    function gen_data=call_gendata

    %   Function 'call_data' returns generator's data
    %   Pmin<P<Pmax  &   Cost=a*P^2+b*P+c

    %   Unit    Pmin    Pmax          a       b       c
    gen_data=[
        1       0       350           0.01    0.3     0.2
        2       0       1145.55       0.01    0.3     0.2
        3       0       750           0.01    0.3     0.2
        4       0       732           0.01    0.3     0.2
        5       0       608           0.01    0.3     0.2
        6       0       750           0.01    0.3     0.2
        7       0       660           0.01    0.3     0.2
        8       0       640           0.01    0.3     0.2
        9       0       930           0.006   0.3     0.2
        10      0       1100          0.006   0.3     0.2   ];
```

e) "costfun" M-file code

```
    function cost=costfun(x)
```

```
% Function 'costfun' calculates total cost for GA & IP methods

% Recall generator's data
gen_data=call_gendata;
x=x';

% Calculate cost
cost=sum(gen_data(:,4).*x.^2+gen_data(:,5).*x+gen_data(:,6));
```

f) "costfunsa" M-file code

```
function objective=costfunsa(x)
% Objective function includes total cost
% and penalty function for SA method

% Recall generator's data
gen_data=call_gendata;
x=x';

% Calculate cost
cost=sum(gen_data(:,4).*x.^2+gen_data(:,5).*x+gen_data(:,6));

objective=cost+1e4*(sum(x)-6254.23).^2;
```

g) "call_objective" M-file code

```
function  objective=...
    call_objective(population,gen_data,pcf,demand)
% Total objective function includes total cost
% and penalty function for DE & PSO method

npop=size(population,2);

% Calculate total cost
for i=1:npop
  tcost(i)=sum(gen_data(:,4).*population(:,i).^2+...
      gen_data(:,5).*population(:,i)+gen_data(:,6));
end

% Load violation penalized by penalty function
    penalty=pcf*(sum(population)-demand).^2;

% Total objective function includes
% total cost and penalty function
    objective=tcost+penalty;
```

h) "cut2lim" M-file code

```
function x=cut2lim(x,xminmat,xmaxmat)

% Limits on decision variables

rmin=x<xminmat;
rmax=x>xmaxmat;
```

```
x(rmin)=xminmat(rmin);
x(rmax)=xmaxmat(rmax);
```

i) "select_individual" M-file code

```
function slind=select_individual(npop)

%  Select three different individuals
%  for making each trial vector for DE method
    slind=zeros(3,npop);

% First individual in mutation operator term
    slind(1,:)=randperm(npop);

% in order to select three different individuals
% for each trial vector, shift elements of first
% row in 'slind'. This method guarantees that all
% individuals will participate in making trial vectors

slind(2,1:npop-1)=slind(1,2:npop);slind(2,npop)=slind(1,1);
slind(3,1:npop-1)=slind(2,2:npop);slind(3,npop)=slind(2,1);
```

j) "discrete_recombination" M-file code

```
function unew=discrete_recombination(population,...
    trial_vectors,RR)%#ok<FNDEF>

[dimnsn,npop]=size(population);

% Those genes that replaced by genes of trial vectors
    genslct=rand(dimnsn,npop)<RR;

% Check at least one gene is replaced
    checknonzero=sum(genslct);

    for i=1:npop
        if ~checknonzero(i)
            genslct(fix(1+dimnsn*rand),i)=1;
        end
    end

    unew=population;

% Discrete recombination
    unew(genslct)= trial_vectors(genslct);
```

References

1. Clerc M, Kennedy J (2002) The particle swarm-explosion, stability, and convergence in a multidimensional complex space. IEEE Trans Evol Comput 6(1):58–73
2. Storn R, Price Price K (1997) Differential evolution – a simple and efficient heuristic for global optimization over continuous spaces. J Global Optim11(4):341–59
3. Zimmerman RD, Murillo-Sanchez CE, Gan D. MATPOWER: A MATLAB power system simulation package 2006. www.pserc.cornell.edu/matpower

Appendix C
AutoRegressive Moving Average (ARMA) Modeling

ARMA models are mathematical models of autocorrelation, in a time series. ARMA models can be used to predict behavior of a time series from past values alone. Such a prediction can be used as a baseline to evaluate possible importance of other variables to the system. An AR model expresses a time series as a linear function of its past values. The order of the AR model tells how many lagged past values are included. The simplest AR model is the first order autoregressive as follows

$$y_t + a_t y_{t-1} = e_t \tag{C.1}$$

or

$$y_t = -a_t y_{t-1} + e_t \tag{C.2}$$

where y_t is the mean-adjusted series in year (or time) t, y_{t-1} is the series in previous year, a_t is the lag-1 autoregressive coefficient and e_t is the noise. We can see that the model has the form of a regression model in which y_t is regressed on its previous value. The name autoregressive refers to the regression on self (auto).

Higher order AR models may also be assumed. A second order case is as follows

$$y_t + a_1 y_{t-1} + a_2 y_{t-2} = e_t \tag{C.3}$$

The Moving Average (MA) model is a form of ARMA model in which time series is regarded as a moving average (unevenly weighted) of a random shock noise e_t. A first order moving average model is given by

$$y_t = e_t + c_1 e_{t-1} \tag{C.4}$$

If we include both AR and MA, we reach at the ARMA model. A first order ARMA model is given by

$$y_t + a_1 y_{t-1} = e_t + c_1 e_{t-1} \tag{C.5}$$

For more details on ARMA modeling, refer to the references at the end of Chap. 4 and vast literature available on the subject.

Appendix D
What is EViews

EViews provides sophisticated data analysis, regression, and forecasting tools on Windows based computers. With EViews you can quickly develop a statistical relation from your data and then use the relation to forecast future values of the data. Areas where EViews can be useful include: scientific data analysis and evaluation, financial analysis, macroeconomic forecasting, simulation, sales forecasting, and cost analysis.

EViews is a new version of a set of tools for manipulating time series data originally developed in the Time Series Processor software for large computers. The immediate predecessor of EViews was MicroTSP, first released in 1981. Though EViews was developed by economists and most of its uses are in economics, there is nothing in its design that limits its usefulness to economic time series. Even quite large cross-section projects can be handled in EViews.

EViews provides convenient visual ways to enter data series from the keyboard or from disk files, to create new series from existing ones, to display and print series, and to carry out statistical analysis of the relationships among series.

EViews takes advantage of the visual features of modern Windows software. You can use your mouse to guide the operation with standard Windows menus and dialogs. Results appear in windows and can be manipulated with standard Windows techniques.

Alternatively, you may use EViews powerful command and batch processing language. You can enter and edit commands in the command window. You can create and store the commands in programs that document your research project for later execution.

Appendix E
The Calculations of the Reliability Indices

The analytical approach in calculating the reliability indices of a generation system may be, briefly, described as follows

- **Generation model.** A Capacity Outage Probability Table (COPT) should be, initially, generated in which various generation capacities as well as their respective probabilities are described. If the generation units are identical, a simple procedure is adopted to generate COPT. If the units are not similar, a recursive approach should be followed.
- **Load model.** The load may be described as Daily Peak Load Variation Curve (DPLVC) or Load Duration Curve (LDC). DPLVC is a cumulative representation of loads; descending order generated from the daily peak loads. LDC is generated from the hourly loads; descending order generated. DPLVC is widely used due to its simplicity. However, LDC shows a more practical representation of the load behavior.
- **Risk model.** The Loss of Load Expectation (LOLE) can be determined from convolving the generation and the load models. If DPLVC (LDC) is used as the load model, LOLE represents the expected days (*hours*) during a specific period in which the daily peak (*hourly*) load exceeds the generation capacity. According to Fig. E.1, for a generation outage of O_k; more than the available reserve, the load is lost for a period of t_k.

Mathematically speaking, *LOLE* is calculated as follows

$$LOLE = \sum_{i=1}^{N} p_k t_k = \sum_{i=1}^{N} P_k (t_k - t_{k-1}) \tag{E.1}$$

where

N The number of cases for which the generation outage is more than the reserve available

p_k The probability of the generation outage O_k

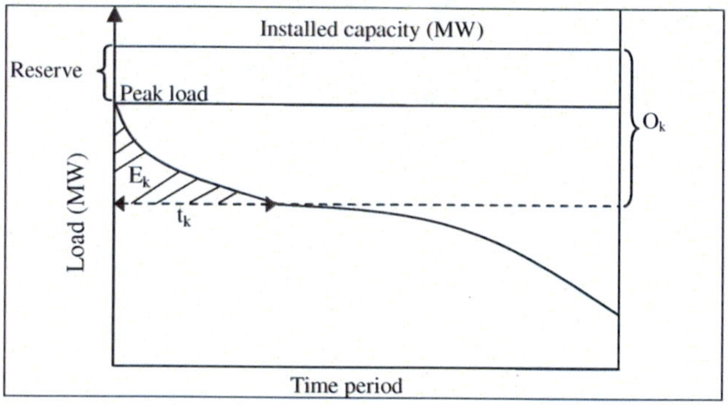

Fig. E.1 Relationship between capacity, load and reserve

t_k The period of lost load in generation outage O_k
P_k The cumulative probability of the generation outage O_k and more

If t_k is represented in per unit, the index calculated from (E.1) is called LOLP (Loss of Load Probability). The LOLP is expressed in terms of the average fraction of total time the system is expected to be in a state of failure. The area under an LDC shows the total energy demand. The Loss Of Expected Energy (LOEE) or the so called *Expected Energy Not Served* (EENS) or *Expected Unserved Energy* (EUE) may be calculated as

$$LOEE = \sum_{i=1}^{n} p_k E_k \quad \text{(E.2)}$$

where E_k is defined in Fig. E.1.

Example E.1 A generation system is composed of three units as follows

Unit 1: 10 MW, $FOR_1 = 1\%$
Unit 2: 20 MW, $FOR_2 = 2\%$
Unit 3: 60 MW, $FOR_3 = 3\%$

COPT is generated as shown in Table E.1. The probability of each capacity being *out* is FOR of its respective unit. Its probability being *in* is 1-FOR of its respective unit. For the LDC as shown in Fig. E.2, p_k, t_k and E_k (see (E.1)), should be determined for each row of Table E.1. Once done, (E.1) and (E.2) may be used to calculate LOLE and LOEE. As the reserve is 40 MW, the first four rows do not result in any lost load. Based on the results shown in Table E.2.

LOLE = 2.0298 (hours/100 hours)
LOLP = 0.020298
LOEE = 21.351 (MWh/100 hours)

Table E.1 The COPT of the generation system of the example

No.	Unit status (0:Out and 1:In)			Capacity (MW)		Probability	
	10 MW	20 MW	60 MW	In	Out	Individual	Cumulative
1	1	1	1	90	0	0.941094	1.000000
2	0	1	1	80	10	0.009506	0.058906
3	1	0	1	70	20	0.019206	0.049400
4	0	0	1	60	30	0.000194	0.030194
5	1	1	0	30	60	0.029106	0.030000
6	0	1	0	20	70	0.000294	0.000894
7	1	0	0	10	80	0.000594	0.000600
8	0	0	0	0	90	0.000006	0.000006

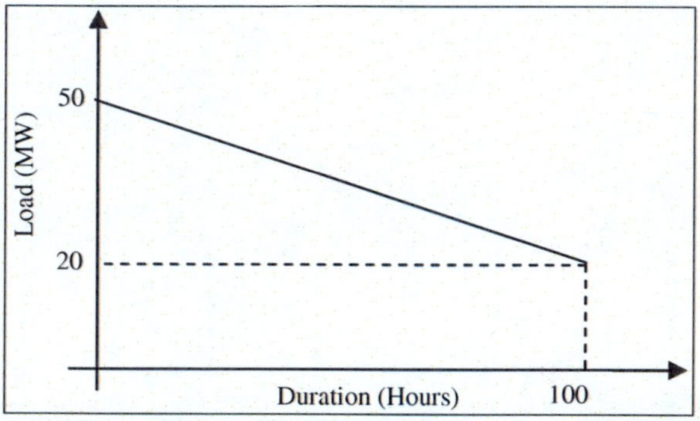

Fig. E.2 The load model (LDC) of the example

Table E.2 The required parameters for reliability indices calculation

No.	t_k	E_k	$p_k \times t_k$	$p_k \times E_k$	t_k/T	$p_k \times t_k/T$	t_k-t_{k-1}	$P_k \times (t_k$-$t_{k-1})$
1	0	0	0	0	0	0	0	0
2	0	0	0	0	0	0	0	0
3	0	0	0	0	0	0	0	0
4	0	0	0	0	0	0	0	0
5	66.67	666.67	1.9404	19.404	0.67	0.019440	66.67	2
6	100	1500	0.0294	0.441	1	0.000294	33.34	0.03
7	100	2500	0.0594	1.485	1	0.000594	0	0
8	100	3500	0.0006	0.021	1	0.000006	0	0

Note that the total energy demand is 3500 MWh; calculated from the area under LDC.

Appendix F
Garver Test System Data

In this book, Garver test system is used in Chaps. 6, 8 and 9 to describe generation and transmission network planning problems. The relevant data of this system are provided in current appendix. The base case, as used in Chaps. 6 and 8, is described in Sect. F.1. The modified case, as used in Chap. 9, is described in Sect. F.2.

F.1 The Base Case

The base Garver test system is shown in Fig. F.1, with the details given in Tables F.1 and F.2.

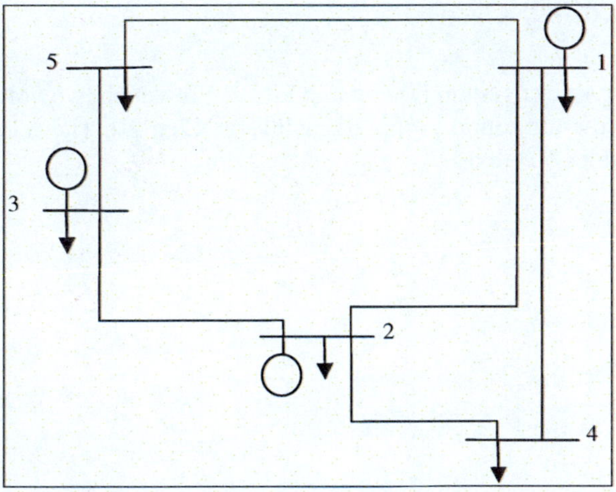

Fig. F.1 Garver test system

Table F.1 Network data[a]

Line no.	Bus		R (p.u.)	X (p.u.)	Capacity limit (p.u.)	Path length (km)
	From	To				
1	1	2	0.1000	0.40	1.0	400.0
2	1	4	0.1500	0.60	0.8	600.0
3	1	5	0.0500	0.20	1.0	200.0
4	2	3	0.0500	0.20	1.0	200.0
5	2	4	0.1000	0.40	1.0	400.0
6	3	5	0.0500	0.20	1.0	200.0
7	1	3	0.0950	0.38	1.0	380.0
8	2	5	0.0775	0.31	1.0	310.0
9	3	4	0.1475	0.59	0.8	590.0
10	4	5	0.1575	0.63	0.8	630.0

[a] It should be mentioned that some lines (7 through 10) are used as candidates in some places; while still some candidates may be considered in the some corridors of existing lines (1 through 6)

Table F.2 Generation and load data

Bus	Load		Generation
	P_D (p.u.)	Q_D (p.u.)	P_G (p.u.)
1	0.240	0.116	1.130
2	0.720	0.348	0.500
3	0.120	0.058	0.650
4	0.480	0.232	-
5	0.720	0.348	-

F.2 The Modified Case

A modified Garver test system is shown in Fig. F.2 in which two voltage levels are used to assess the algorithm proposed mainly in Chap. 9. The relevant data are provided in Tables F.3 and F.4.

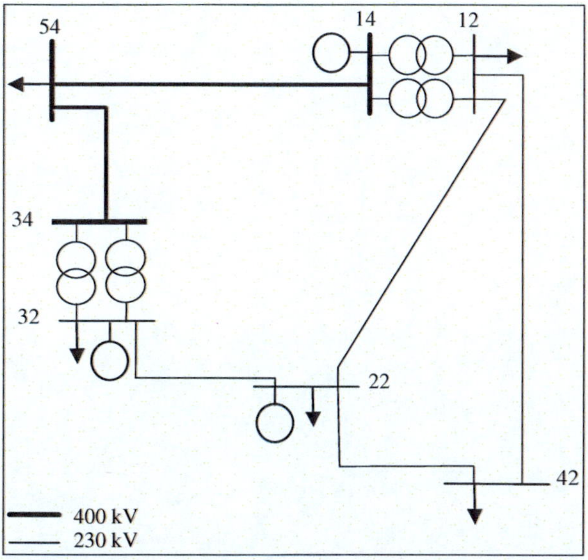

Fig. F.2 Modified Garver test system

Table F.3 Network data

Line no.	Bus		R (p.u.)	X (p.u.)	B (p.u.)	Capacity limit (p.u.)
	From	To				
1	12	22	0.10	0.40	0.8	1.0
2	12	42	0.15	0.60	1.2	0.8
3	14	54	0.05	0.20	0.4	1.0
4	22	32	0.05	0.20	0.4	1.0
5	22	42	0.10	0.40	0.8	1.0
6	34	54	0.05	0.20	0.4	1.0

Table F.4 Generation and load data

Bus	Load		Generation
	P_D (p.u.)	Q_D (p.u.)	P_G (p.u.)
12	0.240	0.116	-
22	0.720	0.348	0.500
32	0.120	0.058	0.650
42	0.480	0.232	-
14	-	-	1.13
54	0.720	0.348	-

Appendix G
Geographical Information System

A Geographical Information System (GIS) is a system of hardware, software and procedures to facilitate the management, manipulation, analysis, modeling, representation and display of *georeferenced* data to solve complex problems regarding planning and management of resources.

The georeferenced data or information is the geographic information identified according to locations (an alternative term is *spatial* data or information). In other words, the information; normally in digital form, is linked to specific places in the earth, using earth coordinates (such as latitude/longitude). In this way, a *layer* (also known as theme) may be formed, consisting of geographic data linked to descriptive, or tabular information. For various types of information, different layers may thus be created. The layers may then be combined as required to perform analyses.

GIS has found widespread use in many decision making activities in various desciplines. It may be used in both daily operation or long term planning of a system in which the decision making is, somehow, related to the geography. The issues referred to in this book, are mainly related to long term planning of a power system. As detailed in some chapters, the geographical information of load points, existing and candidate substations, transmission lines routes, etc. are used in some types of decision makings in GEP, SEP and NEP problems. So, if GIS is used, it can mathematically transform map features from one scale or projection to another, to allow map layers from different sources to be used together. If information is created through a GIS, it is quite simple to update the data on the computer to generate an updated product.

For data manipulation and storage in layers, two models, namely, *raster data model* and *vector data model* may be used. In the former, the region under study is divided into small regular blocks, with each block having a specific value attached to it. In the latter, all objects of interest are described in terms of geometric elements such as points, lines, polygons, etc.

While raster data are best used for representing continuous variables (such as elevations) and all satellite and aerial photograph data come in raster form, the vector data are very widely used in analysis of networks and municipal data bases (containing description of buildings, streets, etc.).

Briefly speaking, the GIS functions are as follows

- Capture
- Store
- Query
- Analyze
- Output

Capturing may be performed using hardcopy maps, Global Positioning Systems (GPS), digital data from some sources such as satellites, aerial photography, etc.

Storing can be carried out using one of the techniques already described (raster and vector).

Query may come in two forms. One is looking to identify or find features of interest of some points on the map. The other tries to identify the features based on some specific conditions (for instance, identifying the stream with the longest length and in the southern province).

Analysis of any type of data involves searching for patterns within one variable and relationships between two or more variables. For example, we can say that census tract A is next to census tract B, and both adjoin tract C; that city A is 100 km northwest of city B; that my house is on the same street as yours, etc.

Output may be in the form of paper/hardeopy files, map digital files, images, etc.

There are vast literatures on GIS. Instead of introducing some to the reader, we encourage him or her to search for the relevant materials in the form of books, tutorials, websites, etc.

Appendix H
84-Bus Test System Data

The relevant data of the 84-bus test system, as reported in Sect. 8.6.2, are provided in current appendix. This is a single voltage level network with detailed information as below

- Bus data are provided in Table H.1.
- Line data are provided in Table H.2.
- Candidate lines data are provided in Table H.3.
- Generation data are provided in Table H.4.

Table H.1 Bus data

No.	Bus name	X[a]	Y[a]	Area no.	P_D (p.u.)	No.	Bus name	X	Y	Area no.	P_D (p.u.)
1	B 1V4	47.65	37.19	1	0.36	19	B 19V4	50.90	35.42	4	0.00
2	B 2V4	46.17	38.08	1	0.00	20	B 20V4	51.38	35.75	1	5.66
3	B 3V4	54.90	36.93	1	6.28	21	B 21V4	51.57	35.75	1	4.50
4	B 4V4	51.20	36.50	1	2.08	22	B 22V4	51.57	35.75	1	4.50
5	B 5V4	52.63	36.35	1	5.40	23	B 23V4	51.65	35.33	1	2.05
6	B 6V4	53.25	36.82	1	6.26	24	B 24V4	51.13	35.75	1	7.86
7	B 7V4	53.43	35.60	2	3.25	25	B 25V4	50.47	36.10	1	6.56
8	B 8V4	54.87	36.42	2	2.23	26	B 26V4	57.40	37.05	2	1.76
9	B 9V4	51.87	35.43	1	0.00	27	B 27V4	59.40	36.42	2	0.00
10	B 10V4	51.30	35.62	4	6.90	28	B 28V4	58.68	36.28	2	1.08
11	B 11V4	51.30	35.62	4	6.83	29	B 29V4	59.02	33.75	2	2.63
12	B 12V4	51.85	35.42	1	0.00	30	B 30V4	58.77	36.20	2	3.71
13	B 13V4	51.28	35.77	1	6.53	31	B 31V4	57.75	36.25	2	0.27
14	B 14V4	50.90	35.42	4	0.00	32	B 32V4	59.08	35.18	2	3.66
15	B 15V4	51.83	35.75	1	2.60	33	B 33V4	57.93	37.40	2	2.27
16	B 16V4	50.32	36.15	1	0.00	34	B 34V4	60.65	35.23	2	2.53
17	B 17V4	50.32	36.15	1	0.00	35	B 35V4	59.40	36.42	2	1.42
18	B 18V4	51.58	35.52	1	8.10	36	B 36V4	54.38	31.81	2	3.47

(continued)

Table H.1 (continued)

No.	Bus name	X[a]	Y[a]	Area no.	P_D (p.u.)	No.	Bus name	X	Y	Area no.	P_D (p.u.)
37	B 37V4	54.17	31.90	2	3.39	61	B 61V4	50.87	32.24	2	3.19
38	B 38V4	48.28	30.45	4	5.22	62	B 62V4	51.22	32.49	2	0.00
39	B 39V4	49.60	32.05	4	0.00	63	B 63V4	51.31	32.41	2	1.05
40	B 40V4	48.82	31.30	4	6.64	64	B 64V4	52.71	27.45	3	1.34
41	B 41V4	48.67	31.45	4	2.77	65	B 65V4	52.61	27.45	3	0.00
42	B 42V4	48.35	32.47	4	3.04	66	B 66V4	50.92	28.83	3	3.36
43	B 43V4	49.37	32.02	4	1.86	67	B 67V4	51.02	28.98	4	0.00
44	B 44V4	48.08	30.37	4	2.20	68	B 68V4	51.02	28.98	4	2.54
45	B 45V4	48.12	32.50	4	0.00	69	B 69V4	53.67	29.08	3	1.90
46	B 46V4	49.98	31.93	3	0.00	70	B 70V4	52.05	27.83	3	2.15
47	B 47V4	49.25	30.58	4	4.08	71	B 71V4	51.72	29.52	3	0.00
48	B 48V4	49.70	30.80	3	8.06	72	B 72V4	54.32	29.20	2	1.59
49	B 49V4	49.68	30.85	4	0.00	73	B 73V4	52.45	29.58	3	2.58
50	B 50V4	48.75	32.15	4	4.85	74	B 74V4	52.83	31.00	3	2.27
51	B 51V4	51.37	30.60	3	1.11	75	B 75V4	45.54	34.74	4	1.38
52	B 52V4	49.83	34.00	4	3.47	76	B 76V4	46.60	34.12	4	2.67
53	B 53V4	48.22	33.43	4	4.35	77	B 77V4	47.35	34.35	4	0.68
54	B 54V4	48.87	35.13	1	0.00	78	B 78V4	56.11	27.15	2	0.00
55	B 55V4	51.47	32.25	2	4.56	79	B 79V4	56.00	28.32	2	0.26
56	B 56V4	51.49	32.80	2	2.29	80	B 80V4	54.30	27.02	2	2.62
57	B 57V4	50.32	33.41	2	2.48	81	B 81V4	48.58	36.65	1	3.97
58	B 58V4	51.47	32.25	2	0.00	82	B 82V4	56.78	30.23	2	2.33
59	B 59V4	51.33	32.59	2	2.97	83	B 83V4	55.75	29.43	2	4.03
60	B 60V4	51.42	32.25	2	8.51	84	B 84V4	49.63	37.18	1	0.00

[a] Geographical characteristics

Table H.2 Line data

No.	From bus	To bus	R (p.u.)	X (p.u.)	$\bar{P}_L$ (p.u.)	No.	From bus	To bus	R (p.u.)	X (p.u.)	$\bar{P}_L$ (p.u.)
1	B 1V4	B 81V4	0.0022	0.0258	15.0	15	B 9V4	B 12V4	0.0000	0.0002	13.9
2	B 2V4	B 16V4	0.0092	0.0927	11.7	16	B 9V4	B 15V4	0.0007	0.0089	18.2
3	B 3V4	B 6V4	0.0029	0.0326	15.0	17	B 9V4	B 15V4	0.0007	0.0089	18.2
4	B 3V4	B 6V4	0.0029	0.0326	15.0	18	B 9V4	B 21V4	0.0008	0.0113	12.5
5	B 3V4	B 26V4	0.0054	0.0567	14.7	19	B 9V4	B 22V4	0.0008	0.0113	12.5
6	B 4V4	B 5V4	0.0024	0.0278	15.0	20	B 10V4	B 14V4	0.0006	0.0083	15.3
7	B 4V4	B 25V4	0.0023	0.0264	11.7	21	B 11V4	B 19V4	0.0006	0.0085	18.4
8	B 5V4	B 6V4	0.0016	0.0181	15.0	22	B 12V4	B 18V4	0.0006	0.0057	15.1
9	B 6V4	B 7V4	0.0027	0.0315	15.0	23	B 12V4	B 19V4	0.0010	0.0164	22.2
10	B 6V4	B 12V4	0.0040	0.0550	9.1	24	B 12V4	B 19V4	0.0010	0.0164	22.2
11	B 7V4	B 8V4	0.0028	0.0319	15.0	25	B 12V4	B 23V4	0.0004	0.0045	15.0
12	B 7V4	B 12V4	0.0034	0.0317	9.0	26	B 12V4	B 57V4	0.0060	0.0633	10.7
13	B 8V4	B 31V4	0.0056	0.0588	14.7	27	B 13V4	B 20V4	0.0002	0.0021	16.6
14	B 9V4	B 12V4	0.0000	0.0002	9.9	28	B 13V4	B 25V4	0.0013	0.0180	11.7

(continued)

Table H.2 (continued)

No.	From bus	To bus	R (p.u.)	X (p.u.)	$\bar{P}_L$ (p.u.)	No.	From bus	To bus	R (p.u.)	X (p.u.)	$\bar{P}_L$ (p.u.)
29	B 14V4	B 19V4	0.0000	0.0002	16.6	73	B 40V4	B 49V4	0.0010	0.0177	27.1
30	B 15V4	B 24V4	0.0009	0.0126	4.8	74	B 41V4	B 50V4	0.0015	0.0167	15.0
31	B 15V4	B 24V4	0.0009	0.0126	4.8	75	B 42V4	B 43V4	0.0023	0.0265	15.0
32	B 16V4	B 17V4	0.0000	0.0002	16.1	76	B 42V4	B 53V4	0.0024	0.0275	15.0
33	B 16V4	B 19V4	0.0013	0.0185	16.8	77	B 43V4	B 46V4	0.0014	0.0165	15.0
34	B 16V4	B 24V4	0.0013	0.0177	18.2	78	B 43V4	B 50V4	0.0012	0.0134	22.0
35	B 16V4	B 25V4	0.0003	0.0044	24.5	79	B 43V4	B 57V4	0.0051	0.0583	15.0
36	B 16V4	B 25V4	0.0003	0.0044	24.5	80	B 45V4	B 50V4	0.0017	0.0196	15.0
37	B 16V4	B 84V4	0.0032	0.0340	10.7	81	B 45V4	B 53V4	0.0025	0.0291	15.0
38	B 17V4	B 19V4	0.0013	0.0185	10.7	82	B 46V4	B 48V4	0.0020	0.0306	22.5
39	B 17V4	B 24V4	0.0013	0.0177	18.2	83	B 46V4	B 55V4	0.0021	0.0319	22.5
40	B 18V4	B 23V4	0.0004	0.0045	15.0	84	B 46V4	B 62V4	0.0019	0.0292	22.5
41	B 19V4	B 52V4	0.0035	0.0395	7.6	85	B 48V4	B 49V4	0.0002	0.0019	14.8
42	B 19V4	B 52V4	0.0035	0.0395	7.6	86	B 49V4	B 67V4	0.0023	0.0402	27.1
43	B 20V4	B 24V4	0.0004	0.0050	16.6	87	B 49V4	B 68V4	0.0032	0.0477	22.5
44	B 24V4	B 25V4	0.0010	0.0147	16.6	88	B 51V4	B 67V4	0.0013	0.0198	22.5
45	B 26V4	B 30V4	0.0031	0.0325	14.7	89	B 52V4	B 57V4	0.0016	0.0181	15.0
46	B 26V4	B 31V4	0.0029	0.0304	14.7	90	B 53V4	B 77V4	0.0026	0.0299	15.0
47	B 26V4	B 33V4	0.0014	0.0151	14.7	91	B 54V4	B 77V4	0.0031	0.0350	15.0
48	B 27V4	B 35V4	0.0000	0.0001	22.7	92	B 54V4	B 81V4	0.0032	0.0371	15.0
49	B 28V4	B 30V4	0.0003	0.0026	14.7	93	B 55V4	B 58V4	0.0000	0.0002	15.0
50	B 29V4	B 32V4	0.0035	0.0370	14.7	94	B 55V4	B 58V4	0.0000	0.0002	15.0
51	B 29V4	B 34V4	0.0048	0.0507	14.7	95	B 55V4	B 60V4	0.0002	0.0023	15.0
52	B 30V4	B 32V4	0.0028	0.0290	14.7	96	B 55V4	B 62V4	0.0009	0.0107	15.0
53	B 30V4	B 35V4	0.0014	0.0144	14.7	97	B 56V4	B 57V4	0.0024	0.0276	15.0
54	B 34V4	B 35V4	0.0039	0.0410	14.7	98	B 56V4	B 62V4	0.0013	0.0148	15.0
55	B 36V4	B 37V4	0.0004	0.0071	27.1	99	B 57V4	B 62V4	0.0026	0.0295	15.0
56	B 36V4	B 37V4	0.0004	0.0071	27.1	100	B 58V4	B 74V4	0.0024	0.0360	22.5
57	B 36V4	B 82V4	0.0036	0.0540	22.0	101	B 59V4	B 62V4	0.0004	0.0049	15.0
58	B 36V4	B 83V4	0.0030	0.0531	27.1	102	B 59V4	B 62V4	0.0004	0.0041	15.0
59	B 37V4	B 56V4	0.0029	0.0517	27.1	103	B 60V4	B 62V4	0.0008	0.0087	15.0
60	B 37V4	B 58V4	0.0032	0.0486	22.5	104	B 61V4	B 62V4	0.0008	0.0093	15.0
61	B 38V4	B 41V4	0.0023	0.0268	15.0	105	B 62V4	B 63V4	0.0003	0.0031	15.0
62	B 38V4	B 44V4	0.0038	0.0436	15.0	106	B 62V4	B 63V4	0.0003	0.0031	15.0
63	B 38V4	B 47V4	0.0016	0.0185	15.0	107	B 64V4	B 65V4	0.0000	0.0002	15.0
64	B 39V4	B 40V4	0.0034	0.0284	12.8	108	B 64V4	B 70V4	0.0029	0.0237	16.0
65	B 39V4	B 46V4	0.0011	0.0122	15.0	109	B 65V4	B 69V4	0.0014	0.0216	22.5
66	B 39V4	B 49V4	0.0037	0.0304	12.8	110	B 65V4	B 80V4	0.0040	0.0328	16.0
67	B 39V4	B 52V4	0.0047	0.0544	15.0	111	B 66V4	B 68V4	0.0003	0.0045	22.5
68	B 39V4	B 57V4	0.0048	0.0550	15.0	112	B 66V4	B 68V4	0.0003	0.0045	22.5
69	B 39V4	B 61V4	0.0025	0.0291	15.0	113	B 67V4	B 68V4	0.0015	0.0225	22.5
70	B 40V4	B 41V4	0.0004	0.0051	18.0	114	B 67V4	B 71V4	0.0007	0.0108	22.5
71	B 40V4	B 43V4	0.0020	0.0227	18.0	115	B 67V4	B 73V4	0.0010	0.0177	27.1
72	B 40V4	B 47V4	0.0008	0.0112	18.0	116	B 68V4	B 70V4	0.0044	0.0364	16.0

(continued)

Table H.2 (continued)

No.	From bus	To bus	R (p.u.)	X (p.u.)	$\bar{P}_L$ (p.u.)	No.	From bus	To bus	R (p.u.)	X (p.u.)	$\bar{P}_L$ (p.u.)
117	B 69V4	B 72V4	0.0008	0.0143	27.1	123	B 78V4	B 79V4	0.0031	0.0340	14.4
118	B 69V4	B 73V4	0.0014	0.0253	27.1	124	B 78V4	B 80V4	0.0048	0.0414	16.0
119	B 71V4	B 73V4	0.0024	0.0200	16.0	125	B 78V4	B 83V4	0.0031	0.0552	27.1
120	B 72V4	B 83V4	0.0016	0.0283	27.1	126	B 79V4	B 83V4	0.0027	0.0302	14.4
121	B 75V4	B 76V4	0.0036	0.0412	15.0	127	B 81V4	B 84V4	0.0023	0.0268	15.0
122	B 76V4	B 77V4	0.0011	0.0171	22.5	128	B 82V4	B 83V4	0.0027	0.0309	15.0

Table H.3 Candidate lines data[a]

No.	From bus	To bus	No.	From bus	To bus	No.	From bus	To bus	No.	From bus	To bus
1	B 14V4	B 19V4	30	B 12V4	B 23V4	59	B 60V4	B 62V4	88	B 56V4	B 63V4
2	B 21V4	B 22V4	31	B 10V4	B 24V4	60	B 15V4	B 18V4	89	B 21V4	B 23V4
3	B 10V4	B 11V4	32	B 11V4	B 24V4	61	B 55V4	B 62V4	90	B 22V4	B 23V4
4	B 27V4	B 35V4	33	B 38V4	B 44V4	62	B 58V4	B 62V4	91	B 18V4	B 24V4
5	B 55V4	B 58V4	34	B 42V4	B 45V4	63	B 9V4	B 15V4	92	B 15V4	B 23V4
6	B 67V4	B 68V4	35	B 40V4	B 41V4	64	B 12V4	B 15V4	93	B 47V4	B 48V4
7	B 16V4	B 17V4	36	B 39V4	B 43V4	65	B 39V4	B 46V4	94	B 13V4	B 15V4
8	B 9V4	B 12V4	37	B 18V4	B 23V4	66	B 59V4	B 60V4	95	B 10V4	B 15V4
9	B 58V4	B 60V4	38	B 36V4	B 37V4	67	B 13V4	B 18V4	96	B 11V4	B 15V4
10	B 55V4	B 60V4	39	B 20V4	B 24V4	68	B 22V4	B 24V4	97	B 47V4	B 49V4
11	B 48V4	B 49V4	40	B 9V4	B 23V4	69	B 21V4	B 24V4	98	B 60V4	B 61V4
12	B 13V4	B 20V4	41	B 55V4	B 63V4	70	B 58V4	B 59V4	99	B 42V4	B 50V4
13	B 64V4	B 65V4	42	B 58V4	B 63V4	71	B 55V4	B 59V4	100	B 13V4	B 19V4
14	B 28V4	B 30V4	43	B 15V4	B 22V4	72	B 15V4	B 20V4	101	B 13V4	B 14V4
15	B 62V4	B 63V4	44	B 15V4	B 21V4	73	B 19V4	B 24V4	102	B 20V4	B 23V4
16	B 13V4	B 24V4	45	B 18V4	B 21V4	74	B 14V4	B 24V4	103	B 11V4	B 12V4
17	B 16V4	B 25V4	46	B 18V4	B 22V4	75	B 11V4	B 14V4	104	B 10V4	B 12V4
18	B 17V4	B 25V4	47	B 13V4	B 22V4	76	B 10V4	B 19V4	105	B 9V4	B 10V4
19	B 59V4	B 62V4	48	B 13V4	B 21V4	77	B 11V4	B 19V4	106	B 9V4	B 11V4
20	B 11V4	B 20V4	49	B 12V4	B 18V4	78	B 10V4	B 14V4	107	B 12V4	B 20V4
21	B 10V4	B 20V4	50	B 10V4	B 18V4	79	B 56V4	B 62V4	108	B 58V4	B 61V4
22	B 10V4	B 13V4	51	B 11V4	B 18V4	80	B 61V4	B 62V4	109	B 55V4	B 61V4
23	B 11V4	B 13V4	52	B 56V4	B 59V4	81	B 12V4	B 22V4	110	B 3V4	B 8V4
24	B 20V4	B 22V4	53	B 9V4	B 18V4	82	B 12V4	B 21V4	111	B 9V4	B 20V4
25	B 20V4	B 21V4	54	B 11V4	B 21V4	83	B 9V4	B 22V4	112	B 19V4	B 20V4
26	B 66V4	B 68V4	55	B 11V4	B 22V4	84	B 9V4	B 21V4	113	B 14V4	B 20V4
27	B 66V4	B 67V4	56	B 10V4	B 21V4	85	B 10V4	B 23V4	114	B 59V4	B 61V4
28	B 59V4	B 63V4	57	B 10V4	B 22V4	86	B 11V4	B 23V4	115	B 43V4	B 46V4
29	B 60V4	B 63V4	58	B 18V4	B 20V4	87	B 61V4	B 63V4	116	B 13V4	B 23V4

(continued)

Table H.3 (continued)

No.	From bus	To bus	No.	From bus	To bus	No.	From bus	To bus	No.	From bus	To bus
117	B 43V4	B 50V4	161	B 16V4	B 24V4	205	B 21V4	B 25V4	249	B 43V4	B 45V4
118	B 26V4	B 33V4	162	B 12V4	B 14V4	206	B 22V4	B 25V4	250	B 17V4	B 84V4
119	B 55V4	B 56V4	163	B 12V4	B 19V4	207	B 42V4	B 53V4	251	B 16V4	B 84V4
120	B 56V4	B 58V4	164	B 9V4	B 19V4	208	B 38V4	B 40V4	252	B 53V4	B 77V4
121	B 56V4	B 60V4	165	B 9V4	B 14V4	209	B 42V4	B 43V4	253	B 40V4	B 46V4
122	B 27V4	B 30V4	166	B 4V4	B 17V4	210	B 66V4	B 71V4	254	B 78V4	B 79V4
123	B 30V4	B 35V4	167	B 4V4	B 16V4	211	B 5V4	B 7V4	255	B 69V4	B 73V4
124	B 14V4	B 18V4	168	B 4V4	B 21V4	212	B 39V4	B 41V4	256	B 5V4	B 20V4
125	B 18V4	B 19V4	169	B 4V4	B 22V4	213	B 81V4	B 84V4	257	B 57V4	B 59V4
126	B 15V4	B 24V4	170	B 40V4	B 47V4	214	B 39V4	B 40V4	258	B 5V4	B 18V4
127	B 69V4	B 72V4	171	B 68V4	B 71V4	215	B 41V4	B 47V4	259	B 46V4	B 62V4
128	B 12V4	B 13V4	172	B 67V4	B 71V4	216	B 4V4	B 18V4	260	B 57V4	B 62V4
129	B 9V4	B 13V4	173	B 46V4	B 61V4	217	B 44V4	B 47V4	261	B 41V4	B 44V4
130	B 23V4	B 24V4	174	B 20V4	B 25V4	218	B 5V4	B 21V4	262	B 82V4	B 83V4
131	B 27V4	B 28V4	175	B 41V4	B 43V4	219	B 5V4	B 22V4	263	B 43V4	B 49V4
132	B 28V4	B 35V4	176	B 30V4	B 31V4	220	B 30V4	B 32V4	264	B 4V4	B 9V4
133	B 14V4	B 23V4	177	B 14V4	B 15V4	221	B 41V4	B 49V4	265	B 16V4	B 18V4
134	B 19V4	B 23V4	178	B 15V4	B 19V4	222	B 38V4	B 41V4	266	B 17V4	B 18V4
135	B 65V4	B 70V4	179	B 10V4	B 25V4	223	B 41V4	B 42V4	267	B 4V4	B 12V4
136	B 14V4	B 21V4	180	B 11V4	B 25V4	224	B 46V4	B 50V4	268	B 39V4	B 49V4
137	B 14V4	B 22V4	181	B 38V4	B 47V4	225	B 18V4	B 25V4	269	B 41V4	B 46V4
138	B 19V4	B 21V4	182	B 26V4	B 31V4	226	B 75V4	B 76V4	270	B 46V4	B 63V4
139	B 19V4	B 22V4	183	B 40V4	B 50V4	227	B 17V4	B 21V4	271	B 4V4	B 23V4
140	B 45V4	B 50V4	184	B 40V4	B 43V4	228	B 17V4	B 22V4	272	B 6V4	B 7V4
141	B 71V4	B 73V4	185	B 40V4	B 49V4	229	B 16V4	B 21V4	273	B 23V4	B 25V4
142	B 24V4	B 25V4	186	B 13V4	B 17V4	230	B 16V4	B 22V4	274	B 5V4	B 13V4
143	B 76V4	B 77V4	187	B 13V4	B 16V4	231	B 39V4	B 61V4	275	B 40V4	B 42V4
144	B 12V4	B 24V4	188	B 16V4	B 19V4	232	B 41V4	B 48V4	276	B 43V4	B 48V4
145	B 9V4	B 24V4	189	B 17V4	B 19V4	233	B 4V4	B 19V4	277	B 39V4	B 48V4
146	B 5V4	B 6V4	190	B 14V4	B 16V4	234	B 4V4	B 14V4	278	B 57V4	B 61V4
147	B 64V4	B 70V4	191	B 14V4	B 17V4	235	B 5V4	B 9V4	279	B 46V4	B 60V4
148	B 41V4	B 50V4	192	B 5V4	B 15V4	236	B 46V4	B 49V4	280	B 27V4	B 32V4
149	B 4V4	B 25V4	193	B 4V4	B 11V4	237	B 51V4	B 71V4	281	B 32V4	B 35V4
150	B 52V4	B 57V4	194	B 4V4	B 10V4	238	B 5V4	B 12V4	282	B 72V4	B 83V4
151	B 39V4	B 50V4	195	B 40V4	B 48V4	239	B 40V4	B 44V4	283	B 38V4	B 49V4
152	B 4V4	B 13V4	196	B 4V4	B 15V4	240	B 79V4	B 83V4	284	B 28V4	B 33V4
153	B 13V4	B 25V4	197	B 1V4	B 81V4	241	B 39V4	B 42V4	285	B 38V4	B 48V4
154	B 28V4	B 31V4	198	B 45V4	B 53V4	242	B 28V4	B 32V4	286	B 25V4	B 84V4
155	B 4V4	B 24V4	199	B 16V4	B 20V4	243	B 41V4	B 45V4	287	B 7V4	B 9V4
156	B 4V4	B 20V4	200	B 17V4	B 20V4	244	B 56V4	B 57V4	288	B 26V4	B 28V4
157	B 19V4	B 25V4	201	B 11V4	B 17V4	245	B 46V4	B 48V4	289	B 32V4	B 34V4
158	B 14V4	B 25V4	202	B 10V4	B 16V4	246	B 15V4	B 25V4	290	B 15V4	B 16V4
159	B 56V4	B 61V4	203	B 10V4	B 17V4	247	B 31V4	B 33V4	291	B 15V4	B 17V4
160	B 17V4	B 24V4	204	B 11V4	B 16V4	248	B 4V4	B 5V4	292	B 43V4	B 61V4

(continued)

Table H.3 (continued)

No.	From bus	To bus	No.	From bus	To bus	No.	From bus	To bus	No.	From bus	To bus
293	B 5V4	B 23V4	338	B 67V4	B 70V4	380	B 43V4	B 57V4	425	B 71V4	B 74V4
294	B 7V4	B 12V4	339	B 1V4	B 2V4	381	B 78V4	B 80V4	426	B 3V4	B 7V4
295	B 57V4	B 63V4	339	B 1V4	B 2V4	382	B 25V4	B 81V4	427	B 6V4	B 9V4
296	B 5V4	B 10V4	339	B 1V4	B 2V4	383	B 25V4	B 54V4	428	B 53V4	B 54V4
297	B 5V4	B 11V4	339	B 1V4	B 2V4	384	B 47V4	B 50V4	429	B 46V4	B 51V4
298	B 46V4	B 55V4	340	B 54V4	B 77V4	385	B 43V4	B 62V4	430	B 70V4	B 73V4
299	B 46V4	B 58V4	341	B 49V4	B 51V4	386	B 51V4	B 67V4	431	B 43V4	B 58V4
300	B 7V4	B 15V4	342	B 7V4	B 23V4	387	B 51V4	B 68V4	432	B 43V4	B 55V4
301	B 12V4	B 25V4	343	B 64V4	B 80V4	388	B 51V4	B 60V4	433	B 50V4	B 61V4
302	B 51V4	B 74V4	344	B 42V4	B 46V4	389	B 51V4	B 58V4	434	B 6V4	B 12V4
303	B 46V4	B 59V4	345	B 57V4	B 60V4	390	B 51V4	B 55V4	435	B 39V4	B 53V4
304	B 9V4	B 25V4	346	B 16V4	B 81V4	391	B 14V4	B 52V4	436	B 5V4	B 8V4
305	B 3V4	B 6V4	347	B 17V4	B 81V4	392	B 19V4	B 52V4	437	B 30V4	B 34V4
306	B 39V4	B 45V4	348	B 46V4	B 47V4	393	B 7V4	B 20V4	438	B 51V4	B 63V4
307	B 40V4	B 45V4	349	B 39V4	B 57V4	394	B 72V4	B 73V4	439	B 51V4	B 66V4
308	B 31V4	B 35V4	350	B 39V4	B 63V4	395	B 45V4	B 46V4	440	B 50V4	B 57V4
309	B 27V4	B 31V4	351	B 39V4	B 47V4	396	B 4V4	B 6V4	441	B 47V4	B 51V4
310	B 5V4	B 24V4	352	B 7V4	B 18V4	397	B 5V4	B 14V4	442	B 38V4	B 43V4
311	B 50V4	B 53V4	353	B 46V4	B 57V4	398	B 5V4	B 19V4	443	B 52V4	B 56V4
312	B 17V4	B 23V4	354	B 55V4	B 57V4	399	B 14V4	B 54V4	444	B 64V4	B 69V4
313	B 16V4	B 23V4	355	B 57V4	B 58V4	400	B 19V4	B 54V4	445	B 69V4	B 83V4
314	B 6V4	B 8V4	356	B 53V4	B 76V4	401	B 43V4	B 63V4	446	B 6V4	B 20V4
315	B 30V4	B 33V4	357	B 7V4	B 21V4	402	B 51V4	B 61V4	447	B 44V4	B 50V4
316	B 52V4	B 54V4	358	B 7V4	B 22V4	403	B 58V4	B 74V4	448	B 24V4	B 84V4
317	B 51V4	B 73V4	359	B 31V4	B 32V4	404	B 55V4	B 74V4	449	B 6V4	B 18V4
318	B 67V4	B 73V4	360	B 49V4	B 50V4	405	B 43V4	B 53V4	450	B 5V4	B 17V4
319	B 68V4	B 73V4	361	B 33V4	B 35V4	406	B 70V4	B 71V4	451	B 5V4	B 16V4
320	B 26V4	B 30V4	362	B 27V4	B 33V4	407	B 72V4	B 79V4	452	B 7V4	B 24V4
321	B 66V4	B 70V4	363	B 66V4	B 73V4	408	B 49V4	B 61V4	453	B 65V4	B 69V4
322	B 7V4	B 8V4	364	B 54V4	B 81V4	409	B 26V4	B 27V4	454	B 52V4	B 59V4
323	B 4V4	B 84V4	365	B 75V4	B 77V4	410	B 26V4	B 35V4	455	B 51V4	B 62V4
324	B 29V4	B 32V4	366	B 46V4	B 56V4	411	B 6V4	B 21V4	456	B 69V4	B 70V4
325	B 39V4	B 62V4	367	B 36V4	B 74V4	412	B 6V4	B 22V4	457	B 42V4	B 57V4
326	B 12V4	B 17V4	368	B 39V4	B 60V4	413	B 7V4	B 10V4	458	B 6V4	B 13V4
327	B 12V4	B 16V4	369	B 39V4	B 59V4	414	B 7V4	B 11V4	459	B 52V4	B 62V4
328	B 43V4	B 47V4	370	B 16V4	B 54V4	415	B 60V4	B 74V4	460	B 3V4	B 5V4
329	B 9V4	B 16V4	371	B 17V4	B 54V4	416	B 38V4	B 50V4	461	B 28V4	B 34V4
330	B 9V4	B 17V4	372	B 65V4	B 80V4	417	B 43V4	B 60V4	462	B 63V4	B 74V4
331	B 48V4	B 51V4	373	B 34V4	B 35V4	418	B 48V4	B 61V4	463	B 13V4	B 84V4
332	B 37V4	B 74V4	374	B 27V4	B 34V4	419	B 43V4	B 59V4	464	B 24V4	B 54V4
333	B 52V4	B 53V4	375	B 6V4	B 15V4	420	B 53V4	B 57V4	465	B 43V4	B 56V4
334	B 73V4	B 74V4	376	B 48V4	B 50V4	421	B 7V4	B 13V4	466	B 38V4	B 39V4
335	B 44V4	B 49V4	377	B 1V4	B 84V4	422	B 69V4	B 71V4	467	B 45V4	B 77V4
336	B 44V4	B 48V4	378	B 39V4	B 58V4	423	B 5V4	B 25V4	468	B 39V4	B 52V4
337	B 68V4	B 70V4	379	B 39V4	B 55V4	424	B 39V4	B 56V4	469	B 52V4	B 61V4

(continued)

Table H.3 (continued)

No.	From bus	To bus	No.	From bus	To bus	No.	From bus	To bus	No.	From bus	To bus
470	B 42V4	B 52V4	513	B 13V4	B 54V4	556	B 42V4	B 61V4	599	B 21V4	B 52V4
471	B 6V4	B 23V4	514	B 19V4	B 57V4	557	B 64V4	B 73V4	600	B 22V4	B 52V4
472	B 42V4	B 49V4	515	B 14V4	B 57V4	558	B 45V4	B 47V4	601	B 50V4	B 60V4
473	B 6V4	B 10V4	516	B 7V4	B 14V4	559	B 48V4	B 67V4	602	B 48V4	B 59V4
474	B 6V4	B 11V4	517	B 7V4	B 19V4	560	B 48V4	B 68V4	603	B 47V4	B 66V4
475	B 40V4	B 61V4	518	B 49V4	B 58V4	561	B 70V4	B 80V4	604	B 15V4	B 84V4
476	B 43V4	B 44V4	519	B 49V4	B 55V4	562	B 47V4	B 61V4	605	B 23V4	B 54V4
477	B 79V4	B 80V4	520	B 45V4	B 52V4	563	B 20V4	B 52V4	606	B 1V4	B 54V4
478	B 51V4	B 59V4	521	B 46V4	B 52V4	564	B 25V4	B 52V4	607	B 18V4	B 84V4
479	B 29V4	B 34V4	522	B 65V4	B 68V4	565	B 48V4	B 71V4	608	B 21V4	B 54V4
480	B 3V4	B 26V4	523	B 65V4	B 67V4	566	B 45V4	B 48V4	609	B 22V4	B 54V4
481	B 23V4	B 52V4	524	B 38V4	B 46V4	567	B 50V4	B 63V4	610	B 49V4	B 66V4
482	B 20V4	B 84V4	525	B 61V4	B 74V4	568	B 72V4	B 80V4	611	B 71V4	B 72V4
483	B 4V4	B 7V4	526	B 52V4	B 77V4	569	B 69V4	B 79V4	612	B 40V4	B 51V4
484	B 52V4	B 63V4	527	B 49V4	B 63V4	570	B 12V4	B 52V4	613	B 78V4	B 83V4
485	B 41V4	B 53V4	528	B 18V4	B 52V4	571	B 17V4	B 52V4	614	B 50V4	B 58V4
486	B 6V4	B 24V4	529	B 48V4	B 55V4	572	B 16V4	B 52V4	615	B 50V4	B 55V4
487	B 43V4	B 52V4	530	B 48V4	B 58V4	573	B 40V4	B 53V4	616	B 65V4	B 72V4
488	B 10V4	B 52V4	531	B 39V4	B 51V4	574	B 52V4	B 60V4	617	B 26V4	B 32V4
489	B 11V4	B 52V4	532	B 54V4	B 57V4	575	B 49V4	B 68V4	618	B 37V4	B 58V4
490	B 38V4	B 42V4	533	B 64V4	B 66V4	576	B 49V4	B 67V4	619	B 37V4	B 55V4
491	B 62V4	B 74V4	534	B 49V4	B 62V4	577	B 51V4	B 56V4	620	B 67V4	B 69V4
492	B 42V4	B 48V4	535	B 46V4	B 53V4	578	B 9V4	B 52V4	621	B 68V4	B 69V4
493	B 79V4	B 82V4	536	B 4V4	B 81V4	579	B 42V4	B 76V4	622	B 8V4	B 31V4
494	B 41V4	B 61V4	537	B 42V4	B 44V4	580	B 49V4	B 71V4	623	B 4V4	B 54V4
495	B 65V4	B 66V4	538	B 48V4	B 63V4	581	B 23V4	B 57V4	624	B 45V4	B 61V4
496	B 19V4	B 84V4	539	B 50V4	B 62V4	582	B 65V4	B 71V4	625	B 6V4	B 25V4
497	B 14V4	B 84V4	540	B 22V4	B 84V4	583	B 72V4	B 74V4	626	B 13V4	B 81V4
498	B 59V4	B 74V4	541	B 21V4	B 84V4	584	B 47V4	B 68V4	627	B 18V4	B 57V4
499	B 49V4	B 60V4	542	B 45V4	B 49V4	585	B 47V4	B 67V4	628	B 10V4	B 57V4
500	B 42V4	B 47V4	543	B 8V4	B 26V4	586	B 43V4	B 51V4	629	B 11V4	B 57V4
501	B 11V4	B 54V4	544	B 54V4	B 76V4	587	B 50V4	B 59V4	630	B 37V4	B 60V4
502	B 10V4	B 54V4	545	B 39V4	B 44V4	588	B 52V4	B 55V4	631	B 6V4	B 19V4
503	B 24V4	B 52V4	546	B 48V4	B 62V4	589	B 52V4	B 58V4	632	B 6V4	B 14V4
504	B 69V4	B 74V4	547	B 56V4	B 74V4	590	B 49V4	B 59V4	633	B 40V4	B 62V4
505	B 38V4	B 45V4	548	B 44V4	B 45V4	591	B 48V4	B 66V4	634	B 72V4	B 82V4
506	B 42V4	B 77V4	549	B 13V4	B 52V4	592	B 19V4	B 81V4	635	B 12V4	B 57V4
507	B 45V4	B 57V4	550	B 69V4	B 80V4	593	B 14V4	B 81V4	636	B 1V4	B 16V4
508	B 45V4	B 76V4	551	B 65V4	B 73V4	594	B 18V4	B 54V4	637	B 1V4	B 17V4
509	B 11V4	B 84V4	552	B 64V4	B 68V4	595	B 64V4	B 71V4	638	B 47V4	B 71V4
510	B 10V4	B 84V4	553	B 64V4	B 67V4	596	B 24V4	B 81V4	639	B 3V4	B 31V4
511	B 50V4	B 52V4	554	B 20V4	B 54V4	597	B 64V4	B 72V4	640	B 40V4	B 63V4
512	B 48V4	B 60V4	555	B 54V4	B 84V4	598	B 44V4	B 46V4	641	B 2V4	B 81V4

(continued)

Table H.3 (continued)

No.	From bus	To bus	No.	From bus	To bus	No.	From bus	To bus	No.	From bus	To bus
641	B 2V4	B 81V4	685	B 41V4	B 55V4	729	B 9V4	B 56V4	773	B 8V4	B 22V4
642	B 9V4	B 57V4	686	B 41V4	B 58V4	730	B 15V4	B 57V4	774	B 8V4	B 21V4
643	B 15V4	B 52V4	687	B 12V4	B 84V4	731	B 36V4	B 83V4	775	B 37V4	B 73V4
644	B 41V4	B 62V4	688	B 1V4	B 25V4	732	B 48V4	B 57V4	776	B 23V4	B 59V4
645	B 50V4	B 56V4	689	B 48V4	B 56V4	733	B 42V4	B 56V4	777	B 54V4	B 75V4
646	B 41V4	B 57V4	690	B 9V4	B 84V4	734	B 14V4	B 56V4	778	B 36V4	B 62V4
647	B 32V4	B 33V4	691	B 42V4	B 59V4	735	B 19V4	B 56V4	779	B 15V4	B 81V4
648	B 40V4	B 60V4	692	B 53V4	B 61V4	736	B 38V4	B 51V4	780	B 36V4	B 73V4
649	B 66V4	B 69V4	693	B 47V4	B 55V4	737	B 36V4	B 63V4	781	B 53V4	B 63V4
650	B 70V4	B 72V4	694	B 47V4	B 58V4	738	B 50V4	B 76V4	782	B 37V4	B 82V4
651	B 42V4	B 62V4	695	B 41V4	B 59V4	739	B 48V4	B 73V4	783	B 38V4	B 67V4
652	B 11V4	B 81V4	696	B 23V4	B 56V4	740	B 18V4	B 81V4	784	B 38V4	B 68V4
653	B 10V4	B 81V4	697	B 47V4	B 63V4	741	B 52V4	B 76V4	785	B 36V4	B 69V4
654	B 20V4	B 81V4	698	B 5V4	B 84V4	742	B 47V4	B 59V4	786	B 45V4	B 60V4
655	B 24V4	B 57V4	699	B 47V4	B 62V4	743	B 53V4	B 62V4	787	B 55V4	B 73V4
656	B 37V4	B 56V4	700	B 28V4	B 29V4	744	B 29V4	B 35V4	788	B 58V4	B 73V4
657	B 41V4	B 63V4	701	B 8V4	B 15V4	745	B 27V4	B 29V4	789	B 38V4	B 66V4
658	B 40V4	B 55V4	702	B 36V4	B 60V4	746	B 25V4	B 57V4	790	B 53V4	B 56V4
659	B 40V4	B 58V4	703	B 21V4	B 57V4	747	B 48V4	B 74V4	791	B 8V4	B 18V4
660	B 6V4	B 16V4	704	B 22V4	B 57V4	748	B 45V4	B 63V4	792	B 60V4	B 73V4
661	B 6V4	B 17V4	705	B 67V4	B 74V4	749	B 42V4	B 54V4	793	B 23V4	B 81V4
662	B 12V4	B 54V4	706	B 68V4	B 74V4	750	B 36V4	B 59V4	794	B 61V4	B 71V4
663	B 7V4	B 25V4	707	B 37V4	B 62V4	751	B 45V4	B 54V4	795	B 37V4	B 61V4
664	B 29V4	B 30V4	708	B 31V4	B 34V4	752	B 37V4	B 72V4	796	B 37V4	B 83V4
665	B 3V4	B 33V4	709	B 21V4	B 81V4	753	B 49V4	B 74V4	797	B 10V4	B 56V4
666	B 40V4	B 57V4	710	B 22V4	B 81V4	754	B 45V4	B 59V4	798	B 11V4	B 56V4
667	B 23V4	B 84V4	711	B 53V4	B 75V4	755	B 29V4	B 31V4	799	B 8V4	B 23V4
668	B 41V4	B 51V4	712	B 7V4	B 17V4	756	B 49V4	B 73V4	800	B 40V4	B 52V4
669	B 9V4	B 54V4	713	B 7V4	B 16V4	757	B 40V4	B 56V4	801	B 51V4	B 70V4
670	B 41V4	B 60V4	714	B 72V4	B 78V4	758	B 18V4	B 56V4	802	B 46V4	B 71V4
671	B 37V4	B 63V4	715	B 36V4	B 82V4	759	B 50V4	B 51V4	803	B 52V4	B 81V4
672	B 49V4	B 56V4	716	B 42V4	B 60V4	760	B 37V4	B 51V4	804	B 45V4	B 55V4
673	B 13V4	B 57V4	717	B 46V4	B 74V4	761	B 80V4	B 83V4	805	B 45V4	B 58V4
674	B 15V4	B 54V4	718	B 36V4	B 72V4	762	B 66V4	B 74V4	806	B 44V4	B 51V4
675	B 40V4	B 59V4	719	B 45V4	B 62V4	763	B 41V4	B 52V4	807	B 36V4	B 51V4
676	B 50V4	B 77V4	720	B 49V4	B 57V4	764	B 58V4	B 71V4	808	B 38V4	B 61V4
677	B 47V4	B 60V4	721	B 8V4	B 9V4	765	B 55V4	B 71V4	809	B 1V4	B 77V4
678	B 20V4	B 57V4	722	B 8V4	B 33V4	766	B 4V4	B 52V4	810	B 19V4	B 59V4
679	B 42V4	B 63V4	723	B 36V4	B 56V4	767	B 53V4	B 59V4	811	B 14V4	B 59V4
680	B 37V4	B 59V4	724	B 12V4	B 56V4	768	B 3V4	B 15V4	812	B 37V4	B 69V4
681	B 36V4	B 58V4	725	B 57V4	B 77V4	769	B 17V4	B 57V4	813	B 45V4	B 56V4
682	B 36V4	B 55V4	726	B 8V4	B 12V4	770	B 16V4	B 57V4	814	B 49V4	B 53V4
683	B 51V4	B 69V4	727	B 42V4	B 58V4	771	B 60V4	B 71V4	815	B 23V4	B 62V4
684	B 77V4	B 81V4	728	B 42V4	B 55V4	772	B 41V4	B 56V4	816	B 12V4	B 59V4

(continued)

Table H.3 (continued)

No.	From bus	To bus	No.	From bus	To bus	No.	From bus	To bus	No.	From bus	To bus
817	B 3V4	B 9V4	863	B 12V4	B 62V4	909	B 8V4	B 24V4	955	B 26V4	B 34V4
818	B 9V4	B 59V4	864	B 38V4	B 53V4	910	B 19V4	B 77V4	956	B 9V4	B 55V4
819	B 73V4	B 83V4	865	B 47V4	B 53V4	911	B 14V4	B 77V4	957	B 9V4	B 58V4
820	B 43V4	B 77V4	866	B 50V4	B 54V4	912	B 3V4	B 28V4	958	B 41V4	B 68V4
821	B 2V4	B 84V4	867	B 61V4	B 73V4	913	B 45V4	B 75V4	959	B 41V4	B 67V4
822	B 3V4	B 12V4	868	B 1V4	B 75V4	914	B 47V4	B 74V4	960	B 19V4	B 60V4
823	B 69V4	B 78V4	869	B 9V4	B 62V4	915	B 41V4	B 77V4	961	B 14V4	B 60V4
824	B 67V4	B 72V4	870	B 63V4	B 73V4	916	B 39V4	B 71V4	962	B 9V4	B 60V4
825	B 68V4	B 72V4	871	B 8V4	B 10V4	917	B 18V4	B 63V4	963	B 46V4	B 66V4
826	B 8V4	B 20V4	872	B 8V4	B 11V4	918	B 65V4	B 79V4	964	B 14V4	B 58V4
827	B 44V4	B 68V4	873	B 66V4	B 72V4	919	B 43V4	B 74V4	965	B 14V4	B 55V4
828	B 44V4	B 67V4	874	B 3V4	B 4V4	920	B 75V4	B 81V4	966	B 19V4	B 58V4
829	B 44V4	B 66V4	875	B 40V4	B 68V4	921	B 38V4	B 71V4	967	B 19V4	B 55V4
830	B 63V4	B 71V4	876	B 40V4	B 67V4	922	B 65V4	B 78V4	968	B 10V4	B 63V4
831	B 6V4	B 84V4	877	B 62V4	B 71V4	923	B 43V4	B 76V4	969	B 11V4	B 63V4
832	B 48V4	B 53V4	878	B 76V4	B 81V4	924	B 11V4	B 62V4	970	B 43V4	B 71V4
833	B 51V4	B 72V4	879	B 36V4	B 61V4	925	B 10V4	B 62V4	971	B 17V4	B 53V4
834	B 12V4	B 81V4	880	B 16V4	B 77V4	926	B 78V4	B 82V4	972	B 16V4	B 53V4
835	B 1V4	B 4V4	881	B 17V4	B 77V4	927	B 3V4	B 13V4	973	B 38V4	B 60V4
836	B 47V4	B 56V4	882	B 3V4	B 18V4	928	B 43V4	B 54V4	974	B 38V4	B 62V4
837	B 9V4	B 81V4	883	B 19V4	B 63V4	929	B 39V4	B 54V4	975	B 53V4	B 81V4
838	B 3V4	B 22V4	884	B 14V4	B 63V4	930	B 1V4	B 24V4	976	B 36V4	B 71V4
839	B 3V4	B 21V4	885	B 44V4	B 61V4	931	B 8V4	B 30V4	977	B 70V4	B 74V4
840	B 23V4	B 63V4	886	B 11V4	B 59V4	932	B 23V4	B 61V4	978	B 38V4	B 63V4
841	B 53V4	B 60V4	887	B 10V4	B 59V4	933	B 59V4	B 73V4	979	B 1V4	B 13V4
842	B 69V4	B 82V4	888	B 73V4	B 80V4	934	B 20V4	B 59V4	980	B 25V4	B 53V4
843	B 18V4	B 59V4	889	B 64V4	B 79V4	935	B 1V4	B 14V4	981	B 39V4	B 76V4
844	B 14V4	B 62V4	890	B 64V4	B 78V4	936	B 1V4	B 19V4	982	B 3V4	B 24V4
845	B 19V4	B 62V4	891	B 12V4	B 63V4	937	B 24V4	B 59V4	983	B 42V4	B 75V4
846	B 47V4	B 73V4	892	B 18V4	B 62V4	938	B 21V4	B 59V4	984	B 41V4	B 71V4
847	B 39V4	B 74V4	893	B 9V4	B 63V4	939	B 22V4	B 59V4	985	B 24V4	B 62V4
848	B 51V4	B 57V4	894	B 44V4	B 53V4	940	B 46V4	B 73V4	986	B 61V4	B 68V4
849	B 22V4	B 56V4	895	B 40V4	B 66V4	941	B 37V4	B 71V4	987	B 61V4	B 67V4
850	B 21V4	B 56V4	896	B 40V4	B 71V4	942	B 19V4	B 61V4	988	B 20V4	B 62V4
851	B 20V4	B 56V4	897	B 3V4	B 20V4	943	B 14V4	B 61V4	989	B 41V4	B 66V4
852	B 4V4	B 8V4	898	B 8V4	B 28V4	944	B 13V4	B 59V4	990	B 38V4	B 55V4
853	B 15V4	B 56V4	899	B 3V4	B 23V4	945	B 42V4	B 51V4	991	B 38V4	B 58V4
854	B 24V4	B 56V4	900	B 46V4	B 67V4	946	B 3V4	B 10V4	992	B 44V4	B 71V4
855	B 74V4	B 83V4	901	B 46V4	B 68V4	947	B 3V4	B 11V4	993	B 5V4	B 81V4
856	B 39V4	B 77V4	902	B 23V4	B 55V4	948	B 12V4	B 55V4	994	B 18V4	B 55V4
857	B 8V4	B 13V4	903	B 23V4	B 58V4	949	B 12V4	B 58V4	995	B 18V4	B 58V4
858	B 19V4	B 53V4	904	B 33V4	B 34V4	950	B 54V4	B 56V4	996	B 46V4	B 77V4
859	B 14V4	B 53V4	905	B 23V4	B 60V4	951	B 1V4	B 76V4	997	B 18V4	B 60V4
860	B 53V4	B 55V4	906	B 59V4	B 71V4	952	B 41V4	B 76V4	998	B 22V4	B 62V4
861	B 53V4	B 58V4	907	B 25V4	B 77V4	953	B 12V4	B 60V4	999	B 21V4	B 62V4
862	B 13V4	B 56V4	908	B 62V4	B 73V4	954	B 3V4	B 30V4	1000	B 12V4	B 61V4

(continued)

Table H.3 (continued)

No.	From bus	To bus	No.	From bus	To bus	No.	From bus	To bus	No.	From bus	To bus
1001	B 5V4	B 54V4	1045	B 24V4	B 77V4	1089	B 1V4	B 18V4	1133	B 36V4	B 79V4
1002	B 60V4	B 67V4	1046	B 65V4	B 83V4	1090	B 46V4	B 76V4	1134	B 3V4	B 16V4
1003	B 60V4	B 68V4	1047	B 10V4	B 61V4	1091	B 20V4	B 77V4	1135	B 3V4	B 17V4
1004	B 40V4	B 77V4	1048	B 11V4	B 61V4	1092	B 41V4	B 74V4	1136	B 6V4	B 33V4
1005	B 55V4	B 68V4	1049	B 23V4	B 53V4	1093	B 43V4	B 73V4	1137	B 4V4	B 77V4
1006	B 58V4	B 67V4	1050	B 39V4	B 66V4	1094	B 48V4	B 70V4	1138	B 3V4	B 32V4
1007	B 58V4	B 68V4	1051	B 44V4	B 60V4	1095	B 70V4	B 83V4	1139	B 73V4	B 82V4
1008	B 55V4	B 67V4	1052	B 13V4	B 53V4	1096	B 8V4	B 32V4	1140	B 50V4	B 66V4
1009	B 9V4	B 61V4	1053	B 7V4	B 84V4	1097	B 3V4	B 35V4	1141	B 1V4	B 12V4
1010	B 39V4	B 68V4	1054	B 40V4	B 74V4	1098	B 3V4	B 27V4	1142	B 12V4	B 46V4
1011	B 39V4	B 67V4	1055	B 60V4	B 66V4	1099	B 8V4	B 35V4	1143	B 1V4	B 9V4
1012	B 6V4	B 26V4	1056	B 55V4	B 66V4	1100	B 8V4	B 27V4	1144	B 18V4	B 46V4
1013	B 64V4	B 83V4	1057	B 58V4	B 66V4	1101	B 3V4	B 25V4	1145	B 40V4	B 54V4
1014	B 1V4	B 20V4	1058	B 44V4	B 55V4	1102	B 1V4	B 15V4	1146	B 2V4	B 16V4
1015	B 1V4	B 10V4	1059	B 44V4	B 58V4	1103	B 17V4	B 76V4	1147	B 2V4	B 17V4
1016	B 1V4	B 11V4	1060	B 43V4	B 66V4	1104	B 16V4	B 76V4	1148	B 10V4	B 39V4
1017	B 46V4	B 54V4	1061	B 1V4	B 21V4	1105	B 49V4	B 70V4	1149	B 11V4	B 39V4
1018	B 51V4	B 65V4	1062	B 1V4	B 22V4	1106	B 18V4	B 77V4	1150	B 18V4	B 39V4
1019	B 18V4	B 61V4	1063	B 18V4	B 53V4	1107	B 6V4	B 31V4	1151	B 74V4	B 79V4
1020	B 24V4	B 53V4	1064	B 10V4	B 77V4	1108	B 2V4	B 54V4	1152	B 12V4	B 77V4
1021	B 54V4	B 61V4	1065	B 11V4	B 77V4	1109	B 50V4	B 74V4	1153	B 48V4	B 69V4
1022	B 20V4	B 63V4	1066	B 39V4	B 73V4	1110	B 23V4	B 77V4	1154	B 2V4	B 77V4
1023	B 24V4	B 63V4	1067	B 20V4	B 53V4	1111	B 50V4	B 71V4	1155	B 12V4	B 39V4
1024	B 22V4	B 63V4	1068	B 74V4	B 82V4	1112	B 70V4	B 78V4	1156	B 9V4	B 77V4
1025	B 21V4	B 63V4	1069	B 66V4	B 80V4	1113	B 23V4	B 46V4	1157	B 44V4	B 73V4
1026	B 10V4	B 53V4	1070	B 68V4	B 80V4	1114	B 8V4	B 16V4	1158	B 7V4	B 36V4
1027	B 11V4	B 53V4	1071	B 67V4	B 80V4	1115	B 8V4	B 17V4	1159	B 80V4	B 82V4
1028	B 73V4	B 79V4	1072	B 71V4	B 83V4	1116	B 47V4	B 70V4	1160	B 49V4	B 69V4
1029	B 43V4	B 68V4	1073	B 7V4	B 26V4	1117	B 41V4	B 54V4	1161	B 5V4	B 26V4
1030	B 43V4	B 67V4	1074	B 13V4	B 77V4	1118	B 23V4	B 39V4	1162	B 37V4	B 79V4
1031	B 45V4	B 51V4	1075	B 70V4	B 79V4	1119	B 38V4	B 73V4	1163	B 15V4	B 77V4
1032	B 51V4	B 64V4	1076	B 19V4	B 39V4	1120	B 1V4	B 23V4	1164	B 76V4	B 84V4
1033	B 8V4	B 14V4	1077	B 14V4	B 39V4	1121	B 50V4	B 68V4	1165	B 71V4	B 79V4
1034	B 8V4	B 19V4	1078	B 64V4	B 74V4	1122	B 50V4	B 67V4	1166	B 38V4	B 76V4
1035	B 10V4	B 60V4	1079	B 26V4	B 29V4	1123	B 50V4	B 75V4	1167	B 11V4	B 43V4
1036	B 11V4	B 60V4	1080	B 65V4	B 74V4	1124	B 21V4	B 77V4	1168	B 10V4	B 43V4
1037	B 10V4	B 55V4	1081	B 37V4	B 46V4	1125	B 22V4	B 77V4	1169	B 38V4	B 74V4
1038	B 11V4	B 58V4	1082	B 7V4	B 31V4	1126	B 25V4	B 76V4	1170	B 44V4	B 76V4
1039	B 11V4	B 55V4	1083	B 8V4	B 25V4	1127	B 36V4	B 46V4	1171	B 2V4	B 25V4
1040	B 10V4	B 58V4	1084	B 3V4	B 19V4	1128	B 6V4	B 81V4	1172	B 51V4	B 83V4
1041	B 71V4	B 80V4	1085	B 3V4	B 14V4	1129	B 7V4	B 54V4	1173	B 37V4	B 48V4
1042	B 2V4	B 75V4	1086	B 40V4	B 73V4	1130	B 7V4	B 37V4	1174	B 37V4	B 49V4
1043	B 77V4	B 84V4	1087	B 14V4	B 46V4	1131	B 41V4	B 73V4	1175	B 2V4	B 76V4
1044	B 40V4	B 76V4	1088	B 19V4	B 46V4	1132	B 29V4	B 33V4	1176	B 12V4	B 43V4

(continued)

Table H.3 (continued)

No.	From bus	To bus	No.	From bus	To bus	No.	From bus	To bus	No.	From bus	To bus
1177	B 38V4	B 77V4	1205	B 24V4	B 45V4	1233	B 48V4	B 65V4	1261	B 21V4	B 45V4
1178	B 37V4	B 67V4	1206	B 13V4	B 42V4	1234	B 16V4	B 50V4	1262	B 49V4	B 72V4
1179	B 37V4	B 68V4	1207	B 24V4	B 50V4	1235	B 17V4	B 50V4	1263	B 47V4	B 64V4
1180	B 9V4	B 43V4	1208	B 75V4	B 84V4	1236	B 41V4	B 75V4	1264	B 29V4	B 36V4
1181	B 29V4	B 82V4	1209	B 47V4	B 69V4	1237	B 38V4	B 70V4	1265	B 7V4	B 30V4
1182	B 24V4	B 43V4	1210	B 36V4	B 48V4	1238	B 20V4	B 50V4	1266	B 5V4	B 33V4
1183	B 24V4	B 42V4	1211	B 5V4	B 31V4	1239	B 3V4	B 84V4	1267	B 22V4	B 76V4
1184	B 49V4	B 77V4	1212	B 36V4	B 49V4	1240	B 36V4	B 66V4	1268	B 21V4	B 76V4
1185	B 9V4	B 37V4	1213	B 42V4	B 71V4	1241	B 20V4	B 45V4	1269	B 6V4	B 28V4
1186	B 17V4	B 42V4	1214	B 20V4	B 42V4	1242	B 48V4	B 76V4	1270	B 40V4	B 75V4
1187	B 16V4	B 42V4	1215	B 44V4	B 74V4	1243	B 25V4	B 75V4	1271	B 51V4	B 80V4
1188	B 44V4	B 77V4	1216	B 16V4	B 75V4	1244	B 20V4	B 76V4	1272	B 37V4	B 47V4
1189	B 25V4	B 42V4	1217	B 17V4	B 75V4	1245	B 49V4	B 65V4	1273	B 22V4	B 37V4
1190	B 73V4	B 78V4	1218	B 67V4	B 83V4	1246	B 48V4	B 64V4	1274	B 21V4	B 37V4
1191	B 7V4	B 33V4	1219	B 68V4	B 83V4	1247	B 66V4	B 83V4	1275	B 68V4	B 79V4
1192	B 36V4	B 67V4	1220	B 9V4	B 42V4	1248	B 8V4	B 84V4	1276	B 67V4	B 79V4
1193	B 36V4	B 68V4	1221	B 37V4	B 66V4	1249	B 45V4	B 74V4	1277	B 4V4	B 76V4
1194	B 24V4	B 76V4	1222	B 45V4	B 81V4	1250	B 45V4	B 71V4	1278	B 71V4	B 82V4
1195	B 7V4	B 81V4	1223	B 13V4	B 76V4	1251	B 47V4	B 65V4	1279	B 36V4	B 70V4
1196	B 48V4	B 77V4	1224	B 49V4	B 76V4	1252	B 44V4	B 70V4	1280	B 40V4	B 70V4
1197	B 13V4	B 43V4	1225	B 42V4	B 67V4	1253	B 2V4	B 4V4	1281	B 37V4	B 70V4
1198	B 37V4	B 43V4	1226	B 42V4	B 68V4	1254	B 7V4	B 28V4	1282	B 15V4	B 36V4
1199	B 1V4	B 5V4	1227	B 74V4	B 80V4	1255	B 49V4	B 64V4	1283	B 29V4	B 37V4
1200	B 17V4	B 45V4	1228	B 13V4	B 50V4	1256	B 48V4	B 72V4	1284	B 6V4	B 30V4
1201	B 16V4	B 45V4	1229	B 42V4	B 81V4	1257	B 15V4	B 37V4	1285	B 1V4	B 6V4
1202	B 42V4	B 74V4	1230	B 13V4	B 45V4	1258	B 45V4	B 67V4	1286	B 66V4	B 79V4
1203	B 50V4	B 73V4	1231	B 47V4	B 76V4	1259	B 45V4	B 68V4	1287	B 20V4	B 37V4
1204	B 25V4	B 45V4	1232	B 25V4	B 50V4	1260	B 8V4	B 29V4			

[a] The length of any candidate line may be readily calculated from geographical characteristics of the sending and receiving buses. For details, see problem 6 of Chap. 7

Table H.4 Generation data

No.	Bus name	P_G (p.u.)	No.	Bus name	P_G (p.u.)
1[a]	B 2V4	1.03	14	B 54V4	2.06
2	B 6V4	14.21	15	B 56V4	7.37
3	B 9V4	13.06	16	B 58V4	7.11
4	B 14V4	6.08	17	B 60V4	2.06
5	B 16V4	9.25	18	B 65V4	4.26
6	B 17V4	7.46	19	B 66V4	8.18
7	B 27V4	7.48	20	B 69V4	2.29
8	B 30V4	8.36	21	B 70V4	0.97
9	B 33V4	7.48	22	B 71V4	3.00
10	B 39V4	8.11	23	B 78V4	0.16
11	B 43V4	16.30	24	B 82V4	3.89
12	B 45V4	3.65	25	B 84V4	2.34
13	B 46V4	16.30			

[a] Slack bus

Appendix I
Numerical Details of the Basic Approach

The details of the proposed approach in Chap. 8 for transmission expansion planning, as discussed and tested on the 84-bus test system (see Chap. 8, Sect. 8.6.2) are given here (as Tables I.1, I.2 and I.3).

Table I.1 The detailed results of the backward stage

No.[a]	From bus	To bus	Length[b] (km)	Voltage level (kV)	No. of lines[c]	Capacity limit (p.u.)	Line flow (p.u.)	Maximum line flow in contingency conditions		
								Flow on line (p.u.)	Relevant contingency	
17	B 16V4	B 25V4	14.56	400	2	6.6	5.725	7.537	B 5V4	B 6V4
18	B 17V4	B 25V4	14.56	400	2	6.6	5.624	7.505	B 5V4	B 6V4

[a] The number shown is taken from the candidate line number given in Table H.3
[b] As X and Y are known for each bus, the line length can be readily calculated. For details, see problem 6 of Chap. 7
[c] Two lines are considered in each corridor

Table I.2 The detailed results of the forward stage

No.	From bus	To bus	Length (km)	Voltage level (kV)	No. of lines	Capacity limit (p.u.)	Line flow (p.u.)	Maximum line flow in contingency conditions		
								Flow on line (p.u.)	Relevant contingency	
2	B 21V4	B 22V4	1	400	2	6.6	0.265	4.5	B 9V4	B 22V4
3	B 10V4	B 11V4	1	400	2	6.6	1.895	6.197	B 11V4	B 19V4
17	B 16V4	B 25V4	14.56	400	2	6.6	4.845	5.782	B 17V4	B 25V4
18	B 17V4	B 25V4	14.56	400	2	6.6	4.868	5.785	B 16V4	B 25V4
21	B 10V4	B 20V4	16.15	400	2	6.6	1.576	4.344	B 20V4	B 24V4
33	B 38V4	B 44V4	21.13	400	2	6.6	2.074	2.2	B 38V4	B 44V4
54	B 11V4	B 21V4	28.33	400	2	6.6	0.533	2.833	B 9V4	B 21V4

(continued)

Table I.2 (continued)

No.	From bus	To bus	Length (km)	Voltage level (kV)	No. of lines	Capacity limit (p.u.)	Line flow (p.u.)	Maximum line flow in contingency conditions		
								Flow on line (p.u.)	Relevant contingency	
284	B 28V4	B 33V4	141.19	400	2	6.6	0.198	2.985	B 26V4	B 33V4
302	B 51V4	B 74V4	146.26	400	2	6.6	1.048	2.272	B 58V4	B 74V4
309	B 27V4	B 31V4	148.89	400	2	6.6	3.805	5.221	B 27V4	B 35V4
339	B 1V4	B 2V4	163.51	400	2	6.6	-3.304	6.357	B 43V4[a]	
374	B 27V4	B 34V4	173.68	400	2	6.6	2.91	5.769	B 30V4	B 32V4
473	B 6V4	B 10V4	219.84	400	2	6.6	4.736	5.721	B 6V4	B 7V4
699	B 47V4	B 62V4	282.55	400	2	6.6	-3.032	5.303	B 40V4	B 47V4
868	B 1V4	B 75V4	331.81	400	2	6.6	2.055	4.053	B 76V4	B 77V4
1113	B 23V4	B 46V4	408.13	400	2	6.6	-3.034	3.411	B 39V4	B 46V4
1161	B 5V4	B 26V4	431.95	400	2	6.6	-2.546	3.869	B 3V4	B 26V4
1197	B 13V4	B 43V4	452.34	400	2	6.6	-2.734	3.258	B 43V4	B 50V4
1253	B 2V4	B 4V4	477.96	400	2	6.6	0.822	3.319	B 6V4[b]	
1266	B 5V4	B 33V4	485.24	400	2	6.6	-3.414	4.841	B 26V4	B 33V4

[a, b] Contingency on generation which is located in this bus

Table I.3 The detailed results of the decrease stage

No.	From bus	To bus	Length (km)	Voltage level (kV)	No. of lines	Capacity limit (p.u.)	Line flow (p.u.)	Maximum line flow in contingency conditions		
								Flow on line (p.u.)	Relevant contingency	
2	B 21V4	B 22V4	1	400	2	6.6	0.312	4.5	B 9V4	B 22V4
3	B 10V4	B 11V4	1	400	2	6.6	1.975	6.501	B 11V4	B 19V4
17	B 16V4	B 25V4	14.56	400	2	6.6	4.874	5.816	B 17V4	B 25V4
18	B 17V4	B 25V4	14.56	400	2	6.6	4.894	5.816	B 16V4	B 25V4
21	B 10V4	B 20V4	16.15	400	2	6.6	1.487	4.108	B 20V4	B 24V4
33	B 38V4	B 44V4	21.13	400	1	3.3	1.962	2.2	B 38V4	B 44V4
54	B 11V4	B 21V4	28.33	400	1	3.3	0.628	2.569	B 9V4	B 21V4
284	B 28V4	B 33V4	141.19	400	2	6.6	-0.196	4.559	B 26V4	B 33V4
302	B 51V4	B 74V4	146.26	400	1	3.3	0.899	2.272	B 58V4	B 74V4
309	B 27V4	B 31V4	148.89	400	2	6.6	4.024	5.548	B 26V4	B 33V4
339	B 1V4	B 2V4	163.51	400	2	6.6	-3.213	6.295	B 43V4[a]	
374	B 27V4	B 34V4	173.68	400	2	6.6	2.895	5.768	B 30V4	B 32V4
473	B 6V4	B 10V4	219.84	400	2	6.6	4.713	5.656	B 6V4	B 7V4
699	B 47V4	B 62V4	282.55	400	2	6.6	-3.018	5.295	B 40V4	B 47V4
868	B 1V4	B 75V4	331.81	400	2	6.6	1.987	4.053	B 76V4	B 77V4
1113	B 23V4	B 46V4	408.13	400	1	3.3	-1.83	2.083	B 39V4	B 46V4
1161	B 5V4	B 26V4	431.95	400	2	6.6	-3.208	4.969	B 3V4	B 26V4
1197	B 13V4	B 43V4	452.34	400	2	6.6	-3.047	3.612	B 42V4	B 43V4
1253	B 2V4	B 4V4	477.96	400	2	6.6	0.898	3.398	B 6V4[b]	
1266	B 5V4	B 33V4	485.24	400	1	3.3	-2.114	3.268	B 26V4	B 33V4

[a, b] Contingency on generation which is located in this bus

Appendix J
77-Bus Test System Data

A 77-bus dual voltage level test system is used in Chap. 9 to assess the capability of the proposed hybrid approach for transmission expansion planning problem. Moreover, this test system is used in Chap. 10 for RPP analysis. The relevant data of this test system are provided as follows

- Bus data are provided in Table J.1.
- Line data are provided in Table J.2.
- Candidate lines data are provided in Table J.3.
- Generation data are provided in Table J.4.

Table J.1 Bus data

No.	Bus name	X	Y	P_D (p.u.)	Q_D (p.u.)	No.	Bus name	X	Y	P_D (p.u.)	Q_D (p.u.)
1	B 1V4	53.43	35.60	0.00	0.00	20	B 20V2	50.00	36.28	0.00	0.00
2	B 2V2	50.07	36.22	0.63	0.58	21	B 21V2	50.87	34.68	1.24	0.54
3	B 3V2	51.52	35.75	2.17	1.05	22	B 22V2	50.75	34.58	1.56	0.76
4	B 4V4	49.83	34.00	0.00	0.00	23	B 23V2	50.95	34.62	1.61	0.54
5	B 5V2	49.83	34.00	0.00	0.00	24	B 24V2	51.43	35.68	4.25	1.98
6	B 6V2	50.15	35.95	1.68	0.66	25	B 25V2	52.15	35.67	0.22	0.11
7	B 7V2	51.32	35.67	3.00	1.28	26	B 26V4	50.32	33.41	0.00	0.00
8	B 8V2	50.95	33.71	0.00	0.00	27	B 27V4	51.20	36.50	0.28	1.43
9	B 9V2	51.43	35.63	1.73	0.58	28	B 28V2	50.77	35.95	1.56	0.71
10	B 10V2	50.10	35.75	0.42	0.44	29	B 29V4	51.85	35.42	0.00	0.00
11	B 11V4	51.87	35.43	0.00	0.00	30	B 30V2	51.77	35.82	0.00	0.00
12	B 12V2	51.50	35.70	3.46	1.75	31	B 31V2	50.83	35.82	3.48	1.68
13	B 13V2	51.27	35.57	2.68	1.30	32	B 32V2	51.28	35.77	2.29	0.80
14	B 14V4	51.30	35.62	0.00	0.00	33	B 33V4	51.28	35.77	0.00	0.00
15	B 15V2	51.30	35.62	0.00	0.00	34	B 34V2	51.02	35.85	1.93	0.93
16	B 16V4	51.30	35.62	0.00	0.00	35	B 35V2	51.33	34.05	0.68	0.23
17	B 17V2	52.73	35.77	0.50	0.24	36	B 36V2	51.67	35.75	0.00	0.00
18	B 18V2	51.12	35.43	0.08	0.01	37	B 37V2	51.25	35.70	2.94	1.23
19	B 19V2	52.33	35.25	0.00	0.00	38	B 38V2	51.05	35.77	0.09	0.02

(continued)

Table J.1 (continued)

No.	Bus name	X	Y	P_D (p.u.)	Q_D (p.u.)	No.	Bus name	X	Y	P_D (p.u.)	Q_D (p.u.)
39	B 39V2	51.02	35.73	0.00	0.00	59	B 59V4	50.90	35.42	0.00	0.00
40	B 40V2	51.00	35.75	3.92	1.86	60	B 60V2	52.97	36.17	0.00	0.00
41	B 41V2	51.42	35.73	3.77	0.70	61	B 61V2	50.37	35.02	0.27	0.44
42	B 42V2	51.48	35.62	2.24	1.08	62	B 62V2	51.07	35.68	2.48	0.98
43	B 43V4	49.63	37.18	0.00	0.00	63	B 63V2	51.38	35.75	0.92	0.36
44	B 44V2	51.02	35.47	0.00	0.00	64	B 64V4	51.38	35.75	0.00	0.00
45	B 45V4	50.90	35.42	0.00	0.00	65	B 65V2	51.43	35.67	2.12	1.02
46	B 46V2	51.38	35.78	3.25	1.15	66	B 66V2	50.55	35.82	0.59	0.23
47	B 47V4	53.25	36.82	0.00	0.00	67	B 67V4	46.17	38.08	0.00	0.00
48	B 48V2	50.57	34.23	0.40	0.19	68	B 68V2	51.35	35.73	3.25	0.83
49	B 49V2	51.52	35.80	1.95	0.67	69	B 69V2	51.43	35.63	0.00	0.00
50	B 50V2	51.02	35.47	2.47	1.67	70	B 70V4	51.57	35.75	0.00	0.00
51	B 51V2	51.83	35.75	1.71	0.83	71	B 71V4	51.57	35.75	0.00	0.00
52	B 52V4	51.83	35.75	0.00	0.00	72	B 72V2	51.57	35.75	1.78	0.73
53	B 53V2	51.43	35.80	1.60	0.56	73	B 73V4	51.65	35.33	1.52	0.74
54	B 54V4	50.32	36.15	0.00	0.00	74	B 74V2	51.13	35.75	2.55	1.11
55	B 55V4	50.32	36.15	0.00	0.00	75	B 75V4	51.13	35.75	0.00	0.00
56	B 56V2	51.40	35.52	3.15	1.52	76	B 76V2	50.47	36.10	0.82	0.40
57	B 57V2	51.58	35.52	2.89	1.19	77	B 77V4	50.47	36.10	0.00	0.00
58	B 58V4	51.58	35.52	0.00	0.00						

Table J.2 Line data

No.	From bus	To bus	R (p.u.)	X (p.u.)	B (p.u.)	$\bar{P}_L$ (p.u.)
1	B 10V2	B 31V2	0.0148	0.0611	-0.3425	3.0
2	B 12V2	B 57V2	0.0027	0.0156	-0.0524	4.9
3	B 12V2	B 72V2	0.0008	0.0046	-0.0155	4.9
4	B 13V2	B 15V2	0.0016	0.0092	-0.0163	2.8
5	B 13V2	B 18V2	0.0031	0.0196	-0.0407	2.8
6	B 13V2	B 44V2	0.0059	0.0267	-0.0472	2.8
7	B 13V2	B 56V2	0.0029	0.0172	-0.0305	2.8
8	B 15V2	B 32V2	0.0024	0.0139	-0.0466	4.8
9	B 15V2	B 39V2	0.0050	0.0220	-0.0569	2.8
10	B 15V2	B 56V2	0.0020	0.0120	-0.0294	3.4
11	B 15V2	B 69V2	0.0023	0.0121	-0.0210	3.4
12	B 15V2	B 69V2	0.0023	0.0121	-0.0210	3.4
13	B 17V2	B 60V2	0.0003	0.0024	-0.0043	2.7
14	B 18V2	B 21V2	0.0150	0.0886	-0.1672	3.4
15	B 18V2	B 44V2	0.0021	0.0128	-0.0189	3.2
16	B 18V2	B 56V2	0.0050	0.0293	-0.0553	3.4
17	B 19V2	B 57V2	0.0090	0.0670	-0.1223	3.8
18	B 21V2	B 22V2	0.0020	0.0120	-0.0213	2.7

(continued)

Table J.2 (continued)

No.	From bus	To bus	R (p.u.)	X (p.u.)	B (p.u.)	$\bar{P}_L$ (p.u.)
19	B 21V2	B 22V2	0.0020	0.0120	-0.0213	2.7
20	B 21V2	B 23V2	0.0026	0.0156	-0.0285	3.2
21	B 22V2	B 23V2	0.0037	0.0218	-0.0401	3.2
22	B 22V2	B 23V2	0.0037	0.0218	-0.0401	6.5
23	B 22V2	B 44V2	0.0296	0.1314	-0.2323	2.8
24	B 23V2	B 35V2	0.0108	0.0665	-0.1202	4.0
25	B 24V2	B 41V2	0.0003	0.0027	-0.5850	5.7
26	B 24V2	B 65V2	0.0002	0.0017	-0.3656	6.4
27	B 17V2	B 25V2	0.0077	0.0556	-0.0687	2.7
28	B 25V2	B 51V2	0.0184	0.1326	-0.1638	2.7
29	B 27V4	B 77V4	0.0023	0.0264	-0.6973	11.7
30	B 28V2	B 31V2	0.0030	0.0178	-0.0329	3.3
31	B 28V2	B 76V2	0.0043	0.0257	-0.0375	3.3
32	B 1V4	B 29V4	0.0034	0.0317	-0.8608	9.0
33	B 11V4	B 29V4	0.0000	0.0002	-0.0062	9.9
34	B 11V4	B 29V4	0.0000	0.0002	-0.0060	9.9
35	B 26V4	B 29V4	0.0060	0.0633	-1.7161	10.7
36	B 29V4	B 47V4	0.0040	0.0550	-1.3729	9.1
37	B 29V4	B 58V4	0.0006	0.0057	-0.0823	12.1
38	B 29V4	B 59V4	0.0010	0.0164	-0.6666	22.2
39	B 29V4	B 59V4	0.0010	0.0164	-0.6733	22.2
40	B 29V4	B 73V4	0.0004	0.0045	-0.1207	15.0
41	B 2V2	B 76V2	0.0064	0.0452	-0.0815	3.2
42	B 31V2	B 40V2	0.0031	0.0188	-0.0259	3.3
43	B 31V2	B 40V2	0.0069	0.0388	-0.1568	2.3
44	B 31V2	B 74V2	0.0040	0.0223	-0.0846	4.9
45	B 32V2	B 37V2	0.0011	0.0063	-0.0216	4.9
46	B 32V2	B 46V2	0.0010	0.0058	-0.0194	4.9
47	B 32V2	B 74V2	0.0016	0.0091	-0.0333	4.9
48	B 33V4	B 77V4	0.0013	0.0180	-0.6305	11.7
49	B 31V2	B 34V2	0.0023	0.0128	-0.0487	4.9
50	B 34V2	B 74V2	0.0010	0.0046	-0.0333	4.9
51	B 30V2	B 36V2	0.0004	0.0058	-0.2063	2.3
52	B 9V2	B 36V2	0.0072	0.0364	-0.3079	2.3
53	B 37V2	B 39V2	0.0039	0.0246	-0.0844	4.9
54	B 38V2	B 39V2	0.0008	0.0046	-0.0089	1.4
55	B 38V2	B 39V2	0.0008	0.0046	-0.0089	1.4
56	B 39V2	B 40V2	0.0006	0.0025	-0.0059	4.7
57	B 39V2	B 40V2	0.0006	0.0025	-0.0059	4.7
58	B 39V2	B 49V2	0.0059	0.0331	-0.1072	4.8
59	B 39V2	B 74V2	0.0038	0.0108	-0.0352	4.8
60	B 3V2	B 49V2	0.0015	0.0086	-0.0287	4.8
61	B 3V2	B 72V2	0.0010	0.0051	-0.0172	4.9
62	B 41V2	B 63V2	0.0003	0.0025	-0.0059	3.5
63	B 41V2	B 63V2	0.0002	0.0021	-0.0076	3.5

(continued)

Table J.2 (continued)

No.	From bus	To bus	R (p.u.)	X (p.u.)	B (p.u.)	$\bar{P}_L$ (p.u.)
64	B 42V2	B 57V2	0.0020	0.0112	-0.0392	2.9
65	B 43V4	B 54V4	0.0032	0.0340	-0.8453	10.7
66	B 44V2	B 61V2	0.0162	0.0733	-0.1294	2.7
67	B 14V4	B 45V4	0.0006	0.0083	-0.2686	8.3
68	B 46V2	B 49V2	0.0017	0.0098	-0.0330	4.8
69	B 1V4	B 47V4	0.0027	0.0315	-0.8335	15.0
70	B 21V2	B 48V2	0.0070	0.0401	-0.0764	3.3
71	B 49V2	B 53V2	0.0011	0.0059	-0.0214	4.8
72	B 4V4	B 26V4	0.0016	0.0181	-0.4794	15.0
73	B 4V4	B 59V4	0.0035	0.0395	-1.0460	7.6
74	B 44V2	B 50V2	0.0010	0.0008	-0.0016	3.2
75	B 44V2	B 50V2	0.0010	0.0008	-0.0016	3.2
76	B 30V2	B 51V2	0.0014	0.0066	-0.0131	2.7
77	B 11V4	B 52V4	0.0007	0.0089	-0.3122	18.2
78	B 11V4	B 52V4	0.0007	0.0089	-0.3122	18.2
79	B 52V4	B 75V4	0.0009	0.0126	-0.4433	4.8
80	B 53V2	B 72V2	0.0023	0.0130	-0.0417	4.9
81	B 54V4	B 55V4	0.0000	0.0002	-0.0124	12.1
82	B 54V4	B 59V4	0.0013	0.0185	-0.6867	16.8
83	B 54V4	B 77V4	0.0003	0.0044	-0.1561	13.5
84	B 54V4	B 77V4	0.0003	0.0044	-0.1561	13.5
85	B 55V4	B 59V4	0.0013	0.0185	-0.6805	10.7
86	B 56V2	B 57V2	0.0023	0.0136	-0.0253	3.4
87	B 56V2	B 57V2	0.0023	0.0136	-0.0253	3.4
88	B 57V2	B 65V2	0.0039	0.0246	-0.0844	4.9
89	B 58V4	B 73V4	0.0004	0.0045	-0.1207	15.0
90	B 16V4	B 59V4	0.0006	0.0085	-0.2846	18.4
91	B 4V4	B 59V4	0.0035	0.0395	-1.0460	7.6
92	B 45V4	B 59V4	0.0000	0.0002	0.0000	16.6
93	B 5V2	B 22V2	0.0260	0.1323	-0.2399	2.7
94	B 40V2	B 62V2	0.0025	0.0140	-0.0529	4.9
95	B 33V4	B 64V4	0.0002	0.0021	-0.0448	16.6
96	B 65V2	B 69V2	0.0001	0.0010	-0.2194	5.5
97	B 15V2	B 66V2	0.0028	0.0016	-0.0544	4.9
98	B 62V2	B 66V2	0.0008	0.0045	-0.0169	4.9
99	B 54V4	B 67V4	0.0092	0.0927	-2.5224	11.7
100	B 32V2	B 68V2	0.0008	0.0046	-0.0157	4.8
101	B 32V2	B 68V2	0.0008	0.0046	-0.0157	4.8
102	B 6V2	B 10V2	0.0033	0.0191	-0.0371	4.0
103	B 11V4	B 70V4	0.0008	0.0113	-0.3871	12.5
104	B 11V4	B 71V4	0.0008	0.0113	-0.3640	12.5
105	B 57V2	B 72V2	0.0027	0.0156	-0.0617	4.9
106	B 52V4	B 75V4	0.0009	0.0126	-0.4433	4.8
107	B 54V4	B 75V4	0.0013	0.0177	-0.6400	18.2
108	B 64V4	B 75V4	0.0004	0.0050	-0.1792	16.6

(continued)

Table J.2 (continued)

No.	From bus	To bus	R (p.u.)	X (p.u.)	B (p.u.)	$\bar{P}_L$ (p.u.)
109	B 75V4	B 77V4	0.0010	0.0147	-0.5073	16.6
110	B 20V2	B 76V2	0.0099	0.0518	-0.0895	3.0
111	B 6V2	B 76V2	0.0054	0.0306	-0.0593	2.8
112	B 55V4	B 77V4	0.0013	0.0177	-0.6400	18.2
113	B 7V2	B 15V2	0.0002	0.0034	-0.0118	9.6
114	B 7V2	B 15V2	0.0002	0.0034	-0.0118	9.6
115	B 7V2	B 24V2	0.0004	0.0039	-0.8555	6.4
116	B 8V2	B 35V2	0.0089	0.0509	-0.0970	3.7
117	B 8V2	B 48V2	0.0097	0.0555	-0.1058	3.3
118	B 9V2	B 42V2	0.0006	0.0040	-0.0137	4.8
119	B 9V2	B 56V2	0.0035	0.0154	-0.0269	2.8
120	B 14V4	B 15V2	0.0013	0.0257	1.0000	5.0
121	B 14V4	B 15V2	0.0013	0.0257	1.0000	5.0
122	B 16V4	B 15V2	0.0013	0.0257	1.0000	5.0
123	B 16V4	B 15V2	0.0013	0.0257	1.0000	5.0
124	B 33V4	B 32V2	0.0012	0.0242	1.0000	5.0
125	B 33V4	B 32V2	0.0012	0.0242	1.0000	5.0
126	B 52V4	B 51V2	0.0013	0.0257	1.0000	5.0
127	B 52V4	B 51V2	0.0013	0.0257	1.0000	5.0
128	B 58V4	B 57V2	0.0012	0.0240	1.0000	5.0
129	B 58V4	B 57V2	0.0012	0.0240	1.0000	5.0
130	B 64V4	B 63V2	0.0004	0.0257	1.0000	5.0
131	B 64V4	B 63V2	0.0004	0.0257	1.0000	5.0
132	B 70V4	B 72V2	0.0012	0.0229	1.0000	5.0
133	B 71V4	B 72V2	0.0012	0.0229	1.0000	5.0
134	B 75V4	B 74V2	0.0012	0.0241	1.0000	5.0
135	B 75V4	B 74V2	0.0012	0.0241	1.0000	5.0
136	B 77V4	B 76V2	0.0013	0.0269	1.0000	5.0
137	B 77V4	B 76V2	0.0013	0.0269	1.0000	5.0

Table J.3 Candidate lines data[a]

No.	From bus	To bus	No.	From bus	To bus	No.	From bus	To bus	No.	From bus	To bus
1	B 44V2	B 50V2	12	B 71V4	B 72V2	38	B 11V4	B 29V4	49	B 41V2	B 64V4
2	B 44V2	B 50V2	13	B 70V4	B 72V2	39	B 39V2	B 40V2	50	B 41V2	B 64V4
3	B 74V2	B 75V4	14	B 9V2	B 69V2	40	B 39V2	B 40V2	51	B 41V2	B 63V2
4	B 45V4	B 59V4	15	B 9V2	B 69V2	41	B 46V2	B 64V4	52	B 41V2	B 63V2
5	B 45V4	B 59V4	16	B 54V4	B 55V4	42	B 46V2	B 64V4	53	B 65V2	B 69V2
6	B 76V2	B 77V4	17	B 54V4	B 55V4	43	B 46V2	B 63V2	54	B 65V2	B 69V2
7	B 57V2	B 58V4	18	B 32V2	B 33V4	44	B 46V2	B 63V2	55	B 9V2	B 65V2
8	B 63V2	B 64V4	19	B 4V4	B 5V2	45	B 64V4	B 68V2	56	B 9V2	B 65V2
9	B 14V4	B 15V2	35	B 24V2	B 65V2	46	B 64V4	B 68V2	57	B 3V2	B 71V4
10	B 15V2	B 16V4	36	B 24V2	B 65V2	47	B 63V2	B 68V2	58	B 3V2	B 71V4
11	B 51V2	B 52V4	37	B 11V4	B 29V4	48	B 63V2	B 68V2	59	B 3V2	B 72V2

(continued)

Table J.3 (continued)

No.	From bus	To bus	No.	From bus	To bus	No.	From bus	To bus	No.	From bus	To bus
60	B 3V2	B 72V2	104	B 41V2	B 65V2	148	B 12V2	B 70V4	192	B 37V2	B 68V2
61	B 3V2	B 70V4	105	B 12V2	B 65V2	149	B 12V2	B 71V4	193	B 65V2	B 68V2
62	B 3V2	B 70V4	106	B 12V2	B 65V2	150	B 12V2	B 71V4	194	B 65V2	B 68V2
63	B 42V2	B 69V2	107	B 7V2	B 37V2	151	B 24V2	B 64V4	195	B 3V2	B 53V2
64	B 42V2	B 69V2	108	B 7V2	B 37V2	152	B 24V2	B 64V4	196	B 3V2	B 53V2
65	B 9V2	B 42V2	109	B 49V2	B 70V4	153	B 24V2	B 63V2	197	B 21V2	B 23V2
66	B 9V2	B 42V2	110	B 49V2	B 70V4	154	B 24V2	B 63V2	198	B 21V2	B 23V2
67	B 46V2	B 53V2	111	B 49V2	B 71V4	155	B 36V2	B 72V2	199	B 7V2	B 65V2
68	B 46V2	B 53V2	112	B 49V2	B 71V4	156	B 36V2	B 72V2	200	B 7V2	B 65V2
69	B 38V2	B 40V2	113	B 49V2	B 72V2	157	B 36V2	B 70V4	201	B 63V2	B 65V2
70	B 38V2	B 40V2	114	B 49V2	B 72V2	158	B 36V2	B 70V4	202	B 63V2	B 65V2
71	B 38V2	B 39V2	115	B 53V2	B 64V4	159	B 36V2	B 71V4	203	B 64V4	B 65V2
72	B 38V2	B 39V2	116	B 53V2	B 64V4	160	B 36V2	B 71V4	204	B 64V4	B 65V2
73	B 3V2	B 49V2	117	B 53V2	B 63V2	161	B 12V2	B 42V2	205	B 14V4	B 37V2
74	B 3V2	B 49V2	118	B 53V2	B 63V2	162	B 12V2	B 42V2	206	B 14V4	B 37V2
75	B 9V2	B 24V2	119	B 39V2	B 62V2	163	B 32V2	B 46V2	207	B 15V2	B 37V2
76	B 9V2	B 24V2	120	B 39V2	B 62V2	164	B 32V2	B 46V2	208	B 15V2	B 37V2
77	B 24V2	B 69V2	121	B 42V2	B 65V2	165	B 33V4	B 46V2	209	B 16V4	B 37V2
78	B 24V2	B 69V2	122	B 42V2	B 65V2	166	B 33V4	B 46V2	210	B 16V4	B 37V2
79	B 24V2	B 41V2	123	B 7V2	B 68V2	167	B 24V2	B 68V2	211	B 7V2	B 24V2
80	B 24V2	B 41V2	124	B 7V2	B 68V2	168	B 24V2	B 68V2	212	B 7V2	B 24V2
81	B 3V2	B 12V2	125	B 38V2	B 74V2	169	B 2V2	B 20V2	213	B 40V2	B 62V2
82	B 3V2	B 12V2	126	B 38V2	B 74V2	170	B 2V2	B 20V2	214	B 40V2	B 62V2
83	B 7V2	B 16V4	127	B 38V2	B 75V4	171	B 33V4	B 63V2	215	B 9V2	B 12V2
84	B 7V2	B 16V4	128	B 38V2	B 75V4	172	B 33V4	B 63V2	216	B 9V2	B 12V2
85	B 7V2	B 14V4	129	B 32V2	B 68V2	173	B 33V4	B 64V4	217	B 12V2	B 69V2
86	B 7V2	B 14V4	130	B 32V2	B 68V2	174	B 33V4	B 64V4	218	B 12V2	B 69V2
87	B 7V2	B 15V2	131	B 33V4	B 68V2	175	B 32V2	B 64V4	219	B 18V2	B 44V2
88	B 7V2	B 15V2	132	B 33V4	B 68V2	176	B 32V2	B 64V4	220	B 18V2	B 44V2
89	B 46V2	B 68V2	133	B 41V2	B 53V2	177	B 32V2	B 63V2	221	B 18V2	B 50V2
90	B 46V2	B 68V2	134	B 41V2	B 53V2	178	B 32V2	B 63V2	222	B 18V2	B 50V2
91	B 13V2	B 15V2	135	B 12V2	B 41V2	179	B 3V2	B 41V2	223	B 38V2	B 62V2
92	B 13V2	B 15V2	136	B 12V2	B 41V2	180	B 3V2	B 41V2	224	B 38V2	B 62V2
93	B 13V2	B 16V4	137	B 24V2	B 42V2	181	B 34V2	B 38V2	225	B 39V2	B 74V2
94	B 13V2	B 16V4	138	B 24V2	B 42V2	182	B 34V2	B 38V2	226	B 39V2	B 74V2
95	B 13V2	B 14V4	139	B 49V2	B 53V2	183	B 30V2	B 52V4	227	B 39V2	B 75V4
96	B 13V2	B 14V4	140	B 49V2	B 53V2	184	B 30V2	B 52V4	228	B 39V2	B 75V4
97	B 41V2	B 68V2	141	B 33V4	B 37V2	185	B 30V2	B 51V2	229	B 7V2	B 63V2
98	B 41V2	B 68V2	142	B 33V4	B 37V2	186	B 30V2	B 51V2	230	B 7V2	B 63V2
99	B 41V2	B 46V2	143	B 32V2	B 37V2	187	B 62V2	B 75V4	231	B 7V2	B 64V4
100	B 41V2	B 46V2	144	B 32V2	B 37V2	188	B 62V2	B 75V4	232	B 7V2	B 64V4
101	B 12V2	B 24V2	145	B 12V2	B 72V2	189	B 62V2	B 74V2	233	B 53V2	B 68V2
102	B 12V2	B 24V2	146	B 12V2	B 72V2	190	B 62V2	B 74V2	234	B 53V2	B 68V2
103	B 41V2	B 65V2	147	B 12V2	B 70V4	191	B 37V2	B 68V2	235	B 7V2	B 69V2

(continued)

Table J.3 (continued)

No.	From bus	To bus	No.	From bus	To bus	No.	From bus	To bus	No.	From bus	To bus
236	B 7V2	B 69V2	279	B 3V2	B 65V2	322	B 46V2	B 65V2	365	B 41V2	B 72V2
237	B 7V2	B 9V2	280	B 3V2	B 65V2	323	B 16V4	B 68V2	366	B 41V2	B 72V2
238	B 7V2	B 9V2	281	B 12V2	B 63V2	324	B 16V4	B 68V2	367	B 41V2	B 70V4
239	B 41V2	B 69V2	282	B 12V2	B 63V2	325	B 15V2	B 68V2	368	B 41V2	B 70V4
240	B 41V2	B 69V2	283	B 12V2	B 64V4	326	B 15V2	B 68V2	369	B 41V2	B 71V4
241	B 9V2	B 41V2	284	B 12V2	B 64V4	327	B 14V4	B 68V2	370	B 41V2	B 71V4
242	B 9V2	B 41V2	285	B 37V2	B 74V2	328	B 14V4	B 68V2	371	B 53V2	B 70V4
243	B 7V2	B 41V2	286	B 37V2	B 74V2	329	B 3V2	B 46V2	372	B 53V2	B 70V4
244	B 7V2	B 41V2	287	B 37V2	B 75V4	330	B 3V2	B 46V2	373	B 53V2	B 71V4
245	B 3V2	B 24V2	288	B 37V2	B 75V4	331	B 68V2	B 69V2	374	B 53V2	B 71V4
246	B 3V2	B 24V2	289	B 45V4	B 50V2	332	B 68V2	B 69V2	375	B 53V2	B 72V2
247	B 34V2	B 40V2	290	B 45V4	B 50V2	333	B 9V2	B 68V2	376	B 53V2	B 72V2
248	B 34V2	B 40V2	291	B 50V2	B 59V4	334	B 9V2	B 68V2	377	B 49V2	B 64V4
249	B 12V2	B 49V2	292	B 50V2	B 59V4	335	B 42V2	B 56V2	378	B 49V2	B 64V4
250	B 12V2	B 49V2	293	B 44V2	B 45V4	336	B 42V2	B 56V2	379	B 49V2	B 63V2
251	B 7V2	B 32V2	294	B 44V2	B 45V4	337	B 24V2	B 53V2	380	B 49V2	B 63V2
252	B 7V2	B 32V2	295	B 44V2	B 59V4	338	B 24V2	B 53V2	381	B 32V2	B 53V2
253	B 7V2	B 33V4	296	B 44V2	B 59V4	339	B 34V2	B 39V2	382	B 32V2	B 53V2
254	B 7V2	B 33V4	297	B 9V2	B 56V2	340	B 34V2	B 39V2	383	B 33V4	B 53V2
255	B 40V2	B 75V4	298	B 9V2	B 56V2	341	B 7V2	B 46V2	384	B 33V4	B 53V2
256	B 40V2	B 75V4	299	B 56V2	B 69V2	342	B 7V2	B 46V2	385	B 12V2	B 68V2
257	B 40V2	B 74V2	300	B 56V2	B 69V2	343	B 41V2	B 42V2	386	B 12V2	B 68V2
258	B 40V2	B 74V2	301	B 3V2	B 63V2	344	B 41V2	B 42V2	387	B 12V2	B 46V2
259	B 9V2	B 16V4	302	B 3V2	B 63V2	345	B 33V4	B 41V2	388	B 12V2	B 46V2
260	B 9V2	B 16V4	303	B 3V2	B 64V4	346	B 33V4	B 41V2	389	B 9V2	B 64V4
261	B 9V2	B 14V4	304	B 3V2	B 64V4	347	B 32V2	B 41V2	390	B 9V2	B 64V4
262	B 9V2	B 14V4	305	B 12V2	B 53V2	348	B 32V2	B 41V2	391	B 9V2	B 63V2
263	B 9V2	B 15V2	306	B 12V2	B 53V2	349	B 14V4	B 24V2	392	B 9V2	B 63V2
264	B 9V2	B 15V2	307	B 46V2	B 49V2	350	B 14V4	B 24V2	393	B 63V2	B 69V2
265	B 16V4	B 69V2	308	B 46V2	B 49V2	351	B 15V2	B 24V2	394	B 63V2	B 69V2
266	B 16V4	B 69V2	309	B 37V2	B 64V4	352	B 15V2	B 24V2	395	B 64V4	B 69V2
267	B 15V2	B 69V2	310	B 37V2	B 64V4	353	B 16V4	B 24V2	396	B 64V4	B 69V2
268	B 15V2	B 69V2	311	B 37V2	B 63V2	354	B 16V4	B 24V2	397	B 42V2	B 58V4
269	B 14V4	B 69V2	312	B 37V2	B 63V2	355	B 3V2	B 36V2	398	B 42V2	B 58V4
270	B 14V4	B 69V2	313	B 15V2	B 65V2	356	B 3V2	B 36V2	399	B 42V2	B 57V2
271	B 30V2	B 36V2	314	B 15V2	B 65V2	357	B 32V2	B 75V4	400	B 42V2	B 57V2
272	B 30V2	B 36V2	315	B 16V4	B 65V2	358	B 32V2	B 75V4	401	B 16V4	B 56V2
273	B 41V2	B 49V2	316	B 16V4	B 65V2	359	B 33V4	B 74V2	402	B 16V4	B 56V2
274	B 41V2	B 49V2	317	B 14V4	B 65V2	360	B 33V4	B 74V2	403	B 15V2	B 56V2
275	B 24V2	B 46V2	318	B 14V4	B 65V2	361	B 33V4	B 75V4	404	B 15V2	B 56V2
276	B 24V2	B 46V2	319	B 13V2	B 56V2	362	B 33V4	B 75V4	405	B 14V4	B 56V2
277	B 7V2	B 13V2	320	B 13V2	B 56V2	363	B 32V2	B 74V2	406	B 14V4	B 56V2
278	B 7V2	B 13V2	321	B 46V2	B 65V2	364	B 32V2	B 74V2	407	B 36V2	B 51V2

(continued)

Table J.3 (continued)

No.	From bus	To bus	No.	From bus	To bus	No.	From bus	To bus	No.	From bus	To bus
408	B 36V2	B 51V2	452	B 24V2	B 49V2	496	B 42V2	B 70V4	540	B 9V2	B 37V2
409	B 36V2	B 52V4	453	B 3V2	B 69V2	497	B 42V2	B 71V4	541	B 7V2	B 56V2
410	B 36V2	B 52V4	454	B 3V2	B 69V2	498	B 42V2	B 71V4	542	B 7V2	B 56V2
411	B 53V2	B 65V2	455	B 3V2	B 9V2	499	B 7V2	B 12V2	543	B 9V2	B 58V4
412	B 53V2	B 65V2	456	B 3V2	B 9V2	500	B 7V2	B 12V2	544	B 9V2	B 58V4
413	B 13V2	B 37V2	457	B 21V2	B 22V2	501	B 14V4	B 33V4	545	B 9V2	B 57V2
414	B 13V2	B 37V2	458	B 21V2	B 22V2	502	B 14V4	B 33V4	546	B 9V2	B 57V2
415	B 55V4	B 77V4	459	B 37V2	B 41V2	503	B 14V4	B 32V2	547	B 57V2	B 69V2
416	B 55V4	B 77V4	460	B 37V2	B 41V2	504	B 14V4	B 32V2	548	B 57V2	B 69V2
417	B 54V4	B 76V2	461	B 13V2	B 69V2	505	B 16V4	B 32V2	549	B 58V4	B 69V2
418	B 54V4	B 76V2	462	B 13V2	B 69V2	506	B 16V4	B 32V2	550	B 58V4	B 69V2
419	B 54V4	B 77V4	463	B 9V2	B 13V2	507	B 16V4	B 33V4	551	B 22V2	B 23V2
420	B 54V4	B 77V4	464	B 9V2	B 13V2	508	B 16V4	B 33V4	552	B 22V2	B 23V2
421	B 55V4	B 76V2	465	B 16V4	B 64V4	509	B 15V2	B 32V2	553	B 16V4	B 46V2
422	B 55V4	B 76V2	466	B 16V4	B 64V4	510	B 15V2	B 32V2	554	B 16V4	B 46V2
423	B 36V2	B 49V2	467	B 15V2	B 63V2	511	B 15V2	B 33V4	555	B 34V2	B 62V2
424	B 36V2	B 49V2	468	B 15V2	B 63V2	512	B 15V2	B 33V4	556	B 34V2	B 62V2
425	B 37V2	B 46V2	469	B 15V2	B 64V4	513	B 42V2	B 68V2	557	B 30V2	B 71V4
426	B 37V2	B 46V2	470	B 15V2	B 64V4	514	B 42V2	B 68V2	558	B 30V2	B 71V4
427	B 24V2	B 72V2	471	B 14V4	B 64V4	515	B 42V2	B 64V4	559	B 30V2	B 72V2
428	B 24V2	B 72V2	472	B 14V4	B 64V4	516	B 42V2	B 64V4	560	B 30V2	B 72V2
429	B 24V2	B 70V4	473	B 14V4	B 63V2	517	B 42V2	B 63V2	561	B 30V2	B 70V4
430	B 24V2	B 70V4	474	B 14V4	B 63V2	518	B 42V2	B 63V2	562	B 30V2	B 70V4
431	B 24V2	B 71V4	475	B 16V4	B 63V2	519	B 31V2	B 40V2	563	B 37V2	B 38V2
432	B 24V2	B 71V4	476	B 16V4	B 63V2	520	B 31V2	B 40V2	564	B 37V2	B 38V2
433	B 3V2	B 42V2	477	B 16V4	B 42V2	521	B 49V2	B 68V2	565	B 37V2	B 53V2
434	B 3V2	B 42V2	478	B 16V4	B 42V2	522	B 49V2	B 68V2	566	B 37V2	B 53V2
435	B 34V2	B 75V4	479	B 14V4	B 42V2	523	B 9V2	B 46V2	567	B 31V2	B 39V2
436	B 34V2	B 75V4	480	B 14V4	B 42V2	524	B 9V2	B 46V2	568	B 31V2	B 39V2
437	B 34V2	B 74V2	481	B 15V2	B 42V2	525	B 46V2	B 69V2	569	B 18V2	B 59V4
438	B 34V2	B 74V2	482	B 15V2	B 42V2	526	B 46V2	B 69V2	570	B 18V2	B 59V4
439	B 28V2	B 31V2	483	B 56V2	B 57V2	527	B 31V2	B 34V2	571	B 18V2	B 45V4
440	B 28V2	B 31V2	484	B 56V2	B 57V2	528	B 31V2	B 34V2	572	B 18V2	B 45V4
441	B 65V2	B 70V4	485	B 56V2	B 58V4	529	B 46V2	B 72V2	573	B 12V2	B 14V4
442	B 65V2	B 70V4	486	B 56V2	B 58V4	530	B 46V2	B 72V2	574	B 12V2	B 14V4
443	B 65V2	B 71V4	487	B 12V2	B 36V2	531	B 46V2	B 70V4	575	B 12V2	B 15V2
444	B 65V2	B 71V4	488	B 12V2	B 36V2	532	B 46V2	B 70V4	576	B 12V2	B 15V2
445	B 65V2	B 72V2	489	B 37V2	B 62V2	533	B 46V2	B 71V4	577	B 31V2	B 38V2
446	B 65V2	B 72V2	490	B 37V2	B 62V2	534	B 46V2	B 71V4	578	B 31V2	B 38V2
447	B 7V2	B 42V2	491	B 49V2	B 65V2	535	B 7V2	B 53V2	579	B 13V2	B 18V2
448	B 7V2	B 42V2	492	B 49V2	B 65V2	536	B 7V2	B 53V2	580	B 13V2	B 18V2
449	B 3V2	B 68V2	493	B 42V2	B 72V2	537	B 37V2	B 69V2	581	B 29V4	B 73V4
450	B 3V2	B 68V2	494	B 42V2	B 72V2	538	B 37V2	B 69V2	582	B 29V4	B 73V4
451	B 24V2	B 49V2	495	B 42V2	B 70V4	539	B 9V2	B 37V2	583	B 32V2	B 38V2

(continued)

Table J.3 (continued)

No.	From bus	To bus	No.	From bus	To bus	No.	From bus	To bus	No.	From bus	To bus
584	B 32V2	B 38V2	628	B 13V2	B 33V4	672	B 13V2	B 44V2	716	B 31V2	B 74V2
585	B 33V4	B 38V2	629	B 36V2	B 53V2	673	B 13V2	B 50V2	717	B 18V2	B 62V2
586	B 33V4	B 38V2	630	B 36V2	B 53V2	674	B 13V2	B 50V2	718	B 18V2	B 62V2
587	B 37V2	B 39V2	631	B 30V2	B 49V2	675	B 31V2	B 66V2	719	B 49V2	B 51V2
588	B 37V2	B 39V2	632	B 30V2	B 49V2	676	B 31V2	B 66V2	720	B 49V2	B 51V2
589	B 14V4	B 75V4	633	B 6V2	B 10V2	677	B 58V4	B 70V4	721	B 49V2	B 52V4
590	B 14V4	B 75V4	634	B 6V2	B 10V2	678	B 58V4	B 70V4	722	B 49V2	B 52V4
591	B 15V2	B 74V2	635	B 11V4	B 73V4	679	B 58V4	B 71V4	723	B 13V2	B 57V2
592	B 15V2	B 74V2	636	B 11V4	B 73V4	680	B 58V4	B 71V4	724	B 13V2	B 57V2
593	B 15V2	B 75V4	637	B 51V2	B 72V2	681	B 58V4	B 72V2	725	B 13V2	B 58V4
594	B 15V2	B 75V4	638	B 51V2	B 72V2	682	B 58V4	B 72V2	726	B 13V2	B 58V4
595	B 14V4	B 74V2	639	B 52V4	B 70V4	683	B 57V2	B 70V4	727	B 13V2	B 39V2
596	B 14V4	B 74V2	640	B 52V4	B 70V4	684	B 57V2	B 70V4	728	B 13V2	B 39V2
597	B 16V4	B 74V2	641	B 52V4	B 71V4	685	B 57V2	B 71V4	729	B 39V2	B 50V2
598	B 16V4	B 74V2	642	B 52V4	B 71V4	686	B 57V2	B 71V4	730	B 39V2	B 50V2
599	B 16V4	B 75V4	643	B 52V4	B 72V2	687	B 57V2	B 72V2	731	B 39V2	B 44V2
600	B 16V4	B 75V4	644	B 52V4	B 72V2	688	B 57V2	B 72V2	732	B 39V2	B 44V2
601	B 32V2	B 62V2	645	B 51V2	B 70V4	689	B 31V2	B 62V2	733	B 13V2	B 38V2
602	B 32V2	B 62V2	646	B 51V2	B 70V4	690	B 31V2	B 62V2	734	B 13V2	B 38V2
603	B 33V4	B 62V2	647	B 51V2	B 71V4	691	B 36V2	B 58V4	735	B 56V2	B 72V2
604	B 33V4	B 62V2	648	B 51V2	B 71V4	692	B 36V2	B 58V4	736	B 56V2	B 72V2
605	B 3V2	B 32V2	649	B 13V2	B 75V4	693	B 36V2	B 57V2	737	B 56V2	B 70V4
606	B 3V2	B 32V2	650	B 13V2	B 75V4	694	B 36V2	B 57V2	738	B 56V2	B 70V4
607	B 3V2	B 33V4	651	B 13V2	B 74V2	695	B 29V4	B 57V2	739	B 56V2	B 71V4
608	B 3V2	B 33V4	652	B 13V2	B 74V2	696	B 29V4	B 57V2	740	B 56V2	B 71V4
609	B 13V2	B 62V2	653	B 2V2	B 55V4	697	B 29V4	B 58V4	741	B 25V2	B 52V4
610	B 13V2	B 62V2	654	B 2V2	B 55V4	698	B 29V4	B 58V4	742	B 25V2	B 52V4
611	B 16V4	B 62V2	655	B 2V2	B 54V4	699	B 6V2	B 54V4	743	B 25V2	B 51V2
612	B 16V4	B 62V2	656	B 2V2	B 54V4	700	B 6V2	B 54V4	744	B 25V2	B 51V2
613	B 15V2	B 62V2	657	B 44V2	B 62V2	701	B 6V2	B 55V4	745	B 28V2	B 40V2
614	B 15V2	B 62V2	658	B 44V2	B 62V2	702	B 6V2	B 55V4	746	B 28V2	B 40V2
615	B 14V4	B 62V2	659	B 50V2	B 62V2	703	B 18V2	B 56V2	747	B 2V2	B 6V2
616	B 14V4	B 62V2	660	B 50V2	B 62V2	704	B 18V2	B 56V2	748	B 2V2	B 6V2
617	B 32V2	B 49V2	661	B 3V2	B 30V2	705	B 53V2	B 74V2	749	B 56V2	B 73V4
618	B 32V2	B 49V2	662	B 3V2	B 30V2	706	B 53V2	B 74V2	750	B 56V2	B 73V4
619	B 33V4	B 49V2	663	B 33V4	B 39V2	707	B 53V2	B 75V4	751	B 40V2	B 44V2
620	B 33V4	B 49V2	664	B 33V4	B 39V2	708	B 53V2	B 75V4	752	B 40V2	B 44V2
621	B 57V2	B 73V4	665	B 32V2	B 39V2	709	B 11V4	B 57V2	753	B 40V2	B 50V2
622	B 57V2	B 73V4	666	B 32V2	B 39V2	710	B 11V4	B 57V2	754	B 40V2	B 50V2
623	B 58V4	B 73V4	667	B 28V2	B 66V2	711	B 11V4	B 58V4	755	B 49V2	B 57V2
624	B 58V4	B 73V4	668	B 28V2	B 66V2	712	B 11V4	B 58V4	756	B 49V2	B 57V2
625	B 13V2	B 32V2	669	B 28V2	B 34V2	713	B 31V2	B 75V4	757	B 28V2	B 76V2
626	B 13V2	B 32V2	670	B 28V2	B 34V2	714	B 31V2	B 75V4	758	B 28V2	B 76V2
627	B 13V2	B 33V4	671	B 13V2	B 44V2	715	B 31V2	B 74V2	759	B 28V2	B 77V4

(continued)

Table J.3 (continued)

No.	From bus	To bus	No.	From bus	To bus	No.	From bus	To bus	No.	From bus	To bus
760	B 28V2	B 77V4	803	B 56V2	B 62V2	846	B 2V2	B 77V4	889	B 18V2	B 57V2
761	B 66V2	B 76V2	804	B 56V2	B 62V2	847	B 18V2	B 38V2	890	B 18V2	B 57V2
762	B 66V2	B 76V2	805	B 44V2	B 56V2	848	B 18V2	B 38V2	891	B 18V2	B 58V4
763	B 66V2	B 77V4	806	B 44V2	B 56V2	849	B 6V2	B 66V2	892	B 18V2	B 58V4
764	B 66V2	B 77V4	807	B 50V2	B 56V2	850	B 6V2	B 66V2	893	B 39V2	B 66V2
765	B 20V2	B 54V4	808	B 50V2	B 56V2	851	B 25V2	B 29V4	894	B 39V2	B 66V2
766	B 20V2	B 54V4	809	B 36V2	B 56V2	852	B 25V2	B 29V4	895	B 25V2	B 36V2
767	B 20V2	B 55V4	810	B 36V2	B 56V2	853	B 6V2	B 20V2	896	B 25V2	B 36V2
768	B 20V2	B 55V4	811	B 56V2	B 75V4	854	B 6V2	B 20V2	897	B 11V4	B 30V2
769	B 28V2	B 38V2	812	B 56V2	B 75V4	855	B 28V2	B 75V4	898	B 11V4	B 30V2
770	B 28V2	B 38V2	813	B 18V2	B 74V2	856	B 28V2	B 75V4	899	B 29V4	B 72V2
771	B 44V2	B 74V2	814	B 18V2	B 74V2	857	B 28V2	B 74V2	900	B 29V4	B 72V2
772	B 44V2	B 74V2	815	B 18V2	B 75V4	858	B 28V2	B 74V2	901	B 29V4	B 70V4
773	B 50V2	B 74V2	816	B 18V2	B 75V4	859	B 71V4	B 74V2	902	B 29V4	B 70V4
774	B 50V2	B 74V2	817	B 11V4	B 52V4	860	B 71V4	B 74V2	903	B 29V4	B 71V4
775	B 50V2	B 75V4	818	B 11V4	B 52V4	861	B 11V4	B 36V2	904	B 29V4	B 71V4
776	B 50V2	B 75V4	819	B 11V4	B 51V2	862	B 11V4	B 36V2	905	B 11V4	B 70V4
777	B 44V2	B 75V4	820	B 11V4	B 51V2	863	B 29V4	B 36V2	906	B 11V4	B 70V4
778	B 44V2	B 75V4	821	B 39V2	B 45V4	864	B 29V4	B 36V2	907	B 11V4	B 72V2
779	B 45V4	B 62V2	822	B 39V2	B 45V4	865	B 38V2	B 45V4	908	B 11V4	B 72V2
780	B 45V4	B 62V2	823	B 39V2	B 59V4	866	B 38V2	B 45V4	909	B 31V2	B 59V4
781	B 59V4	B 62V2	824	B 39V2	B 59V4	867	B 38V2	B 59V4	910	B 31V2	B 59V4
782	B 59V4	B 62V2	825	B 29V4	B 51V2	868	B 38V2	B 59V4	911	B 31V2	B 45V4
783	B 28V2	B 39V2	826	B 29V4	B 51V2	869	B 40V2	B 66V2	912	B 31V2	B 45V4
784	B 28V2	B 39V2	827	B 29V4	B 52V4	870	B 40V2	B 66V2	913	B 31V2	B 77V4
785	B 6V2	B 77V4	828	B 29V4	B 52V4	871	B 10V2	B 66V2	914	B 31V2	B 77V4
786	B 6V2	B 77V4	829	B 11V4	B 25V2	872	B 10V2	B 66V2	915	B 31V2	B 76V2
787	B 6V2	B 76V2	830	B 11V4	B 25V2	873	B 54V4	B 66V2	916	B 31V2	B 76V2
788	B 6V2	B 76V2	831	B 18V2	B 40V2	874	B 54V4	B 66V2	917	B 29V4	B 30V2
789	B 38V2	B 44V2	832	B 18V2	B 40V2	875	B 55V4	B 66V2	918	B 29V4	B 30V2
790	B 38V2	B 44V2	833	B 30V2	B 58V4	876	B 55V4	B 66V2	919	B 38V2	B 66V2
791	B 38V2	B 50V2	834	B 30V2	B 58V4	877	B 34V2	B 50V2	920	B 38V2	B 66V2
792	B 38V2	B 50V2	835	B 30V2	B 57V2	878	B 34V2	B 50V2	921	B 28V2	B 54V4
793	B 51V2	B 57V2	836	B 30V2	B 57V2	879	B 34V2	B 44V2	922	B 28V2	B 54V4
794	B 51V2	B 57V2	837	B 40V2	B 45V4	880	B 34V2	B 44V2	923	B 28V2	B 55V4
795	B 52V4	B 57V2	838	B 40V2	B 45V4	881	B 22V2	B 48V2	924	B 28V2	B 55V4
796	B 52V4	B 57V2	839	B 40V2	B 59V4	882	B 22V2	B 48V2	925	B 11V4	B 19V2
797	B 52V4	B 58V4	840	B 40V2	B 59V4	883	B 34V2	B 66V2	926	B 11V4	B 19V2
798	B 52V4	B 58V4	841	B 25V2	B 30V2	884	B 34V2	B 66V2	927	B 20V2	B 77V4
799	B 51V2	B 58V4	842	B 25V2	B 30V2	885	B 31V2	B 50V2	928	B 20V2	B 77V4
800	B 51V2	B 58V4	843	B 2V2	B 76V2	886	B 31V2	B 50V2	929	B 20V2	B 76V2
801	B 18V2	B 39V2	844	B 2V2	B 76V2	887	B 31V2	B 44V2	930	B 20V2	B 76V2
802	B 18V2	B 39V2	845	B 2V2	B 77V4	888	B 31V2	B 44V2	931	B 36V2	B 73V4

(continued)

Table J.3 (continued)

No.	From bus	To bus	No.	From bus	To bus	No.	From bus	To bus	No.	From bus	To bus
932	B 36V2	B 73V4	976	B 25V2	B 58V4	1020	B 34V2	B 36V2	1064	B 17V2	B 19V2
933	B 19V2	B 29V4	977	B 59V4	B 66V2	1021	B 22V2	B 61V2	1065	B 45V4	B 73V4
934	B 19V2	B 29V4	978	B 59V4	B 66V2	1022	B 22V2	B 61V2	1066	B 45V4	B 73V4
935	B 18V2	B 34V2	979	B 45V4	B 66V2	1023	B 28V2	B 45V4	1067	B 59V4	B 73V4
936	B 18V2	B 34V2	980	B 45V4	B 66V2	1024	B 28V2	B 45V4	1068	B 59V4	B 73V4
937	B 10V2	B 55V4	981	B 30V2	B 73V4	1025	B 28V2	B 59V4	1069	B 23V2	B 61V2
938	B 10V2	B 55V4	982	B 30V2	B 73V4	1026	B 28V2	B 59V4	1070	B 23V2	B 61V2
939	B 10V2	B 54V4	983	B 23V2	B 48V2	1027	B 40V2	B 77V4	1071	B 2V2	B 28V2
940	B 10V2	B 54V4	984	B 23V2	B 48V2	1028	B 40V2	B 77V4	1072	B 2V2	B 28V2
941	B 34V2	B 59V4	985	B 6V2	B 28V2	1029	B 40V2	B 76V2	1073	B 30V2	B 40V2
942	B 34V2	B 59V4	986	B 6V2	B 28V2	1030	B 40V2	B 76V2	1074	B 30V2	B 40V2
943	B 34V2	B 45V4	987	B 36V2	B 38V2	1031	B 2V2	B 66V2	1075	B 20V2	B 66V2
944	B 34V2	B 45V4	988	B 36V2	B 38V2	1032	B 2V2	B 66V2	1076	B 20V2	B 66V2
945	B 18V2	B 73V4	989	B 34V2	B 77V4	1033	B 19V2	B 73V4	1077	B 34V2	B 55V4
946	B 18V2	B 73V4	990	B 34V2	B 77V4	1034	B 19V2	B 73V4	1078	B 34V2	B 55V4
947	B 17V2	B 60V2	991	B 34V2	B 76V2	1035	B 6V2	B 31V2	1079	B 34V2	B 54V4
948	B 17V2	B 60V2	992	B 34V2	B 76V2	1036	B 6V2	B 31V2	1080	B 34V2	B 54V4
949	B 51V2	B 73V4	993	B 21V2	B 48V2	1037	B 38V2	B 76V2	1081	B 19V2	B 51V2
950	B 51V2	B 73V4	994	B 21V2	B 48V2	1038	B 38V2	B 76V2	1082	B 19V2	B 51V2
951	B 52V4	B 73V4	995	B 50V2	B 66V2	1039	B 10V2	B 28V2	1083	B 19V2	B 52V4
952	B 52V4	B 73V4	996	B 50V2	B 66V2	1040	B 10V2	B 28V2	1084	B 19V2	B 52V4
953	B 19V2	B 25V2	997	B 44V2	B 66V2	1041	B 39V2	B 77V4	1085	B 23V2	B 35V2
954	B 19V2	B 25V2	998	B 44V2	B 66V2	1042	B 39V2	B 77V4	1086	B 23V2	B 35V2
955	B 18V2	B 31V2	999	B 28V2	B 44V2	1043	B 59V4	B 61V2	1087	B 35V2	B 48V2
956	B 18V2	B 31V2	1000	B 28V2	B 44V2	1044	B 59V4	B 61V2	1088	B 35V2	B 48V2
957	B 44V2	B 57V2	1001	B 28V2	B 50V2	1045	B 45V4	B 61V2	1089	B 5V2	B 48V2
958	B 44V2	B 57V2	1002	B 28V2	B 50V2	1046	B 45V4	B 61V2	1090	B 5V2	B 48V2
959	B 44V2	B 58V4	1003	B 36V2	B 39V2	1047	B 18V2	B 28V2	1091	B 4V4	B 48V2
960	B 44V2	B 58V4	1004	B 36V2	B 39V2	1048	B 18V2	B 28V2	1092	B 4V4	B 48V2
961	B 50V2	B 57V2	1005	B 31V2	B 55V4	1049	B 1V4	B 17V2	1093	B 40V2	B 51V2
962	B 50V2	B 57V2	1006	B 31V2	B 55V4	1050	B 1V4	B 17V2	1094	B 40V2	B 51V2
963	B 50V2	B 58V4	1007	B 31V2	B 54V4	1051	B 18V2	B 29V4	1095	B 40V2	B 52V4
964	B 50V2	B 58V4	1008	B 31V2	B 54V4	1052	B 18V2	B 29V4	1096	B 40V2	B 52V4
965	B 10V2	B 77V4	1009	B 25V2	B 73V4	1053	B 10V2	B 31V2	1097	B 40V2	B 73V4
966	B 10V2	B 77V4	1010	B 25V2	B 73V4	1054	B 10V2	B 31V2	1098	B 40V2	B 73V4
967	B 10V2	B 76V2	1011	B 50V2	B 73V4	1055	B 8V2	B 26V4	1099	B 29V4	B 50V2
968	B 10V2	B 76V2	1012	B 50V2	B 73V4	1056	B 8V2	B 26V4	1100	B 29V4	B 50V2
969	B 8V2	B 35V2	1013	B 44V2	B 73V4	1057	B 18V2	B 66V2	1101	B 29V4	B 44V2
970	B 8V2	B 35V2	1014	B 44V2	B 73V4	1058	B 18V2	B 66V2	1102	B 29V4	B 44V2
971	B 2V2	B 10V2	1015	B 21V2	B 61V2	1059	B 8V2	B 48V2	1103	B 1V4	B 60V2
972	B 2V2	B 10V2	1016	B 21V2	B 61V2	1060	B 8V2	B 48V2	1104	B 1V4	B 60V2
973	B 17V2	B 25V2	1017	B 10V2	B 20V2	1061	B 30V2	B 34V2	1105	B 47V4	B 60V2
974	B 17V2	B 25V2	1018	B 10V2	B 20V2	1062	B 30V2	B 34V2	1106	B 47V4	B 60V2
975	B 25V2	B 58V4	1019	B 34V2	B 36V2	1063	B 17V2	B 19V2	1107	B 27V4	B 77V4

(continued)

Table J.3 (continued)

No.	From bus	To bus	No.	From bus	To bus	No.	From bus	To bus	No.	From bus	To bus
1108	B 27V4	B 77V4	1152	B 11V4	B 45V4	1196	B 8V2	B 21V2	1240	B 5V2	B 23V2
1109	B 27V4	B 76V2	1153	B 11V4	B 59V4	1197	B 5V2	B 8V2	1241	B 4V4	B 23V2
1110	B 27V4	B 76V2	1154	B 11V4	B 59V4	1198	B 5V2	B 8V2	1242	B 4V4	B 23V2
1111	B 22V2	B 35V2	1155	B 27V4	B 55V4	1199	B 4V4	B 8V2	1243	B 17V2	B 47V4
1112	B 22V2	B 35V2	1156	B 27V4	B 55V4	1200	B 4V4	B 8V2	1244	B 17V2	B 47V4
1113	B 4V4	B 26V4	1157	B 27V4	B 54V4	1201	B 17V2	B 73V4	1245	B 25V2	B 27V4
1114	B 4V4	B 26V4	1158	B 27V4	B 54V4	1202	B 17V2	B 73V4	1246	B 25V2	B 27V4
1115	B 5V2	B 26V4	1159	B 17V2	B 29V4	1203	B 20V2	B 27V4	1247	B 55V4	B 61V2
1116	B 5V2	B 26V4	1160	B 17V2	B 29V4	1204	B 20V2	B 27V4	1248	B 55V4	B 61V2
1117	B 10V2	B 45V4	1161	B 48V2	B 61V2	1205	B 6V2	B 27V4	1249	B 54V4	B 61V2
1118	B 10V2	B 45V4	1162	B 48V2	B 61V2	1206	B 6V2	B 27V4	1250	B 54V4	B 61V2
1119	B 10V2	B 59V4	1163	B 61V2	B 66V2	1207	B 51V2	B 60V2	1251	B 51V2	B 76V2
1120	B 10V2	B 59V4	1164	B 61V2	B 66V2	1208	B 51V2	B 60V2	1252	B 51V2	B 76V2
1121	B 19V2	B 30V2	1165	B 27V4	B 30V2	1209	B 52V4	B 60V2	1253	B 52V4	B 76V2
1122	B 19V2	B 30V2	1166	B 27V4	B 30V2	1210	B 52V4	B 60V2	1254	B 52V4	B 76V2
1123	B 17V2	B 51V2	1167	B 25V2	B 60V2	1211	B 2V2	B 43V4	1255	B 52V4	B 77V4
1124	B 17V2	B 51V2	1168	B 25V2	B 60V2	1212	B 2V2	B 43V4	1256	B 52V4	B 77V4
1125	B 17V2	B 52V4	1169	B 26V4	B 48V2	1213	B 30V2	B 60V2	1257	B 51V2	B 77V4
1126	B 17V2	B 52V4	1170	B 26V4	B 48V2	1214	B 30V2	B 60V2	1258	B 51V2	B 77V4
1127	B 21V2	B 35V2	1171	B 8V2	B 22V2	1215	B 1V4	B 25V2	1259	B 11V4	B 60V2
1128	B 21V2	B 35V2	1172	B 8V2	B 22V2	1216	B 1V4	B 25V2	1260	B 11V4	B 60V2
1129	B 21V2	B 59V4	1173	B 27V4	B 52V4	1217	B 26V4	B 35V2	1261	B 10V2	B 27V4
1130	B 21V2	B 59V4	1174	B 27V4	B 52V4	1218	B 26V4	B 35V2	1262	B 10V2	B 27V4
1131	B 21V2	B 45V4	1175	B 27V4	B 51V2	1219	B 19V2	B 60V2	1263	B 43V4	B 55V4
1132	B 21V2	B 45V4	1176	B 27V4	B 51V2	1220	B 19V2	B 60V2	1264	B 43V4	B 55V4
1133	B 10V2	B 61V2	1177	B 8V2	B 23V2	1221	B 61V2	B 77V4	1265	B 43V4	B 54V4
1134	B 10V2	B 61V2	1178	B 8V2	B 23V2	1222	B 61V2	B 77V4	1266	B 43V4	B 54V4
1135	B 59V4	B 77V4	1179	B 21V2	B 73V4	1223	B 61V2	B 76V2	1267	B 11V4	B 27V4
1136	B 59V4	B 77V4	1180	B 21V2	B 73V4	1224	B 61V2	B 76V2	1268	B 11V4	B 27V4
1137	B 45V4	B 76V2	1181	B 23V2	B 73V4	1225	B 23V2	B 29V4	1269	B 22V2	B 26V4
1138	B 45V4	B 76V2	1182	B 23V2	B 73V4	1226	B 23V2	B 29V4	1270	B 22V2	B 26V4
1139	B 45V4	B 77V4	1183	B 6V2	B 61V2	1227	B 5V2	B 21V2	1271	B 2V2	B 61V2
1140	B 45V4	B 77V4	1184	B 6V2	B 61V2	1228	B 5V2	B 21V2	1272	B 2V2	B 61V2
1141	B 59V4	B 76V2	1185	B 20V2	B 43V4	1229	B 4V4	B 21V2	1273	B 1V4	B 47V4
1142	B 59V4	B 76V2	1186	B 20V2	B 43V4	1230	B 4V4	B 21V2	1274	B 1V4	B 47V4
1143	B 29V4	B 59V4	1187	B 2V2	B 27V4	1231	B 11V4	B 23V2	1275	B 10V2	B 21V2
1144	B 29V4	B 59V4	1188	B 2V2	B 27V4	1232	B 11V4	B 23V2	1276	B 10V2	B 21V2
1145	B 29V4	B 45V4	1189	B 5V2	B 22V2	1233	B 11V4	B 21V2	1277	B 5V2	B 35V2
1146	B 29V4	B 45V4	1190	B 5V2	B 22V2	1234	B 11V4	B 21V2	1278	B 5V2	B 35V2
1147	B 11V4	B 17V2	1191	B 4V4	B 22V2	1235	B 4V4	B 61V2	1279	B 4V4	B 35V2
1148	B 11V4	B 17V2	1192	B 4V4	B 22V2	1236	B 4V4	B 61V2	1280	B 4V4	B 35V2
1149	B 17V2	B 30V2	1193	B 1V4	B 19V2	1237	B 5V2	B 61V2	1281	B 11V4	B 22V2
1150	B 17V2	B 30V2	1194	B 1V4	B 19V2	1238	B 5V2	B 61V2	1282	B 11V4	B 22V2
1151	B 11V4	B 45V4	1195	B 8V2	B 21V2	1239	B 5V2	B 23V2	1283	B 35V2	B 61V2

(continued)

Table J.3 (continued)

No.	From bus	To bus	No.	From bus	To bus	No.	From bus	To bus	No.	From bus	To bus
1284	B 35V2	B 61V2	1307	B 6V2	B 21V2	1330	B 6V2	B 23V2	1353	B 23V2	B 54V4
1285	B 43V4	B 77V4	1308	B 6V2	B 21V2	1331	B 21V2	B 54V4	1354	B 23V2	B 54V4
1286	B 43V4	B 77V4	1309	B 27V4	B 43V4	1332	B 21V2	B 54V4	1355	B 2V2	B 21V2
1287	B 43V4	B 76V2	1310	B 27V4	B 43V4	1333	B 21V2	B 55V4	1356	B 2V2	B 21V2
1288	B 43V4	B 76V2	1311	B 17V2	B 27V4	1334	B 21V2	B 55V4	1357	B 27V4	B 47V4
1289	B 10V2	B 22V2	1312	B 17V2	B 27V4	1335	B 19V2	B 27V4	1358	B 27V4	B 47V4
1290	B 10V2	B 22V2	1313	B 23V2	B 25V2	1336	B 19V2	B 27V4	1359	B 2V2	B 22V2
1291	B 19V2	B 23V2	1314	B 23V2	B 25V2	1337	B 25V2	B 54V4	1360	B 2V2	B 22V2
1292	B 19V2	B 23V2	1315	B 21V2	B 25V2	1338	B 25V2	B 54V4	1361	B 19V2	B 47V4
1293	B 20V2	B 61V2	1316	B 21V2	B 25V2	1339	B 25V2	B 55V4	1362	B 19V2	B 47V4
1294	B 20V2	B 61V2	1317	B 25V2	B 47V4	1340	B 25V2	B 55V4	1363	B 2V2	B 23V2
1295	B 6V2	B 43V4	1318	B 25V2	B 47V4	1341	B 10V2	B 48V2	1364	B 2V2	B 23V2
1296	B 6V2	B 43V4	1319	B 19V2	B 35V2	1342	B 10V2	B 48V2	1365	B 6V2	B 48V2
1297	B 23V2	B 26V4	1320	B 19V2	B 35V2	1343	B 22V2	B 25V2	1366	B 6V2	B 48V2
1298	B 23V2	B 26V4	1321	B 6V2	B 22V2	1344	B 22V2	B 25V2	1367	B 4V4	B 10V2
1299	B 19V2	B 21V2	1322	B 6V2	B 22V2	1345	B 25V2	B 61V2	1368	B 4V4	B 10V2
1300	B 19V2	B 21V2	1323	B 19V2	B 22V2	1346	B 25V2	B 61V2	1369	B 19V2	B 48V2
1301	B 10V2	B 23V2	1324	B 19V2	B 22V2	1347	B 22V2	B 55V4	1370	B 19V2	B 48V2
1302	B 10V2	B 23V2	1325	B 27V4	B 60V2	1348	B 22V2	B 55V4	1371	B 67V4	B 71V4
1303	B 21V2	B 26V4	1326	B 27V4	B 60V2	1349	B 22V2	B 54V4	1372	B 67V4	B 71V4
1304	B 21V2	B 26V4	1327	B 10V2	B 43V4	1350	B 22V2	B 54V4			
1305	B 8V2	B 61V2	1328	B 10V2	B 43V4	1351	B 23V2	B 55V4			
1306	B 8V2	B 61V2	1329	B 6V2	B 23V2	1352	B 23V2	B 55V4			

[a] The length of any candidate line may be readily calculated from geographical characteristics of the sending and receiving buses. For details, see problem 6 of Chap. 7

Table J.4 Generation data

No.	Bus name	P_G (p.u.)	V_{set} (p.u.)	$\bar{P}_G$ (p.u.)	$\underline{Q}$ (p.u.)	$\overline{Q}$ (p.u.)
1[a]	B 1V4	1.66	1.0103	3.75	-1.15	2.81
2	B 4V4	3.37	1.0292	4.00	-1.22	3.00
3	B 5V2	0.43	1.0134	0.80	-0.24	0.60
4	B 8V2	0.01	0.9826	0.80	-0.25	0.60
5	B 9V2	2.00	0.9653	3.03	-0.88	0.50
6[a]	B 11V4	13.05	1.0092	24.00	-5.70	9.00
7	B 19V2	0.40	0.9615	0.75	-0.23	0.56
8	B 20V2	0.09	0.9558	0.75	-0.23	0.56
9	B 22V2	5.60	1.0097	7.09	-1.86	3.50
10	B 26V4	2.06	1.0405	2.50	-0.75	1.88
11	B 30V2	0.90	0.9925	1.36	-0.72	0.47
12	B 34V2	0.70	0.9477	0.84	-0.40	0.20
13	B 36V2	0.41	0.9918	0.56	-0.39	0.39
14	B 39V2	7.34	0.9684	12.63	-3.25	5.40
15	B 40V2	4.77	0.9642	7.08	-1.10	1.31
16	B 43V4	0.14	1.0041	2.50	-0.75	1.88
17	B 44V2	6.88	1.0057	12.00	-2.55	4.50
18	B 45V4	6.08	1.0107	9.60	-1.53	3.06
19	B 47V4	2.36	1.0165	3.00	-0.95	2.25
20	B 54V4	9.25	1.0146	12.52	-3.69	6.80
21	B 55V4	7.46	1.0150	12.99	-2.79	3.90
22	B 56V2	5.50	0.9758	12.85	-3.40	3.35
23	B 60V2	0.42	0.9035	0.75	-0.23	0.56
24	B 67V4	1.01	1.0192	5.00	-1.50	3.75
25	B 68V2	0.38	0.9455	0.63	-0.18	0.12
26	B 72V2	0.63	0.9533	0.90	-0.20	0.42

[a] Slack bus

Appendix K
Numerical Details of the Hybrid Approach

The details of the hybrid approach, as discussed and tested on the 77-bus dual voltage level test system (see Chap. 9, Sect 9.6, Table 9.9) are given here (as Tables K.1, K.2, K.3, K.4, K.5 and K.6).

Table K.1 The detailed results of the backward stage

No.[a]	From bus	To bus	Length[b] (km)	Voltage level (kV)	No. of lines[c]	Capacity limit (p.u.)	Line flow (p.u.)	Maximum line flow in contingency conditions	
								Flow on line (p.u.)	Relevant contingency
1	B 44V2	B 50V2	1	230	2	2.2	1.908	2.628	B 44V2 B 50V2
35	B 24V2	B 65V2	1.11	230	2	2.2	-1.482	2.027	B 7V2 B 24V2
38	B 11V4	B 29V4	2.13	400	2	6.6	0.93	1.454	B 11V4 B 29V4
39	B 39V2	B 40V2	2.86	230	2	2.2	1.124	1.477	B 28V2 B 76V2
41	B 46V2	B 64V4	3.33	230	2	2.2	-0.639	0.985	B 46V2 B 64V4
42	B 46V2	B 64V4	3.33	400	2	6.6	-2.555	2.971	B 11V4 B 70V4
43	B 46V2	B 63V2	3.33	230	2	2.2	-0.213	0.87	B 33V4 B 64V4
47	B 63V2	B 68V2	3.5	230	2	2.2	-0.677	1.044	B 63V2 B 68V2
48	B 63V2	B 68V2	3.5	400	2	6.6	-2.709	3.62	B 33V4 B 64V4
49	B 41V2	B 64V4	4.24	230	2	2.2	-1.579	2.004	B 41V2 B 63V2
55	B 9V2	B 65V2	4.44	230	2	2.2	1.428	2.335	B 65V2 B 69V2
58	B 3V2	B 71V4	4.51	400	2	6.6	-2.749	3.919	B 11V4 B 70V4
62	B 3V2	B 70V4	4.51	400	2	6.6	-2.749	3.919	B 11V4 B 71V4
63	B 42V2	B 69V2	4.65	230	2	2.2	1.451	1.713	B 11V4 B 70V4
65	B 9V2	B 42V2	4.65	230	2	2.2	-1.188	2.044	B 65V2 B 69V2
67	B 46V2	B 53V2	5.03	230	2	2.2	0.839	1.374	B 11V4 B 70V4
69	B 38V2	B 40V2	5.03	230	2	2.2	-0.754	0.974	B 38V2 B 39V2
71	B 38V2	B 39V2	5.2	230	2	2.2	-1.347	1.621	B 38V2 B 39V2
73	B 3V2	B 49V2	5.55	230	2	2.2	1.599	1.986	B 3V2 B 49V2
75	B 9V2	B 24V2	5.55	230	2	2.2	1.439	1.93	B 65V2 B 69V2
84	B 7V2	B 16V4	5.84	400	2	6.6	-4.314	5.71	B 14V4 B 45V4

(continued)

Table K.1 (continued)

No.[a]	From bus	To bus	Length[b] (km)	Voltage level (kV)	No. of lines[c]	Capacity limit (p.u.)	Line flow (p.u.)	Maximum line flow in contingency conditions: Flow on line (p.u.)	Maximum line flow in contingency conditions: Relevant contingency
86	B 7V2	B 14V4	5.84	400	2	6.6	-4.439	5.842	B 45V4 B 59V4
101	B 12V2	B 24V2	6.7	230	2	2.2	-1.415	2.641	B 12V2 B 72V2
119	B 39V2	B 62V2	7.16	230	2	2.2	0.696	0.958	B 44V2 B 62V2
124	B 7V2	B 68V2	7.2	400	2	6.6	4.662	5.823	B 7V2 B 24V2
125	B 38V2	B 74V2	7.55	230	2	2.2	1.98	2.456	B 39V2 B 74V2
131	B 33V4	B 68V2	7.72	230	2	2.2	0.925	2.209	B 33V4 B 64V4
143	B 32V2	B 37V2	8.23	230	2	2.2	-1.331	2.164	B 64V4 B 75V4
155	B 36V2	B 72V2	9.02	230	2	2.2	0.917	1.189	B 11V4 B 70V4
181	B 34V2	B 38V2	9.29	230	2	2.2	-1.553	1.787	B 38V2 B 74V2
288	B 37V2	B 75V4	12.17	400	2	6.6	-4.618	6.108	B 64V4 B 75V4
296	B 44V2	B 59V4	12.2	400	2	6.6	-3.289	4.904	B 14V4 B 45V4
398	B 42V2	B 58V4	14.32	400	2	6.6	-5.901	6.619	B 11V4 B 70V4
417	B 54V4	B 76V2	14.56	230	2	2.2	1.775	2.17	B 55V4 B 76V2
421	B 55V4	B 76V2	14.56	230	2	2.2	1.817	2.477	B 54V4 B 55V4
519	B 31V2	B 40V2	17.18	230	2	2.2	-1.453	2.053	B 28V2 B 76V2
657	B 44V2	B 62V2	23.77	230	2	2.2	1.904	2.297	B 14V4 B 45V4
659	B 50V2	B 62V2	23.77	230	2	2.2	1.824	2.21	B 14V4 B 45V4
701	B 6V2	B 55V4	26.96	230	2	2.2	-2.016	2.408	B 6V2 B 76V2
731	B 39V2	B 44V2	28.89	230	2	2.2	-1.394	1.748	B 14V4 B 45V4

[a] The number shown is taken from the candidate line number given in Table J.3
[b] As X and Y are known for each bus, the line length can be readily calculated. For details, see problem 6 of Chap. 7
[c] Two lines are considered in each corridor

Table K.2 The detailed results of the backward stage (transformers)[a]

No.	Bus name	Voltage Level	Transformer capacity (p.u.)
1	B 44V2	400 kV:230 kV	5.50
2	B 37V2	400 kV:230 kV	8.25
3	B 46V2	400 kV:230 kV	5.50
4	B 33V4	400 kV:230 kV	2.75
5	B 3V2	400 kV:230 kV	8.25
6	B 63V2	400 kV:230 kV	5.50
7	B 42V2	400 kV:230 kV	8.25
8	B 54V4	400 kV:230 kV	2.75
9	B 55V4	400 kV:230 kV	5.50
10	B 64V4	400 kV:230 kV	2.75
11	B 68V2	400 kV:230 kV	2.75
12	B 7V2	400 kV:230 kV	5.50

[a] It should be mentioned that the transformers as detailed in Tables K.2, K.4 and K.6 are justified based on the following steps (The steps are described for a typical row 1, Table K.2)

- From Table K.1, a 400 kV line no. 296 (from B 44V2 to B 59V4) is justified. As the former bus is a 230 kV bus, while the latter is a 400 kV one, a 400 kV:230 kV substation is required
- In terms of the transformer (substation) capacity, it is determined based on the maximum flow (for both normal and contingency conditions) through the above mentioned line (line no. 296). This flow is 5.50 p.u

Table K.3 The detailed results of the forward stage

No.	From bus	To bus	Length (km)	Voltage level (kV)	No. of lines	Capacity limit (p.u.)	Line flow (p.u.)	Maximum line flow in contingency conditions	
								Flow on line (p.u.)	Relevant contingency
1	B 44V2	B 50V2	1	230	2	2.2	1.521	2.094	B 44V2 B 50V2
35	B 24V2	B 65V2	1.11	230	2	2.2	-1.282	1.758	B 12V2 B 72V2
38	B 11V4	B 29V4	2.13	400	2	6.6	0.99	1.541	B 11V4 B 29V4
39	B 39V2	B 40V2	2.86	230	2	2.2	0.911	1.17	B 28V2 B 76V2
41	B 46V2	B 64V4	3.33	230	2	2.2	-0.45	0.67	B 46V2 B 64V4
42	B 46V2	B 64V4	3.33	400	2	6.6	-1.798	2.268	B 32V2 B 46V2
43	B 46V2	B 63V2	3.33	230	2	2.2	0.162	0.341	B 63V2 B 68V2
46	B 64V4	B 68V2	3.5	400	2	6.6	0.071	1.031	B 64V4 B 75V4
47	B 63V2	B 68V2	3.5	230	2	2.2	-0.565	0.826	B 63V2 B 68V2
48	B 63V2	B 68V2	3.5	400	2	6.6	-2.258	2.622	B 7V2 B 24V2
49	B 41V2	B 64V4	4.24	230	2	2.2	-1.637	2.086	B 41V2 B 63V2
55	B 9V2	B 65V2	4.44	230	2	2.2	1.28	1.996	B 65V2 B 69V2
58	B 3V2	B 71V4	4.51	400	2	6.6	-2.567	3.079	B 11V4 B 70V4
62	B 3V2	B 70V4	4.51	400	2	6.6	-1.588	2.064	B 11V4 B 71V4
63	B 42V2	B 69V2	4.65	230	2	2.2	1.103	1.315	B 12V2 B 72V2
65	B 9V2	B 42V2	4.65	230	2	2.2	-0.757	1.43	B 65V2 B 69V2
67	B 46V2	B 53V2	5.03	230	2	2.2	-0.06	0.46	B 12V2 B 72V2
69	B 38V2	B 40V2	5.03	230	2	2.2	-0.407	0.547	B 38V2 B 39V2
71	B 38V2	B 39V2	5.2	230	2	2.2	-0.894	1.068	B 38V2 B 39V2
73	B 3V2	B 49V2	5.55	230	2	2.2	1.536	1.904	B 3V2 B 49V2
75	B 9V2	B 24V2	5.55	230	2	2.2	1.281	1.666	B 65V2 B 69V2
84	B 7V2	B 16V4	5.84	400	2	6.6	-2.933	3.631	B 14V4 B 45V4
86	B 7V2	B 14V4	5.84	400	2	6.6	-3.002	3.687	B 16V4 B 59V4
101	B 12V2	B 24V2	6.7	230	2	2.2	-0.288	2.066	B 12V2 B 72V2
119	B 39V2	B 62V2	7.16	230	2	2.2	1.599	1.782	B 40V2 B 62V2
124	B 7V2	B 68V2	7.2	400	2	6.6	2.818	3.688	B 7V2 B 24V2
125	B 38V2	B 74V2	7.55	230	2	2.2	1.502	1.738	B 39V2 B 74V2
131	B 33V4	B 68V2	7.72	230	2	2.2	0.912	2.047	B 33V4 B 64V4
143	B 32V2	B 37V2	8.23	230	2	2.2	-0.573	1.224	B 64V4 B 75V4
155	B 36V2	B 72V2	9.02	230	2	2.2	0.355	0.996	B 30V2 B 51V2
169	B 2V2	B 20V2	9.15	230	2	2.2	-0.324	0.627	B 2V2 B 76V2
172	B 67V4	B 71V4	90.58	230	2	2.2	1.052	1.129	B 54V4 B 75V4
181	B 34V2	B 38V2	9.29	230	2	2.2	-0.721	0.951	B 34V2 B 45V4
288	B 37V2	B 75V4	12.17	400	2	6.6	-3.413	4.594	B 64V4 B 75V4
296	B 44V2	B 59V4	12.2	400	2	6.6	-3.122	4.185	B 16V4 B 59V4
364	B 32V2	B 74V2	13.71	400	2	6.6	-3.476	4.021	B 14V4 B 45V4
398	B 42V2	B 58V4	14.32	400	2	6.6	-3.864	4.269	B 12V2 B 72V2
417	B 54V4	B 76V2	14.56	230	2	2.2	1.495	1.823	B 55V4 B 76V2
421	B 55V4	B 76V2	14.56	230	2	2.2	1.526	1.99	B 54V4 B 55V4
438	B 34V2	B 74V2	14.89	400	2	6.6	1.247	1.574	B 34V2 B 74V2
519	B 31V2	B 40V2	17.18	230	2	2.2	-1.238	1.714	B 28V2 B 76V2

(continued)

Table K.3 (continued)

No.	From bus	To bus	Length (km)	Voltage level (kV)	No. of lines	Capacity limit (p.u.)	Line flow (p.u.)	Maximum line flow in contingency conditions	
								Flow on line (p.u.)	Relevant contingency
657	B 44V2	B 62V2	23.77	230	2	2.2	1.017	1.16	B 44V2 B 74V2
659	B 50V2	B 62V2	23.77	230	2	2.2	0.953	1.093	B 44V2 B 74V2
671	B 13V2	B 44V2	25.19	230	2	2.2	-1.347	1.569	B 13V2 B 44V2
701	B 6V2	B 55V4	26.96	230	2	2.2	-1.173	1.426	B 6V2 B 10V2
705	B 53V2	B 74V2	27.61	230	2	2.2	-0.787	0.892	B 64V4 B 75V4
710	B 11V4	B 57V2	28.08	400	2	6.6	4.432	5.193	B 29V4 B 58V4
731	B 39V2	B 44V2	28.89	230	2	2.2	-0.441	0.605	B 44V2 B 74V2
772	B 44V2	B 74V2	32.66	400	2	6.6	3.518	3.914	B 16V4 B 59V4
900	B 29V4	B 72V2	44.55	400	2	6.6	3.348	3.761	B 11V4 B 70V4
938	B 10V2	B 55V4	48.65	400	2	6.6	-1.484	1.688	B 6V2 B 55V4
944	B 34V2	B 45V4	48.99	400	2	6.6	-2.754	3.263	B 14V4 B 45V4
1219	B 19V2	B 60V2	117.4	230	2	2.2	0.24	0.42	B 17V2 B 60V2
1230	B 4V4	B 21V2	121.71	400	2	6.6	1.044	1.507	B 4V4 B 59V4
1237	B 5V2	B 61V2	123.65	230	2	2.2	0.518	0.595	B 4V4 B 59V4
1310	B 27V4	B 43V4	158.74	400	2	6.6	-0.274	0.341	B 54V4 B 77V4

Table K.4 The detailed results of the forward stage (transformers)

No.	Bus name	Voltage level	Transformer capacity (p.u.)
1	B 10V2	400 kV:230 kV	2.75
2	B 57V2	400 kV:230 kV	5.5
3	B 72V2	400 kV:230 kV	5.5
4	B 44V2	400 kV:230 kV	2.75
5	B 32V2	400 kV:230 kV	5.5
6	B 21V2	400 kV:230 kV	2.75
7	B 74V2	400 kV:230 kV	2.75
8	B 37V2	400 kV:230 kV	5.5
9	B 46V2	400 kV:230 kV	2.75
10	B 33V4	400 kV:230 kV	2.75
11	B 34V2	400 kV:230 kV	2.75
12	B 3V2	400 kV:230 kV	5.5
13	B 63V2	400 kV:230 kV	2.75
14	B 42V2	400 kV:230 kV	5.5
15	B 54V4	400 kV:230 kV	2.75
16	B 55V4	400 kV:230 kV	5.5
17	B 64V4	400 kV:230 kV	2.75
18	B 67V4	400 kV:230 kV	2.75
19	B 68V2	400 kV:230 kV	2.75
20	B 71V4	400 kV:230 kV	2.75
21	B 7V2	400 kV:230 kV	5.5

Table K.5 The detailed results of the decrease stage

No.	From bus	To bus	Length (km)	Voltage level (kV)	No. of lines	Capacity limit (p.u.)	Line flow (p.u.)	Maximum line flow in contingency conditions	
								Flow on line (p.u.)	Relevant contingency
1	B 44V2	B 50V2	1	230	2	2.2	1.565	2.155	B 44V2 B 50V2
35	B 24V2	B 65V2	1.11	230	2	2.2	-1.321	1.79	B 12V2 B 72V2
39	B 39V2	B 40V2	2.86	230	2	2.2	0.877	1.172	B 28V2 B 76V2
41	B 46V2	B 64V4	3.33	230	2	2.2	-0.54	0.827	B 46V2 B 64V4
42	B 46V2	B 64V4	3.33	400	2	6.6	-2.158	2.713	B 32V2 B 46V2
43	B 46V2	B 63V2	3.33	230	1	1.1	0.031	0.144	B 63V2 B 68V2
46	B 64V4	B 68V2	3.5	400	2	6.6	-0.072	1.35	B 64V4 B 75V4
47	B 63V2	B 68V2	3.5	230	2	2.2	-0.59	0.88	B 63V2 B 68V2
48	B 63V2	B 68V2	3.5	400	2	6.6	-2.36	2.741	B 7V2 B 24V2
49	B 41V2	B 64V4	4.24	230	2	2.2	-1.625	2.067	B 41V2 B 63V2
55	B 9V2	B 65V2	4.44	230	2	2.2	1.211	1.924	B 65V2 B 69V2
58	B 3V2	B 71V4	4.51	400	1	3.3	-2.353	2.862	B 11V4 B 70V4
62	B 3V2	B 70V4	4.51	400	1	3.3	-1.652	2.332	B 3V2 B 71V4
63	B 42V2	B 69V2	4.65	230	2	2.2	1.208	1.444	B 12V2 B 72V2
65	B 9V2	B 42V2	4.65	230	1	1.1	-0.489	0.96	B 65V2 B 69V2
67	B 46V2	B 53V2	5.03	230	1	1.1	0.408	0.773	B 49V2 B 53V2
69	B 38V2	B 40V2	5.03	230	1	1.1	-0.308	0.445	B 38V2 B 39V2
71	B 38V2	B 39V2	5.2	230	1	1.1	-0.539	0.692	B 38V2 B 39V2
73	B 3V2	B 49V2	5.55	230	2	2.2	1.536	1.908	B 3V2 B 49V2
75	B 9V2	B 24V2	5.55	230	2	2.2	1.233	1.601	B 65V2 B 69V2
84	B 7V2	B 16V4	5.84	400	2	6.6	-3.158	3.973	B 14V4 B 45V4
86	B 7V2	B 14V4	5.84	400	2	6.6	-3.249	4.043	B 16V4 B 59V4
101	B 12V2	B 24V2	6.7	230	2	2.2	-0.349	2.053	B 12V2 B 72V2
119	B 39V2	B 62V2	7.16	230	2	2.2	1.534	1.716	B 15V2 B 39V2
124	B 7V2	B 68V2	7.2	400	2	6.6	3.154	4.066	B 7V2 B 24V2
125	B 38V2	B 74V2	7.55	230	2	2.2	1.55	1.85	B 39V2 B 74V2
131	B 33V4	B 68V2	7.72	230	2	2.2	0.929	2.108	B 33V4 B 64V4
155	B 36V2	B 72V2	9.02	230	1	1.1	0.315	0.911	B 30V2 B 51V2
169	B 2V2	B 20V2	9.15	230	1	1.1	-0.31	0.627	B 2V2 B 76V2
172	B 67V4	B 71V4	90.58	230	1	1.1	0.805	1.013	B 54V4 B 67V4
181	B 34V2	B 38V2	9.29	230	1	1.1	-0.427	0.851	B 34V2 B 45V4
288	B 37V2	B 75V4	12.17	400	2	6.6	-3.142	4.126	B 64V4 B 75V4
296	B 44V2	B 59V4	12.2	400	2	6.6	-3.396	4.633	B 14V4 B 45V4
364	B 32V2	B 74V2	13.71	400	2	6.6	-3.621	4.21	B 14V4 B 45V4
398	B 42V2	B 58V4	14.32	400	2	6.6	-3.845	4.247	B 12V2 B 72V2
417	B 54V4	B 76V2	14.56	230	2	2.2	1.519	1.852	B 55V4 B 76V2
421	B 55V4	B 76V2	14.56	230	2	2.2	1.549	1.991	B 54V4 B 55V4
519	B 31V2	B 40V2	17.18	230	2	2.2	-1.26	1.751	B 28V2 B 76V2
657	B 44V2	B 62V2	23.77	230	2	2.2	1.117	1.304	B 44V2 B 74V2
659	B 50V2	B 62V2	23.77	230	2	2.2	1.052	1.234	B 44V2 B 74V2
671	B 13V2	B 44V2	25.19	230	1	1.1	-0.82	0.985	B 13V2 B 44V2

(continued)

Table K.5 (continued)

No.	From bus	To bus	Length (km)	Voltage level (kV)	No. of lines	Capacity limit (p.u.)	Line flow (p.u.)	Maximum line flow in contingency conditions		
								Flow on line (p.u.)	Relevant contingency	
701	B 6V2	B 55V4	26.96	230	2	2.2	-1.179	1.43	B 6V2	B 10V2
710	B 11V4	B 57V2	28.08	400	2	6.6	4.492	5.258	B 29V4	B 58V4
731	B 39V2	B 44V2	28.89	230	1	1.1	-0.27	0.378	B 44V2	B 74V2
772	B 44V2	B 74V2	32.66	400	2	6.6	4.027	4.553	B 16V4	B 59V4
900	B 29V4	B 72V2	44.55	400	2	6.6	3.514	3.971	B 11V4	B 70V4
938	B 10V2	B 55V4	48.65	400	2	6.6	-1.506	1.711	B 6V2	B 55V4
944	B 34V2	B 45V4	48.99	400	1	3.3	-1.631	2.092	B 45V4	B 59V4
1219	B 19V2	B 60V2	117.4	230	1	1.1	0.202	0.42	B 17V2	B 60V2
1230	B 4V4	B 21V2	121.71	400	2	6.6	1.054	1.512	B 4V4	B 59V4
1237	B 5V2	B 61V2	123.65	230	2	2.2	0.517	0.593	B 4V4	B 59V4
1310	B 27V4	B 43V4	158.74	400	1	3.3	-0.224	0.279	B 54V4	B 77V4

Table K.6 The detailed results of the decrease stage (transformers)

No.	Bus name	Voltage level	Transformer capacity (p.u.)
1	B 10V2	400 kV:230 kV	2.75
2	B 57V2	400 kV:230 kV	5.50
3	B 72V2	400 kV:230 kV	5.50
4	B 44V2	400 kV:230 kV	2.75
5	B 32V2	400 kV:230 kV	5.50
6	B 21V2	400 kV:230 kV	2.75
7	B 74V2	400 kV:230 kV	2.75
8	B 37V2	400 kV:230 kV	5.50
9	B 46V2	400 kV:230 kV	2.75
10	B 33V4	400 kV:230 kV	2.75
11	B 34V2	400 kV:230 kV	2.75
12	B 3V2	400 kV:230 kV	5.50
13	B 63V2	400 kV:230 kV	2.75
14	B 42V2	400 kV:230 kV	5.50
15	B 54V4	400 kV:230 kV	2.75
16	B 55V4	400 kV:230 kV	5.50
17	B 64V4	400 kV:230 kV	2.75
18	B 67V4	400 kV:230 kV	2.75
19	B 68V2	400 kV:230 kV	2.75
20	B 71V4	400 kV:230 kV	2.75
21	B 7V2	400 kV:230 kV	5.50

Appendix L
Generated Matlab M-files Codes

L.1 GEP1.m

a) "GEP1" M-file code

```
clear
clc
%% Required Input data
%% Required load nodes data:
Gen_Data = xlsread('Gep.xls', 'Gen-Data');
%% Required substations data:
Add_Data = xlsread('Gep.xls', 'Add-Data');
%% Data retrieval from input data
No_Gen = Gen_Data(:,1); % Generator type number
Capaci_Gen = Gen_Data(:,2); % Capacity type plants
% Investment cost type plants:
Invest_Gen = Gen_Data(:,3)*1000;
Life_Gen = Gen_Data(:,4); % Life type plants
FuelCost_Gen = Gen_Data(:,5); % Fuel cost type plants
% Operation and maintenance cost type plants:
O_MCost_Gen = Gen_Data(:,6)*1000*12;
Load = Add_Data(1,1); % Maximum network load (MW)
Reserv = Add_Data(1,2)/100; % Reserve ratio
% Coefficient of annual interest:
Interest_rate = Add_Data(1,3)/100;
Exist_Cap = Add_Data(1,4); % Capacity of existing plants
% Existing power plants, fuel costs:
Exist_FuelCost = Add_Data(1,5);
if isempty(Capaci_Gen)
  fprintf('Input argument "Capaci_Gen" determining');
  fprintf(' capacity type plants.\n');
  error('"Capaci_Gen" is undefined and must be determined.');
end
if isempty(Invest_Gen)
  fprintf('Input argument "Invest_Gen" determining');
  fprintf(' investment cost type plants.\n');
  error('"Invest_Gen" is undefined and must be determined.');
end
if isempty(Life_Gen)
  fprintf('Input argument "Life_Gen" determining');
  fprintf(' life type plants.\n');
  error('"Life_Gen" is undefined and must be determined.');
```

```
end
if isempty(FuelCost_Gen)
  fprintf('Input argument "FuelCost_Gen" determining');
  fprintf('fuel cost type plants.\n');
  error('"FuelCost_Gen" is undefined and must be deter-
mined.');
end
if isempty(O_MCost_Gen)
  fprintf('Input argument "O_MCost_Gen" determining');
  fprintf(' operation and maintenance cost type plants.\n');
  error('"O_MCost_Gen" is undefined and must be determined.');
end
if isempty(Load)
  fprintf('Input argument "Load" determining');
  fprintf(' maximum network load(MW).\n');
  error('"Load" is undefined and must be determined.');
end
if isempty(Reserv)
  fprintf('Input argument "Reserv" determining');
  fprintf(' reserve ratio.\n');
  error('"Reserv" is undefined and must be determined.');
end
if isempty(Interest_rate)
  fprintf('Input argument "Interest_rate" determining');
  fprintf(' coefficient of annual interest.\n');
  error('"Interest_rate" is undefined & must be determined.');
end
if isempty(Exist_Cap)
  fprintf('Input argument "Exist_Cap" determining');
  fprintf(' capacity of existing plants.\n');
  error('"Exist_Cap" is undefined and must be determined.');
end
if isempty(Exist_FuelCost)
  fprintf('Input argument "Exist_FuelCost" determining');
  fprintf(' existing power plants, fuel costs.\n');
  error('"Exist_FuelCost" is undefined & must be determined');
end
%
if (Capaci_Gen==0)
  fprintf('Input argument "Capaci_Gen" determining');
  fprintf(' capacity type plants.\n');
  error('"Capaci_Gen" should not be zero.');
end
if (find(Life_Gen==0))
  fprintf('Input argument "Life_Gen" determining');
  fprintf(' life type plants.\n');
  error('"Life_Gen" should not be zero.');
end
if (Load==0)
  fprintf('Input argument "Load" determining');
  fprintf(' maximum network load(MW).\n');
  error('"Load" should not be zero.');
end
if (Reserv<0)
```

```
    fprintf('Input argument "Reserv" determining');
    fprintf(' reserve ratio.\n');
    error('"Reserv" should not be less than zero.');
  end
  if (Interest_rate<=0)
    fprintf('Input argument "Interest_rate" determining');
    fprintf(' coefficient of annual interest.\n');
    error('"Interest_rate" should be greater than zero.');
  end
  %%
  Gepp;
  %% Print obtained results in command window and results.txt
  Print_GEPP;
```

b) "Gepp" M-file code

```
%% Problem outputs
%Best_Gen: The best units selected
%% Problem inputs
%Capaci_Gen; Capacity type plants
%Invest_Gen; Investment cost type Plants
%Life_Gen; Life type plants
%FuelCost_Gen; Fuel cost type plants
%O_MCost_Gen; Operation and maintenance cost type plants
%Load; Maximum network load(MW)
%Energy; Annual energy consumption(MWh)
%Reserv; Reserve ratio
%Interest_rate; Coefficient of annual interest
%Exist_Cap; Capacity of existing plants
%Exist_FuelCost; Existing power plants, fuel costs

%%Choose the cheapest power plants to produce
CheapFuel = FuelCost_Gen;
CheapFuel(4) = Exist_FuelCost;
[CheapFuel,ICheapFuel] = sort(CheapFuel);
A = (1+Interest_rate);
for i = 1:3
   A_P(i) = (A^Life_Gen(i,1))*Interest_rate;
   A_P(i) = A_P(i)/(A^Life_Gen(i,1)-1);
end
B = zeros (3,1331);
m = 0;
%Create all the solution space
for i = 0:10
    for j = 0:10
        for k = 0:10
            m = m+1;
            B(1,m) = i;
            B(2,m) = j;
            B(3,m) = k;
        end
    end
end
%Calculate the cost of each choice
```

```
for i = 1:1331
    Total_Cap = Exist_Cap+B(1,i)*Capaci_Gen(1)+B(2,i)*...
        Capaci_Gen(2)+B(3,i)*Capaci_Gen(3);
    if Total_Cap < Load*(1+Reserv)
        Total_Cost(i) = 1.0e12;
    else
        Total_Cost(i) = 0.0;
%Calculate the energy production plant
        Energy = Load*8760;
        Energy1 = Energy;
        for j = 1:3
           Energy_Gen(j) = B(j,i) * Capaci_Gen(j) * 8760;
        end
        Energy_Gen(4)=Exist_Cap*8760;
        for j = 1:4
           ii = ICheapFuel(j);
           Energy1 = Energy1-Energy_Gen(ii);
           if Energy1<0.0
              Energy_Gen(ii) = Energy1+Energy_Gen(ii);
              if Energy_Gen(ii)<0.0
                 Energy_Gen(ii) = 0.0;
              end
           end
        end
        if Energy1<=0.0
           for j = 1:3
               Total_Cost(i) = Total_Cost(i)+...
                B(j,i)*Capaci_Gen(j)*...
                (Invest_Gen(j)*A_P(j)+O_MCost_Gen(j))...
                +FuelCost_Gen(j)*Energy_Gen(j);
           end
           Total_Cost(i) = Total_Cost(i)+...
               Exist_FuelCost*Energy_Gen(4);
        else
           Total_Cost(i) = 1.0e12;
        end
    end
end
%Choose the best option
[Solution,II] = min(Total_Cost);
Best_Gen(1) = B(1,II);
Best_Gen(2) = B(2,II);
Best_Gen(3) = B(3,II);
Energy1 = Energy;
for j = 1:3
   Energy_Gen(j) = Best_Gen(j)*Capaci_Gen(j)*8760;
end
Energy_Gen(4) = Exist_Cap*8760;
for j = 1:4
  ii = ICheapFuel(j);
  Energy1 = Energy1-Energy_Gen(ii);
  if Energy1<0.0
     Energy_Gen(ii) = Energy1+Energy_Gen(ii);
     if Energy_Gen(ii)<0.0
```

```
            Energy_Gen(ii) = 0.0;
         end
      end
   end
```

c) "Print_GEPP" M-file code

```
%% Print different costs and optimal capacity of each plant
clc
fprintf('\n Optimal Capacity_Plant1 = %4i',Best_Gen(1));
fprintf('\n Optimal Capacity_Plant1 = %4i',Best_Gen(2));
fprintf('\n Optimal Capacity_Plant1 = %4i',Best_Gen(3));
InvestCost = 0.0;
FuelCost = 0.0;
O_MCost = 0.0;
for i = 1:3
   InvestCost = InvestCost+A_P(i)*Best_Gen(i)*...
       Capaci_Gen(i)*Invest_Gen(i);
   FuelCost = FuelCost+FuelCost_Gen(i)*Energy_Gen(i);
   O_MCost = O_MCost+Best_Gen(i)*Capaci_Gen(i)*...
       O_MCost_Gen(i);
end
FuelCost = FuelCost+Exist_FuelCost*Energy_Gen(4);
Total_Cost1 = InvestCost+FuelCost+O_MCost;
fprintf('\n\n******************************************');
fprintf('**Result***************************************');
fprintf('************\n');
fprintf('| Capacity_Plant1 | Capacity_Plant2 | Capacity');
fprintf('_Plant3 | Investment cost | Fuel cost | Fixed O');
fprintf('&M cost |\n');
fprintf('|       (Mw)       |       (Mw)       |       (Mw) ');
fprintf('      |      (R/yr)      |  (R/yr)    |    (R/yr) ');
fprintf('     |');
fprintf('\n|     %6.2f       |   %6.2f          |      %6.2f',...
    Best_Gen(1)*Capaci_Gen(1),Best_Gen(2)*Capaci_Gen(2),...
    Best_Gen(3)*Capaci_Gen(3));
fprintf('      |   %10.2E    |%10.2E |  %10.2E    |\n',...
    InvestCost,FuelCost,O_MCost);
fprintf('**********************************************');
fprintf('**********************************************');
fprintf('*********\n');
fprintf('\n Total Cost(R) = %10.2E \n',Total_Cost1);

fid = fopen('result.txt', 'wt');
fprintf(fid,'\n Optimal Capacity_Plant1 = %4i',Best_Gen(1));
fprintf(fid,'\n Optimal Capacity_Plant2 = %4i',Best_Gen(2));
fprintf(fid,'\n Optimal Capacity_Plant3 = %4i',Best_Gen(3));
fprintf(fid,'\n\n***************************************');
fprintf(fid,'***Result*********************************');
fprintf(fid,'***********************\n');
fprintf(fid,'\n| Capacity_Plant1 | Capacity_Plant2 | ');
fprintf(fid,'Capacity_Plant3 | Investment cost | Fuel');
fprintf(fid,' cost| Fixed O&M cost |\n');
fprintf(fid,'|       (Mw)       |       (Mw)       |       ');
```

```
fprintf(fid,'(Mw)          |      (R/yr)        |  (R/yr)    | ');
fprintf(fid,'    (R/yr)        |');
fprintf(fid,'\n|     %6.2f        |     %6.2f         |      '...
    ,Best_Gen(1)*Capaci_Gen(1),Best_Gen(2)*Capaci_Gen(2));
fprintf(fid,'%6.2f        |    %10.2E    |%10.2E |  %10.2E'...
    ,Best_Gen(3)*Capaci_Gen(3),InvestCost,FuelCost,O_MCost);
fprintf(fid,''     |\n');
fprintf(fid,'\n****************************************');
fprintf(fid,'*****************************************');
fprintf(fid,'******************\n');
fprintf(fid,'\n Total cost(R) = %10.2E \n',Total_Cost1);
fclose(fid);
```

L.2 GEP2.m

a) "GEP2" M-file code

```
clear
clc
%%Required input data
Busdata = xlsread('Gepdata.xls', 'Busdata');
Linedata = xlsread('Gepdata.xls', 'Linedata');
Candidatesdata = xlsread('Gepdata.xls', 'Candidatesdata');
%Maximum capacity that line i can be enhanced:
Biu = xlsread('Gepdata.xls', 'Biu');
%Investment cost for transmission lines enhancement(R/p.u.km)
Ga = xlsread('Gepdata.xls', 'Gama');
%% Data retrieval from input data
%Candidate buses for generation expansion:
Candidates = Candidatesdata(:,1);
%Beta(i):Investment factor cost of generation expansion
    %in bus i
Beta = Candidatesdata(:,2);
%PGmax(i):Maximum generation expansion limit of bus i
PGmax = Candidatesdata(:,3);
%PGmin(i):Minimum generation expansion limit of bus i
PGmin = Busdata(:,3);
Nlin = Linedata(:,1); %Line number
Nl = Linedata(:,2); %Nl:From bus
Nr = Linedata(:,3); %Nr:To bus
R = Linedata(:,4); %R(i):Line resistance
X = Linedata(:,5); %X(i):Line reactance
%Smax(i):Maximum thermal rating of line i
Smax = Linedata(:,6);
%Length(i):Path length of line i
Length = Linedata(:,7);
Busn = Busdata(:,1); %Bus number
Btype = Busdata(:,2); %Type of bus 1-Slack, 2-PV, 3-PQ
Pg = Busdata(:,3); %Pg(i):Generation of bus i
Pl = Busdata(:,4); %Pl(i):Load of bus i
Nc = setxor(Busn,Candidates); %Nc:Non-candidate buses
```

```
[Ybus] = ybus(Busdata, Linedata); %Computing Ybus
%%
[Gi,Ol,To,Ef] = GEPP(Candidates,Nc,Beta,PGmax,...
PGmin,X,Btype,Nl,Nr,Smax,Length,Biu,Ga,Pg,Pl,Ybus);
%Gi:Generation of candidate buses after expansion
%Ol:Overloaded lines after expansion
%To:Total overload after expansion
%%
if Ef==1
    Print_Gep
else
    fprintf('There is no feasible solution.\n');
end
```

b) "ybus" M-file code

```
function [Ybus] = ybus (Busdata, Linedata)

nbus = size(Busdata,1);
nl = Linedata(:,2);
nr = Linedata(:,3);
Ld = Linedata;
%%
j = sqrt(-1);
X = Ld(:,5);
nbr = length(Ld(:,1));
%Z = R + j*X;
Z = (j*X);
y = ones(nbr,1)./Z;     %Branch admittance
%for n = 1:nbr
Ybus = zeros(nbus,nbus);%Initialize Ybus to zero
%%
%Formation of the off diagonal elements
for k = 1:nbr;
       Ybus(nl(k),nr(k)) = Ybus(nl(k),nr(k))-y(k);
       Ybus(nr(k),nl(k)) = Ybus(nl(k),nr(k));
end
%%
%Formation of the diagonal elements
for n = 1:nbus
    for m = (n+1):nbus
        Ybus(n,n) = Ybus(n,n)-Ybus(n,m);
    end
    for m = 1:n-1
        Ybus(n,n) = Ybus(n,n)-Ybus(n,m);
    end
end
```

c) "GEPP" M-file code

```
function [Gi, Ol, To, Ef]= GEPP (Candidates, Nc, Beta,...
    PGmax,PGmin, X, Btype, Nl, Nr, Smax, Length, Biu,...
    Ga, Pg, Pl, Ybus)
if isempty(Ybus)
  error('Input argument "Ybus" is undefined.');
```

```
end
if isempty(Pg)
  fprintf('Input argument "Pl" determining');
  fprintf(' load demand of buses.\n');
  error('"Pl" is undefined and must be determined.');
end
if isempty(Pg)
  fprintf('Input argument "Pg" determining');
  fprintf(' generation of buses.\n');
  error('"Pg" is undefined and must be determined.');
end
if isempty(Ga)
  fprintf('Input argument "Ga" determining Investment ');
  fprintf('cost of transmission lines enhancement.\n');
  warning('"Ga" is undefined and is set to a default value.');
  Ga = 20;
end
if isempty(Biu)
  fprintf('Input argument "Biu" determining ');
  fprintf('maximum capacity of lines enhancement.\n');
  warning('"Biu" is undefined & is set to a default value.');
  Biu = 1.1;
end
if isempty(Length)
  fprintf('Input argument "Length" determining');
  fprintf(' path length of lines.\n');
  error('"Length" is undefined and must be determined.');
end
if isempty(Smax)
  fprintf('Input argument "Smax" defining');
  fprintf(' lines thermal loading before expansion.\n');
  error('"Smax" is undefined and must be determined.');
end
if isempty(Nr) || isempty(Nl)
  fprintf('Input argument "NL" & "Nr" defining');
  fprintf(' lines sending and ending buses.\n');
  error('"NL" & "Nr" are undefined and must be determined.');
end
if isempty(Btype)
  fprintf('Input argument "Btype" defining');
  fprintf(' information of bus types.\n');
  error('"Btype" is undefined and must be determined.');
end
if isempty(X)
  fprintf('Input argument "X" containing');
  fprintf(' data of lines reactance.\n');
  error('"X" is undefined and must be determined.');
end
if isempty(PGmin)
  fprintf('Input argument "PGmin" defining minimum ');
  fprintf('generation expansion limit of candidate buses.\n');
  error('"PGmin" is undefined and must be determined.');
end
if isempty(PGmax)
```

```
    fprintf('Input argument "PGmax" defining maximum ');
    fprintf('generation expansion limit of candidate buses.\n');
    error('"PGmax" is undefined and must be determined.');
  end
  if isempty(Beta)
    fprintf('Input argument "Beta" defining investment cost');
    fprintf('of generation expansion in candidate buses.\n');
    error('"Beta" is undefined and must be determined.');
  end
  if isempty(Candidates)
    fprintf('Input argument "Candidates" defining');
    fprintf('candidate buses.\n');
    error('"Candidates" is undefined and must be determined.');
  end
  %% Problem outputs
  %Gi:Generation of candidate buses after expansion
  %Ol:Overloaded lines after expansion
  %To:Total overload after expansion
  %Ef:Exit flag, integer identifying the reason the algorithm
      %is terminated. Ef is 1, if there is a feasible solution
  %% Problem Inputs
  %Candidates:Candidate buses for generation expansion
  %Beta(i):investment cost of generation expansion in bus i
  %PGmax(i):Maximum generation expansion limit of bus i
  %PGmin(i):Minimum generation expansion limit of bus i
  %Nlin:Line number
  %Nl:Line from bus
  %Nr:Line to bus
  %R(i):Line resistance
  %X(i):Line reactance
  %Smax(i):Maximum thermal rating of line i
  %Length(i):Path Length of Line i
  %Busn:Bus number
  %Btype:Type of bus 1-Slack, 2-PV, 3-PQ
  %Pg(i):Generation of bus i
  %Pl(i):load of bus i
  %Nc:Non-candidate buses
  %%Obtaining Ybus matrix
  %%
  Ps = (Pg-Pl);
  Na = size (Pg, 1);
  M = size (X, 1);
  %%%%
  [Nons] = find(Btype~=1);
  Nx = length(Nons);
  B = zeros (Nx,Nx);
  for k = 1:Nx
      for j = 1:Nx
          Ymn = Ybus(Nons(k),Nons(j));
          B(k,j) = -imag(Ymn);
      end
  end
  E = inv (B);
  Binv = zeros (Na,Na);
```

```
for k = 1:Nx
    an = Nons(k);
    for j = 1:Nx
        am = Nons(j);
        Binv(an,am) = E(k,j);
    end
end
%% Computing branch admittance calculation (b)
%The admittance matrix in which bii is the admittance
    % of line i and non-diagonal elements are zero
jay = sqrt(-1);
Z = (jay*X);
Y = ones(M,1)./Z;
b = zeros (M,M);
for i = 1:M
    b(i,i) = -imag(Y(i));
end
%% Computing connection matrix (A)
% The connection matrix (M*N) in which aij is 1, if a
    % line exists from bus i to bus j; otherwise zero.
A = zeros (M, Na);
for i = 1:M
    nl = Nl(i);
    nr = Nr(i);
    A(i, nl) = 1;
    A(i, nr) = -1;
end
%% Computing sensitivity matrix (a)
theta = Binv*Ps;
a = b*A*Binv;
%% The line flows are calculated as follows:
Pli = zeros (M,1);
for i = 1:M
    for k = 1:Na
        Pli(i,1) = Pli(i,1)+(a(i,k)*(Pg(k,1)-Pl(k,1)));
    end
end
%% Generation expansion cost of each bus
Pmax = zeros (Na,1);
beta = zeros (Na,1);
for j = 1:length (Nc)
    Inc = Nc(j);
    beta(Inc) = 10^10;
    Pmax(Inc) = 0.000001;
end
for j = 1:length (Candidates)
    Ica = Candidates(j);
    beta(Ica) = Beta(j);
    Pmax(Ica) = PGmax(j);
end
Beta = beta;
PGmax = Pmax;
%% Investment cost for transmission lines enhancement (R/MW)
Gama = Ga*Length;
```

```
%% Maximum possible capacity expansion of each line
Biu = Biu.*ones(M,1);
%% Thermal rating of each line
Pcu = Smax; %Upper bound of thermal rating of each line
Pcl = -Pcu; %Lower bound of thermal rating of each line
%% Defining objective function
for k = 1:Na
    OF(k) = Beta(k);
end
for i = 1:M
    I = i+Na;
    OF(I) = Gama (i);
end
%% First set of inequality constraints: determining
    %% minimum permissible thermal rating of each line
for i = 1:M
    C(i) = (-a(i,:)*Pg)+Pli(i);
end
GH1 = zeros (M, M+Na);
bGH1 = zeros (M,1);
for i = 1:M
    for k = 1:Na
        GH1(i,k) = -a(i,k);
    end
    I = i+Na;
    GH1(i,I) = Pcl(i);
    bGH1(i,1) = C(i);
end
%% Second set of inequality constraints: determining
    %% maximum permissible thermal rating of each line
GH2 = zeros (M, M+Na);
bGH2 = zeros (M,1);
for i = 1:M
    for k = 1:Na
        GH2(i,k) = a(i,k);
    end
    I = i+Na;
    GH2(i,I) = -Pcu(i);
    bGH2(i,1) = -C(i);
end
%% Integrating all inequality constraints
    %% to one matrix, called An & bn here
for i = 1:M
    An(i,:) = GH1(i,:);
    bn(i) = bGH1(i);
    I = i+M;
    An(I,:) = GH2(i,:);
    bn(I) = bGH2(i);
end
%% Determining upper and lower bounds of
    %% decision variables, called lb & ub here
lb = zeros (M+Na,1);
ub = zeros (M+Na,1);
for k = 1:Na
```

```
        lb(k,1) = PGmin(k);
        ub(k,1) = PGmax(k);
    end
    for i = 1:M
        I = i+Na;
        lb(I,1) = 1;
        ub(I,1) = Biu(i);
    end
    %% Defining equality constraint
        %% (Total generation = Total demand)
    Aeq = zeros (1, Na+M);
    for k = 1:Na
        Aeq(1,k) = 1;
    end
    beq = sum (Pl);
    %% Solving the problem and finding the optimal point
    [Dv, Fval, Ef] = linprog(OF,An,bn,Aeq,beq,lb,ub);
    To = 0;
    if Ef~=1
        fprintf('\nWARNING: No feasible solution was found.')
        Gi = zeros(size(Candidates,1),1);
        Ol = zeros(M,1);
    else
        for k = 1:size(Candidates,1)
            Gi (k,1) = Candidates(k,1);
            Gi (k,2) = Dv(k);
        end
        for i = 1:M
            I = i+Na;
            Ol (i,1) = Nl(i);
            Ol (i,2) = Nr(i);
            Ol (i,3) = Dv(I,1)-1;
            To = To+(Dv(I)-1);
        end
    end
```

d) "Print_Gep" M-file code

```
clc
fprintf('**************************************************');
fprintf('***************\n');
fprintf('Generation of each candidate bus after expansion');
fprintf(' is as follows: \n');
fprintf('**************************************************');
fprintf('***************\n');
fprintf('                |Bus number|             |Gi (p.u.)|');
for i = 1:size(Gi,1)
    fprintf('\n %18.0f % 22.2f', Gi(i,1), Gi(i,2));
end
fprintf('\n\n*********************************************');
fprintf('********************\n');
fprintf('      Total overload value and enhanced lines ');
fprintf('are as follows\n');
fprintf('*************************************************');
```

```
fprintf('****************\n');
fprintf('                      |Total overload| \n');
fprintf('%31.2f \n', To);
if To>=0.0001
    El = find (Ol(:,3)>=0.0001);
    Sel = length(El);
    fprintf('*********************************************');
    fprintf('********************\n');
    fprintf('                        |Enhanced lines|       ');
    fprintf('                 \n');

    fprintf('     |From bus|           |To bus|          ');
    fprintf('|Enhancement(%%)|');
    for i = 1:Sel
        fprintf('\n %10i  %18i  % 19.2f \n',...
            Ol(El(i),1), Ol(El(i),2), Ol(El(i),3)*100);
    end
    fprintf('\n*********************************************');
    fprintf('********************\n');
else
    fprintf('\n                         No enhanced line    ');
    fprintf('                    \n');
end

%% Printing the results in results.txt
fid = fopen('results.txt', 'wt');
fprintf(fid,'**********************************************');
fprintf(fid,'*******************\n');
fprintf(fid,...
    'Generation of each candidate bus after expansion');
fprintf(fid,'is as follows: \n');
fprintf(fid,'***********************************************');
fprintf(fid,'******************\n');
fprintf(fid,'             |Bus number|          |Gi (p.u.)|');
for i = 1:size(Gi,1)
    fprintf(fid,'\n %18.0f % 22.2f', Gi(i,1), Gi(i,2));
end
fprintf(fid,'\n\n******************************************');
fprintf(fid,'************************\n');
fprintf(fid,'      Total overload value and enhanced lines');
fprintf(fid,' are as follows\n');
fprintf(fid,'***********************************************');
fprintf(fid,'********************\n');
fprintf(fid,'                      |Total overload| \n');
fprintf(fid,'%31.2f \n', To);
if To>=0.0001
    El = find (Ol(:,3)>=0.0001);
    Sel = length(El);
    fprintf(fid,'*******************************************');
    fprintf(fid,'**********************\n');
    fprintf(fid,'                        |Enhanced lines|   ');
    fprintf(fid,'                   \n');

    fprintf(fid,'     |From bus|           |To bus|       ');
```

```
    fprintf(fid,'   |Enhancement(%%)|');
    for i = 1:Sel
        fprintf(fid,'\n %10i  %18i  % 19.2f \n',...
            Ol(El(i),1), Ol(El(i),2), Ol(El(i),3)*100);
    end
    fprintf(fid,'\n****************************************');
    fprintf(fid,'*************************\n');
else
    fprintf(fid,'\n                              No enhanced line ');
    fprintf(fid,'                              \n');
end
```

L.3 SEP.m

a) "SEP.m" M-file code

```
clear
clc
%% Required Input data
%Required load nodes data:
Ln = xlsread('Sepdata.xls', 'Load nodes');
%Required substations data:
Sub = xlsread('Sepdata.xls', 'Substations');
%Maximum possible distance between load nodes and substations:
Dmax = xlsread('Sepdata.xls', 'Dmax');
%% Data retrieval from input data
Iln = Ln (:,1);%Load node number
%Geographical position of load nodes
    %in terms of X (Lx) and Y (Ly):
Lx = Ln(:,2); Ly = Ln(:,3);
Sl = Ln(:,4); %Sl(i):The load i magnitude in MVA
%The cost of downward feeder unit length (e.g. 1 km)
    %per one unit power transfer capability (e.g. 1 MVA):
Gl = Ln(:,5);
Isub = Sub (:,1); %Substation number
%Geographical position of substations
    %in terms of X (Sx) and Y (Sy):
Sx = Sub(:,2); Sy = Sub(:,3);
Cexis = Sub(:,4); %Existing capacity of substations
%Smax(j):Maximum capacity of the jth substation
Smax = Sub(:,5);
%Gsf(j):The fixed cost of a substation (land cost)
    %for the jth candidate location
Gsf = Sub(:,6);
%Gsv(j):The variable cost of jth substation per MVA
Gsv = Sub(:,7);
%%
[Soc, Cstat_var, Cdown_line, Cstat_fix, Ctotal, XX, Ef]=...
    SEPP (Iln, Lx, Ly, Sl, Gl, Isub, Sx, Sy,...
    Cexis, Smax, Gsf, Gsv, Dmax);
%SOC:Geographical position and optimal capacity of
```

```
    %HV substations after expansion
%Cstat_var:Variable cost of HV substations
%Cstat_fix:Fixed cost of HV substations
%Cdown_line:Downward grid cost
%XX(i,j):1 means the jth load center is
    %connected to the ith substaion
if Ef == 1
    %% Printing the obtained results in the
        %% command window and results.txt
    Print_SEPP
    %% Plotting the expansion results
    Plot_SEPP
else
    fprintf...
      ('\nThere is no feasible solution for this case.\n');
End
```

b) "SEPP" M-file code

```
function [Soc, Cstat_var, Cdown_line, Cstat_fix, Ctotal,...
    XX, Ef]=SEPP (Iln, Lx, Ly, Sl, Gl, Isub, Sx, Sy, ...
    Cexis, Smax, Gsf, Gsv, Dmax);
if isempty(Dmax), Dmax=50; end
if isempty(Gsv)
  fprintf('Input argument "Gsv" containing');
  fprintf(' the variable cost of substations');
  warning('"Gsv" is undefined and is set to a default value');
  Gsv = 2500*ones(size(Sub,1),1);
end
if isempty(Gsf)
  fprintf('Input argument "Gsf" containing');
  fprintf(' the fixed cost of substations');
  warning('"Gsf" is undefined and is set to a default value');
  Gsf = 1700000*ones(size(Sub,1),1);
end
if isempty(Smax)
  fprintf('Input argument "Smax" containing');
  fprintf(' the maximum capacity of substations');
  warning('"Smax" is undefined & is set to a default value');
  Smax = 100*ones(size(Sub,1),1);
end
if isempty(Cexis)
  fprintf('Input argument "Cexis" containing');
  fprintf(' the existing capacity of substations');
  error('"Cexis" is undefined and must be determined');
end
if isempty(Sx) || isempty(Sy)
  fprintf('Input arguments "Sx" & "Sy" containing');
  fprintf(' the geographical position of substations');
  error('"Sx" & "Sy" are undefined and must be determined');
end
if isempty(Isub), Isub=1:size(Sub,1); end
if isempty(Gl)
  fprintf('Input argument "Gl" containing');
```

```
  fprintf(' the cost of downward feeder');
  warning('"Gl" is undefined and is set to a default value');
  Gl = 80*ones(size(Ln,1),1);
end
if isempty(Sl)
  fprintf('Input argument "Sl" containing');
  fprintf(' the load magnitude of each load node');
  error('"Sl" is undefined and must be determined');
end
if isempty(Lx) || isempty(Ly)
  fprintf('Input arguments "Lx" & "Ly" containing');
  fprintf(' the geographical position of load nodes');
  error('"Lx" & "Ly" are undefined and must be determined');
end
if isempty(Iln), Iln=1:size(Ln,1); end
%% Problem outputs
%SOC:Geographical position and optimal capacity
    % of HV substations after expansion
%Cstat_var:Variable cost of HV substations
%Cstat_fix:Fixed cost of HV substations
%Cdown_line:Downward grid cost
%% Problem Inputs
%Iln:Load node number
%Lx & Ly:geographical position of load nodes
    % in terms of X  and Y
%Sl(i)=The load i magnitude in MVA
%Gl:The cost of downward feeder unit length (e.g. 1 km)
    %per one unit power transfer capability (e.g. 1 MVA)
%Isub:Substation number
%Sx & Sy:Geographical position of substations
    % in terms of X (Sx) and Y (Sy)
%Cexis:Current capacity of substations
%Smax(j):Maximum capacity of the jth substation
%Gsf(j):The fixed cost of a substation (land cost)
    % for the jth candidate location
%Gsv(j):The variable cost of jth substation per MVA
%Dmax:Maximum permissible distance between
    % load nodes and substations
%%
Nl = size(Iln,1); % Number of load nodes
Ns = size(Isub,1); % Number of substations
Nls = Nl*Ns;
%% Distance matrix (computing distances between
    %%  the load nodes and the substations)
for i = 1:Nl
    for j = 1:Ns
        D(i,j) = sqrt(((Sx(j)-Lx(i))^2)+((Sy(j)-Ly(i))^2));
        if D(i,j)>Dmax
            D(i,j) = 100000000000000;
        end
    end
end
%% Objective function (forming the objective
    %% function of sep problem)
```

```
for i = 1:Nl
    for j = 1:Ns
        b = ((i-1)*Ns)+j;
        bb = (j)+(Nls);
        fc_total(b) = (Gsv(j)*Sl(i))+(Gl(i)*D(i,j));
        %       fc_total(bb) = (Gsf(j))-(Cexis(j)*Gsv(j));
        fc_total(bb) = (Gsf(j));
        fcstat_var(b) = (Gsv(j)*Sl(i));
        fcdown_line(b) = (Gl(i)*D(i,j));
    end
end
%% Forming constraints
%% Forming equality constraints
Aeq = zeros(Nl,((Nls)+Ns));
for i = 1:Nl
    for j = 1:Ns
        p = ((i-1)*Ns)+(j);
        Aeq(i,p) = 1;
    end
end
Beq=ones(Nl,1);
%% Defining different components of inequality constraints
%% Defining constraints corresponding
    %% with maximum capacity of each substaion
A2 = zeros(Ns,((Nls)+Ns));
for j = 1:Ns
    for i = 1:Nl
        bb = ((i-1)*Ns)+j;
        A2(j,bb) = ((Sl(i)));
    end
    b2(j,1) = Smax(j,1);
end
%% Defining constraints corresponding
    %% with presence of candidate substation
A3 = zeros(Ns,(Nls+Ns));
for j = 1:Ns
    for i = 1:Nl
        bb = ((i-1)*Ns)+j;
        A3(j,bb) = 1;
    end
    A3(j,(Nls+j)) = -Nl;
    b3(j,1) = 0;
end
%% Integrating all inequality constraints
    %% to one matrix, called A & B here
A = zeros((2*Ns), (Nls+Ns));
B = zeros((2*Ns), 1);
for M = 1:Ns
    A(M,:) = A2(M,:);
    B(M,1) = b2(M,1);
end
for m = 1:Ns
    MM = m+Ns;
    A(MM,:) = A3(m,:);
```

```
    B(MM,1) = b3(m,1);
end
%% Solving the problem by branch and bound solver
[x, Fval, Ef] = bintprog(fc_total, A, B, Aeq, Beq);
if Ef~=1
    fprintf('\nWARNING: No feasible solution was found ')
    Soc(:,1) = Isub(:,1); Soc(:,2) = Sx;
    Soc(:,3) = Sy; Soc(:,4) = zeros(Ns,1);
    Cstat_var = 0; Cstat_fix = 0; Cdown_line = 0;
    Ctotal = 0; XX = zeros(Nl,Ns);
else
    %% Calculating the optimal capacity of substations
        %% based on the obtained decision variables in 'x'
    for i = 1:Nl
        for j = 1:Ns
            xx(i,j) = x((((i-1)*Ns)+j),1);
        end
    end
    xx = xx'; XX = xx; %Decision variables
    clear m n
    for m = 1:Nls
        xls(m,1) = x(m,1);
    end
    for n = 1:Ns
        xs(n,1) = x(n+Nls,1);
    end
   %Computing optimal capacity of substations after expansion:
    oc = xx*Sl;
    Soc(:,1) = Isub(:,1); Soc(:,2) = Sx;
    Soc(:,3) = Sy; Soc(:,4) = oc(:,1);
    %% Calculating different components of total cost
    [iaab] = find(Cexis|0);
    iq = 0;
    for q = 1:length(iaab)
        if oc(iaab(q))<Cexis(iaab(q))
            iq = iq+1;
            ip(iq) = iaab(q);
        end
    end
    for jj = 1:iq
        for ii = 1:Nl
            bq = (((ip(jj)-1)*Nl)+ii);
            Cstat_var(bq) = 0;
        end
    end

    for jjj = 1:length(iaab)
        Cstat_fix(iaab(jjj)) = (0);
    End
    %Variable cost of installed substations:
    Cstat_var = (fcstat_var*xls)-((Cexis')*Gsv);
    %Fixed cost of installed substations:
    Cstat_fix = ((Gsf')*xs);
    %Variable cost of lines:
```

```
        Cdown_line = (fcdown_line*xls);
        Ctotal = Cstat_var+Cstat_fix+Cdown_line;
    end
    %%
```

c) "Print_SEPP" M-file code

```
%% Printing different costs
%% Printing optimal capacity of each substation
clc
Nl = size(Iln,1); Ns = size(Isub,1);
fprintf('******************************Costs*************');
fprintf('******************************\n');
fprintf('||Cstat_var       ||  Cstat_fix        ||    ');
fprintf('Cdown_line     ||    Ctotal     ||\n');
fprintf('||    (R)         ||       (R)         ||        ');
fprintf('(R)        ||         (R)       ||');
fprintf('\n %10.1f  %18.1f %19.1f % 19.1f \n',...
    Cstat_var, Cstat_fix, Cdown_line, Ctotal);

%% Printing the optimal capacity of substations
%% Printing the locations of substations
fprintf('\n');
fprintf('**********************************************');
fprintf('******************************\n');
fprintf('***The position and optimal capacity of installed');
fprintf(' substations after expansion***\n');
fprintf('**********************************************');
fprintf('******************************\n');
fprintf(' |Sub_number|     |X|           |Y|     |Optimal ');
fprintf('capacity|\n');
for i = 1:Ns
    if Soc(i,4)~=0
        fprintf('%8.f', Soc(i,1));
        fprintf('    %8.f', Soc(i,2));
        fprintf('    %8.f', Soc(i,3));
        fprintf('     %8.1f',Soc(i,4));
        fprintf('\n');
    end
end
fprintf('**********************************************');
fprintf('********\n');

%% Connected loads nodes
    %% to the selected substation after expansion
for i = 1:Ns
    if Soc(i,4)~=0
        Cln = find (XX(i,:)~=0);
        fprintf('****************************************');
        fprintf('****************\n');
        fprintf(' Connected load nodes to the substation');
        fprintf('%3.0f  are:  \n', Soc(i,1));
        fprintf('|Load_node|     |X|        |Y|     |');
        fprintf('Magnitude(MVA)|\n');
```

```
        for i = 1:length(Cln)
            fprintf('\n %6.0f % 10.0f % 8.0f %10.1f',...
                Cln(i), Lx(Cln(i)), Ly(Cln(i)), Sl(Cln(i)));
        end
        fprintf('\n************************************');
        fprintf('****************\n');
    end
end
%% Printing the results in results.txt
fid = fopen('results.txt', 'wt');
Nl = size(Iln,1); Ns = size(Isub,1); %
fprintf(fid,'*****************************Costs********');
fprintf(fid,'************************************\n');
fprintf(fid,'||Cstat_var        ||  Cstat_fix         ||     ');
fprintf(fid,'Cdown_line     ||    Ctotal     ||\n');
fprintf(fid,'||     (R)         ||       (R)          ||     ');
fprintf(fid,'     (R)       ||        (R)        ||');
fprintf(fid,'\n %10.1f  %18.1f %19.1f % 19.1f \n',...
    Cstat_var, Cstat_fix, Cdown_line, Ctotal);

fprintf(fid,'\n');
fprintf(fid,'*****************************************');
fprintf(fid,'**************************************\n');
fprintf(fid,'***The position and optimal capacity of ');
fprintf(fid,'installed substations after expansion***\n');
fprintf(fid,'*****************************************');
fprintf(fid,'**************************************\n');
fprintf(fid,' |Sub_number|     |X|            |Y|    |Optimal');
fprintf(fid,' capacity| \n');
for i = 1:Ns
    if Soc(i,4)~=0
        fprintf(fid,'%8.f', Soc(i,1));
        fprintf(fid,'    %8.f', Soc(i,2));
        fprintf(fid,'    %8.f', Soc(i,3));
        fprintf(fid,'     %8.1f',Soc(i,4));
        fprintf(fid,'\n');
    end
end
fprintf(fid,'****************************************');
fprintf(fid,'*************\n');
for i = 1:Ns
    if Soc(i,4)~=0
        Cln = find (XX(i,:)~=0);
        fprintf(fid,'********************************');
        fprintf(fid,'********************\n');
        fprintf(fid,' Connected load nodes to the ');
        fprintf(fid,'substation %3.0f  are:  \n', Soc(i,1));
        fprintf(fid,'|Load_node|     |X|       |Y|    ');
        fprintf(fid,'|Magnitude(MVA)|\n');
        for i = 1:length(Cln)
            fprintf(fid,'\n %6.0f % 10.0f % 8.0f %10.1f',...
                Cln(i), Lx(Cln(i)), Ly(Cln(i)), Sl(Cln(i)));
        end
        fprintf(fid,'\n*******************************');
```

```
        fprintf(fid,'**********************\n');
    end
end
%%
```

d) "Plot_SEPP" M-file code

```
hold off
format short
xx = XX;
[jjj,iii] = find(xx==1);
for I = 1:Ns
    [II] = find(jjj==I);
    for J = 1:length(II)
        S_LC(J,I) = iii(II(J));
    end
    clear II
end
z = sum(xx,2);
iz = find(z|0);
izn = find(z==0);
niz = length(iz);
nizn = length(izn);
for bb = 1:niz
subposx(1,bb) = Sx((iz(bb)),1);
subposy(1,bb) = Sy((iz(bb)),1);
end
for bb1 = 1:nizn
  nsubposx(1,bb1) = Sx((izn(bb1)),1);
  nsubposy(1,bb1) = Sy((izn(bb1)),1);
end
for ba = 1:Nl
    loadposx(1,ba) = Lx(ba,1);
    loadposy(1,ba) = Ly(ba,1);
end
Aa = cell(niz,2);
for ia = 1:niz
    Aax = xx(iz(ia),:);
    [iax] = find(Aax==1);
    niax = length(iax);
    for ja = 1:niax
        jab = (2*ja)-1;
        Aaa(1,jab) = Sx(iz(ia));
        Bbb(1,jab) = Sy(iz(ia));
        jaa = (2*ja);
        Aaa(1,jaa) = Lx(iax(ja));
        Bbb(1,jaa) = Ly(iax(ja));
        Aa{ia,1} = Aaa;
        Aa{ia,2} = Bbb;
    end
     clear Aaa Bbb
end
%% Plotting the location of installed & current substations
figure(1)
```

```
subplot(2,2,1)
plot(subposx,subposy,'sb')
xlabel('X Axis')
ylabel('Y Axis')
axis([0 100 0 100])
title('Location of selected candidate substations')
grid on
%% Plotting the location of uninstalled candidate substations
subplot(2,2,2)
plot(nsubposx,nsubposy,'sr')
xlabel('X Axis')
ylabel('Y Axis')
axis([0 100 0 100])
title('Location of unselected candidate substations')
grid on
%% Plotting the location of load nodes
subplot(2,2,3)
plot (loadposx,loadposy,'ok')
xlabel('X Axis')
ylabel('Y Axis')
axis([0 100 0 100])
title('Location of load nodes')
grid on
%% Plotting the location of selected substations, load nodes
    %% & downward lines
subplot(2,2,4)
plot(nsubposx,nsubposy,'sr')
xlabel('X Axis')
ylabel('Y Axis')
axis([0 100 0 100])
title('Selected substations, load nodes & downward lines')
hold on
plot(subposx,subposy,'sb')
hold on
plot(nsubposx,nsubposy,'sr')
hold on
plot (loadposx,loadposy,'ok')
grid on
%% Plotting the Position of selected substations, load nodes
    %% & downward lines
hold on
for iia = 1:niz
    Aab = Aa{iia,1};
    Bba = Aa{iia,2};
    plot(Aab,Bba,'m')
    hold on
end
figure (2)
plot(nsubposx,nsubposy,'sr')
xlabel('X Axis')
ylabel('Y Axis')
axis([0 100 0 100])
title('Selected substations, load nodes & downward lines')
hold on
```

```
plot(subposx,subposy,'sb')
hold on
plot(nsubposx,nsubposy,'sr')
hold on
plot (loadposx,loadposy,'ok')
hold on
for iia = 1:niz
    Aab = Aa{iia,1};
    Bba = Aa{iia,2};
    plot(Aab,Bba,'m')
    hold on
end
grid on
```

L.4 NEP.m

a) "Hybridsearch" M-file code

```
clear
clc
%% Reading the input data %%
%% Reading data of the network buses:
Busdata = xlsread('Nepdata.xls', 'Busdata');
%% Reading data of the network lines:
Linedata = xlsread('Nepdata.xls', 'Linedata');
%% Reading data of the candidate lines:
Candid = xlsread('Nepdata.xls', 'CandidateLinedata');
%% Reading the information of defined line types:
Linetype = xlsread('Nepdata.xls', 'LineType');
inputs = xlsread('Nepdata.xls', 'Otherinputs');
%% Lg: load growth rate (Lg=1 means 100% load growth):
Lg = inputs(1,1);
%% Mof: minimum fitness, which is kept at high value for
    % the first iteration of the forward search algorithm
Mof = inputs(1,2);
Solution = ones (size(Candid,1),1);
%% Calling the hybrid search algorithm to solve the NEP prob-
lem
[Os, Adline, Noll, Coll, Angle, Mof]=...
    HS(Busdata, Linedata, Candid, Linetype, Solution, Lg,
Mof);
%% Os: optimal solution of the NEP problem
%% Adline: final set of selected candidate lines
    % among all candidates
%% Noll: overload of the existing and selected candidate
    % lines in normal condition after adding optimal candidate
    % line in each iteration (or in order of priority)
%% Coll: overload of the existing and selected candidate lines
    % in N-1 condition after adding optimal candidate line
    % in each iteration
%% Angle: voltage phase of all buses for adding the best
```

```
    % candidate line to the network
%%  Printing and saving the obtained results in result.txt
    % in the corresponding directory
Print_Nep
```

b) "Forwardsearch" M-file code

```
clear
clc
%% Reading the Input Data %%
%% Reading data of the network buses
Busdata = xlsread('Nepdata.xls', 'Busdata');
%% Reading data of the network lines
Linedata = xlsread('Nepdata.xls', 'Linedata');
%% Reading data of the candidate lines
Candid = xlsread('Nepdata.xls', 'CandidateLinedata');
%% Reading the information of defined line types
Linetype = xlsread('NEPdata.xls', 'LineType');
inputs = xlsread('Nepdata.xls', 'Otherinputs');
%% Lg: load growth rate Lg=1 means 100% load growth
Lg = inputs(1,1);
%% Mof: minimum fitness, which is kept at high value for
    % the first iteration of the forward search algorithm
Mof = inputs(1,2);
%% Contingency=1 means the problem is solved,
    % considering N-1 condition
Contingency = inputs(1,3);
%% Forward search starts with base network (no candidate line
    % is added to the base network at the beginning)
Solution=zeros (size(Candid,1),1);
%% Calling the forward search algorithm
    % to solve the NEP problem
[Os, Adline, Noll, Coll, Angle] = FS(Busdata, Linedata, ...
    Candid, Linetype, Solution, Contingency, Lg, Mof);
%% Os: optimal solution of the NEP problem
%% Adline: final set of selected candidate lines
    % among all candidates
%% Noll: overload of the existing and selected candidate lines
    % in normal condition after adding optimal candidate line
    % in each iteration (or in order of priority)
%% Coll: overload of the existing and selected candidate lines
    % in N-1 condition after adding optimal candidate line
    % in each iteration
%% Angle: voltage phase of all buses for adding the best
    % candidate line to the network

%%  Printing and saving the obtained results in result.txt in
    % the corresponding directory and in the command window
Print_NEP
```

c) "Backwardsearch" M-file code

```
clear
clc
%% Reading the input data %%
```

```
%% Reading data of the network buses
Busdata = xlsread('Nepdata.xls', 'Busdata');
%% Reading data of the network lines
Linedata = xlsread('Nepdata.xls', 'Linedata');
%% Reading data of the candidate lines
Candid = xlsread('Nepdata.xls', 'CandidateLinedata');
%% Reading the information of defined line types
Linetype = xlsread('NEPdata.xls', 'LineType');
inputs=xlsread('Nepdata.xls', 'Otherinputs');
%% Lg: load growth rate, Lg=1 means 100% load growth
Lg = inputs(1,1);
%% Mof: minimum fitness, which is kept at high value for
    % the first iteration of the forward search algorithm
Mof = inputs(1,2);
%% Contingency=1 means the problem is solved, considering
    % N-1 condition.
Contingency = inputs(1,3);
%% Backward search starts with considering all candidate
    % lines added to the base network at the beginning)
Solution = ones (size(Candid,1),1);
%% Calling the backward search algorithm to
    % solve the NEP problem
[Os, Adline, Noll, Coll, Angle, Mof] = BS(Busdata,...
    Linedata, Candid, Linetype, Solution, ...
    Contingency, Lg, Mof);
%% Os: optimal solution of the NEP problem
%% Adline: final set of selected candidate lines
    % among all candidates
%% Noll: overload of the existing and selected candidate lines
    % in normal condition after adding optimal candidate line
    % in each iteration (or in order of priority)
%% Coll: overload of the existing and selected candidate lines
    % in N-1 condition after adding optimal candidate line
    % in each iteration
%% Angle: voltage phase of all buses for adding the best
    % candidate line to the network

%% Printing and saving the obtained results in result.txt
    % in the corresponding directory
Print_NEP
```

d) "HS" M-file code

```
function [Os, Adline, Noll, Coll, Angle,Mof] = ...
    HS(Busdata, Linedata, Candid, Linetype, Solution, Lg, Mof)
if nargin<7 | isempty(Mof), Mof = 10^20;          end
if nargin<6 | isempty(Lg),  Lg = 0; Mof = 10^20; end
if nargin<5 | isempty(Solution)
    Solution = ones(size(Candid,1),1); Lg = 0; Mof = 10^20;
end
if nargin<4 | isempty(Linetype)
    fprintf('Input argument "Linetype" containing the');
    fprintf(' information of different types of lines.');
    error('"Linetype" is undefined.');
```

```
end
if nargin<3 | isempty(Candid)
    fprintf('Input argument "Candid" containing');
    fprintf(' the information of candidate lines.');
    error('"Candid" is undefined.');
end
if nargin<2 | isempty(Linedata)
    fprintf('Input argument "Linedata" containing');
    fprintf(' the information of existing lines.');
    error('"Linedata" is undefined.');
end
%% Problem outputs:

%% Os: Optimal solution of the NEP problem
%% Adline: final set of selected candidate lines among all
    % candidates.
%% Noll: overload of the existing and selected candidate lines
    % in normal condition after adding optimal candidate line
    % in each iteration (or in order of priority)
%% Coll: overload of the existing and selected candidate lines
    % in N-1 condition after adding optimal candidate line
    % in each iteration
%% Angle: voltage phase of all buses for adding the best
    % candidate line to the network

%% Problem inputs:

%% Busdata: data of the network buses
%% Linedata: data of the network lines
%% Candid: data of candidate lines
%% Linetype: data of different line types
%% Solution: the initial solution, which is a zero vector
    % for hybrid search algorithm
%% Contingency: if contingency=1, the problem is solved by
    % considering N-1 condition.
%% Lg: load growth rate
%% Mof: minimum fitness, which is kept at high value for
    % the first iteration of the forward search algorithm

contingency = 0;
[OSB, added_lineB, NOLLB, COLLB, AngleB, MOFB] = ...
    BS(Busdata, Linedata, Candid, Linetype, Solution,...
        contingency, Lg, Mof);
contingency = 1;
[Os, Adline, Noll, Coll, Angle, Mof] = FS(Busdata, ...
    Linedata, Candid, Linetype, OSB, contingency, Lg, Mof);
if sum(Os-OSB) == 0
    Angle = AngleB;
    Noll = NOLLB;
    Coll = COLLB;
    Mof = MOFB;
    Adline = added_lineB
    Os = OSB;
End
```

e) "BS" M-file code

```
function[Os, Adline, Noll, Coll, Angle, Mof] = BS ...
    (Busdata, Linedata, Candid, Linetype, Solution, ...
    Contingency, Lg, Mof);
if nargin<8 | isempty(Mof), Mof = 10^20;          end
if nargin<7 | isempty(Lg),  Lg = 0; Mof = 10^20; end
if nargin<6 | isempty(Contingency)
    Contingency = 0; Lg = 0; Mof = 10^20;
end
if nargin<5 | isempty(Solution)
    Solution = ones (size(Candid,1),1);
    Contingency = 0; Lg = 0; Mof = 10^20;
end
if nargin<4 | isempty(Linetype)
    fprintf('Input argument "Linetype" containing the');
    fprintf(' information of different types of lines.');
    error('"Linetype" is undefined.');
end
if nargin<3 | isempty(Candid)
    fprintf('Input argument "Candid" containing the');
    fprintf(' information of candidate lines.');
    error('"Candid" is undefined.');
end
if nargin<2 | isempty(Linedata)
    fprintf('Input argument "Linedata" containing');
    fprintf(' the information of existing lines.');
    error('"Linedata" is undefined.');
end

%% Problem outputs:

%% Os: optimal solution of the NEP problem
%% Adline: final set of selected candidate lines among
    % all candidates.
%% Noll: overload of the existing and selected candidate
    % lines in normal condition after adding optimal candidate
    % line in each iteration (or in order of priority)
%% Coll: overload of the existing and selected candidate lines
    % in N-1 condition after adding optimal candidate line
    % in each iteration
%% Angle: voltage phase of all buses for adding the best
    % candidate line to the network

%% Problem inputs:

%% Busdata: data of the network buses
%% Linedata: data of the network lines
%% Candid: data of candidate lines
%% Linetype: data of different line types
%% Solution: the initial solution, which is a zero vector
    % for hybrid search algorithm
%% Contingency: if contingency = 1, the problem is solved
    % by considering N-1 condition
```

```
%% Lg: load growth rate
%% Mof: minimum fitness, which is kept at high value for
    % the first iteration of the forward search algorithm

nc = size (find(Solution ~= 0),1);

%% Backward search algorithm %%
%% Initialization
diff = 1; SID = 0; j = 1;
ii = 0; jj = 0; kk = 0;
Noll = null(1); Coll = null(1);

while diff>0 | j<=2^nc
    Solution1 = Solution;
    [isol] = find(Solution1 ~= 0);
    best_sol = null(1);

%% Adding all candidate lines to the present set of lines and
    % finding the best possible candidate to be eliminated
    % from the set of present and added candidate lines.
    % This step is iterated untill the the obtained fitness
    % function doesn't decreas.
    for i = 1:length (isol)
        Isol = isol(i);
        Solution1 (Isol) = 0;
%% Updating corresponding line data and bus data according
    % to the eliminated candidate line; constructing ybus;
    % computing number of islands
    % after each candidate is eliminated from the network.
        [Ybus, linedata, busdata, nIs, nbus, bus_number] ...
            = ybus_calculation(Busdata, Linedata, ...
            Solution1, Candid, Linetype, Lg);
%% busdata:Updated bus data after considering new candidates
%% linedata:Updated line data after considering new andidates
%% Running DC Power flow for updated line data and bus data
    % to obtain total overload in the normal condition
        [angle_r, angle_d, PF, OL, SOL] = ...
            dcpf(busdata, linedata, Ybus);
%% NOL{i,1}: total overload in case of eliminating the i-th
    % candidate line among the added candidates
        NOL{i,1} = OL;
        angle{i,1} = angle_r;
%% Computing the total cost (TC) after eliminating
    % each candidate line
        Isoln = find(Solution1~=0);
%% TC: Total Cost
        [TC] = Total_Cost(Isoln, Solution1, Candid, Linetype);
%% Computing total overload in N-1 condition after eliminating
    % each candidate line

%% If N-1 condition is considered in the algorithm and there
    % is no island in normal condition
        if Contingency == 1 & nIs == 0
            [COL, CnIs, OLF] = contingency(linedata, busdata);
```

```
%% OOLF{i,1}: total overload in N-1 condition, in case of
    % eliminating the i-th candidate line among not
    % selected candidates
            OOLF{i,1} = OLF;
        else
            COL = 0; CnIs = 0;
        end
        nline = size (linedata,1);
%% Formation of fitness function (OF: NEP Objective Function)
        OF = TC+(10^9*((SOL)+COL))+(10^12*((nIs)+(CnIs)));

        if OF < Mof
            diff = (Mof-OF);
            Mof = OF;
            best_sol = Isol;
            j = j+1;
        else
            j = j+1;
        end

%%  Eliminating the worst candidate line from the set of
    % candidate lines; retrieval the power flow and
    % overloaddata corresponding with the selected candidate
    % of each iteration
        Solution1(Isol) = Candid(Isol,6);
    end
    best_sol_index = isempty(best_sol);
    if best_sol_index == 1;
        break
    else
        Solution(best_sol) = 0;
        ii = ii+1;
        best(ii,1) = best_sol;
        best(ii,2) = Mof;
        if Contingency == 1
            jj = jj+1;
            bsol = find (isol == best_sol);
            Coll{jj,1} = OOLF{bsol,1};
            kk = kk+1;
            Noll{kk,1} = NOL{bsol,1};
            Angle{kk,1} = angle{bsol,1};
            clear angle NOL
        else
            kk = kk+1;
            bsol = find(isol == best_sol);
            Noll{kk,1} = NOL{bsol,1};
            Angle{kk,1} = angle{bsol,1};
            clear angle NOL
        end
    end
end
%% Adline: final set of selected candidate lines
    % among all candidates
Os = Solution; % Optimal solution
```

```
al = find(Os~=0);
if length(al)~=0;
    lb = length(best);
    for i = 1:length(al)
        Adline(i,1) = Candid(al(i),2);
        Adline(i,2) = Candid(al(i),3);
    end
    for i = 1:lb
        removed_line(i,1) = Candid(best(i),2);
        removed_line(i,2) = Candid(best(i),3);
        removed_line(i,3) = (best(i,2)/10^7);
    end
else
    Adline = null(1);
end
```

f) "FS" M-file code

```
function[Os, Adline, Noll, Coll, Angle, Mof] = FS...
    (Busdata, Linedata, Candid, Linetype, Solution, ...
    Contingency, Lg, Mof)
if nargin<8 | isempty(Mof), Mof = 10^9;          end
if nargin<7 | isempty(Lg),  Lg = 0; Mof = 10^9; end
if nargin<6 | isempty(Contingency)
    Contingency = 0; Lg = 0; Mof = 10^9;
end
if nargin<5 | isempty(Solution)
    Solution = zeros(size(Candid,1),1);
    Contingency = 0; Lg = 0; Mof = 10^9;
end
if nargin<4 | isempty(Linetype)
    fprintf('Input argument "Linetype" containing the');
    fprintf(' information of different types of lines.');
    error('"Linetype" is undefined.');
end
if nargin<3 | isempty(Candid)
    fprintf('Input argument "Candid" containing the');
    fprintf(' information of candidate lines.');
    error('"Candid" is undefined.');
end
if nargin<2 | isempty(Linedata)
    fprintf('Input argument "Linedata" containing the');
    fprintf(' information of existing lines.');
    error('"Linedata" is undefined.');
end

%% Problem outputs:

%% Os: optimal solution of the NEP problem
%% Adline: final set of selected candidate lines
    % among all candidates
%% Noll:overload of the existing and selected candidate lines
    % in normal condition after adding optimal candidate line
    % in each iteration (or in order of priority)
```

```
%% Coll:overload of the existing and selected candidate lines
    % in N-1 condition after adding optimal candidate line
    % in each iteration
%% Angle: voltage phase of all buses for adding the best
    % candidate line to the network

%% Problem inputs:

%% Busdata: data of the network buses
%% Linedata: data of the network lines
%% Candid: data of candidate lines
%% Linetype: data of different line types
%% Solution: the initial solution, which is a zero vector for
    % hybrid search algorithm
%% Contingency: if contingency=1, the problem is solved by
    % considering N-1 condition
%% Lg: load growth rate
%% Mof: minimum fitness, which is kept at high value for
    % the first iteration of the forward search algorithm

ncr = length (find(Solution == 0));
%% Forward search algorithm %%
%% Initialization
diff = 1; j = 1; ii = 0; jj = 0;
kk = 0; Noll = null(1); Coll = null(1);
best = null(1); Angle = null(1);
%%
while diff>0 |j<=2^ncr
    Solution1 = Solution;
    %% Finding not selected candidate lines
    [isol] = find(Solution1 == 0);
    best_sol = null(1);

%% Adding each candidate among non-selected candidates to
    % the previously selected lines and finding the best
    % possible candidate for joining to the set of current
    % and previously selected lines.
    % This step is iterated untill the the obtained fitness
    % function doesn't decreas.
    for i = 1:length (isol)
        Isol = isol(i); % Selecting a candidate
        Solution1(Isol) = Candid(Isol,6);
%% Updating corresponding line data and bus data according
    % to the added candidate line; constructing ybus;
    % computing number of islands after each candidate is
    % added to the network

        [Ybus, linedata, busdata, nIs, nbus, bus_number]...
            = ybus_calculation (Busdata, Linedata, ...
            Solution1, Candid, Linetype, Lg);
%% busdata:updated bus data after considering new candidates
%% linedata:updated line data after considering new andidates
%% Running DC Power flow for updated line data and bus data
    % to obtain total overload in the normal condition
```

```
        [angle_r,angle_d, PF, OL, SOL] = ...
            dcpf(busdata, linedata, Ybus);
        NOL{i,1} = OL;
        angle{i,1} = angle_r;
%% Computing Total Cost (TC) for adding each candidate line
        Isoln = find(Solution1~=0);
%% TC: Total Cost
        [TC] = Total_Cost(Isoln,Solution1,Candid,Linetype);
%% Computing total overload in N-1 condition for
    % adding each candidate line
%% If N-1 condition is considered in the algorithm and there
    % is no island in normal condition
if Contingency == 1 & nIs == 0
            [COL,CnIs,OLF] = contingency(linedata,busdata);
%% OOLF{i,1}: total overload in N-1 condition, in case of
    % adding the i-th candidate line among not
    % selected candidates
            OOLF{i,1} = OLF;
        else
            COL = 0; CnIs = 0;
        end
%% Formation of fitness function
%% OF: NEP Objective Function
        OF = TC+(10^9*((SOL)+COL))+(10^12*((nIs)+(CnIs)));
%% Finding the best candidate (which has the least fitness)
    % by comparing OF (fitness of the i-th candidate) with
    % Mof (minimum fitness)
        if OF < Mof
            diff = (Mof-OF);
            Mof = OF; %%
            best_sol = Isol;
            j = j+1;
        else
            j = j+1;
        end
        Solution1(Isol) = 0;
    end
%% Adding the best candidate line to the set of present and
    % previously selected lines by eliminating the selected
    % candidate from the set of candidate lines retrieval
    % the power flow and overload data corresponding with
    % the selected candidate of each iteration
    best_sol_index = isempty(best_sol);
    if best_sol_index == 1;
        break
    else
        Solution(best_sol) = Candid(best_sol,6);
        ii = ii+1;
        best(ii,1) = best_sol;
        best(ii,2) = Mof;
        if Contingency == 1
            jj = jj+1;
            bsol = find (isol == best_sol);
            Coll{jj,1} = OOLF{bsol,1};
```

```
                kk = kk+1;
                Noll{kk,1} = NOL{bsol,1};
                Angle{kk,1} = angle{bsol,1};
                clear angle NOL
            else
                kk = kk+1;
                bsol = find (isol == best_sol);
                Noll{kk,1} = NOL{bsol,1};
                Angle{kk,1} = angle{bsol,1};
                clear angle NOL
            end
        end
    end
%%%%%%%%%%%%%%%%%%%%%%%%%%%%%%%%%%%%%%%%%%%%%%%%%%%%%%%%%%%%
%% added_line: final set of selected candidate lines
    % among all candidates.
Os = Solution;  % Optimal solution
al = find(Os~=0);
if length(al)~=0;
    lb = length(best);
    for i = 1:length(al)
        Adline(i,1) = Candid(al(i),2);
        Adline(i,2) = Candid(al(i),3);
    end
else
    Adline = null(1);
end
```

g) "print_NEP" M-file code

```
fid = fopen('results.txt', 'wt');
fprintf(fid, '---------------------------------------');
fprintf(fid,'\n Added candidate lines are as follows:\n');
fprintf(fid, '--------------------------------------\n');
fprintf(fid, '    From bus     To bus\n');
fprintf(fid, '    --------        -------');
fprintf(fid, '\n %10.0f %15.0f', Adline');
fprintf(fid, '\n\n*******************************************');
fprintf(fid, '******************************');
fprintf(fid,'\n                             Normal');
fprintf(fid,' condition\n');
fprintf(fid, '*********************************************');
fprintf(fid, '*************************\n');
if (isempty(Noll) == 1)
    fprintf(fid, 'No overloaded in normal condition\n');
else
    NNOLL = Noll{size (Noll,1),1};
    NL = size (Linedata,1);
    NS = length (find (Os ~= 0));
    fprintf(fid, '\Overload at normal');
    fprintf(fid,'Total overload of normal condition is');
    fprintf(fid,'3.5f pu\n',sum(NNOLL(:,4)));
    fprintf(fid, '*****************************************');
    fprintf(fid, '******************************');
```

```
      fprintf(fid, '\n************************* Candidate');
      fprintf(fid, 'branches ***************************\n');
      fprintf(fid, '********************************************');
      fprintf(fid, '*********************************\n');
      fprintf(fid, '\n  From bus      To bus        Line flow');
      fprintf(fid, ' (pu)       Overload (pu)\n');
      for i = 1:NS
          fprintf(fid, '\n %10.0f %15.0f %20.5f %20.5f\n',...
              NNOLL(i+NL,:));
      end
      fprintf(fid, '*******************************************');
      fprintf(fid, '***********************************');
      fprintf(fid, '\n************************** Existing');
      fprintf(fid, ' branches ****************************\n');
      fprintf(fid, '*******************************************');
      fprintf(fid, '***********************************\n');
      fprintf(fid, '\n  From bus      To bus        Line flow');
      fprintf(fid, ' (pu)       Overload (pu)\n');
      for i=1:NL
          fprintf(fid, '\n %10.0f %15.0f %20.5f %20.5f\n',...
              NNOLL(i,:));
      end
              fprintf(fid, '\n********************************');
              fprintf(fid, '**********************************');
              fprintf(fid, '**********\n');
  end
  nco = size (Coll,1);
  oc = Coll{nco,1};
  nc = size (oc,1);
  ocl = 0;
  for i = 1:nc
      occ = oc{i,1};
      ocl = ocl+occ(1,3);
  end
  if (ocl == 0)
      fprintf(fid, '\n No overload in N-1 condition\n');
      fprintf(fid, '\n*************************************\n');
  else
              fprintf(fid, '\n********************************');
              fprintf(fid, '**********************************');
              fprintf(fid, '**********\n');
      LCOLL = Coll{size (Coll,1),1};
      fprintf(fid, '                Overloaded lines in');
      fprintf(fid, ' N-1 condition');
      for i = 1: size (LCOLL,1)
          iLCOLL = LCOLL{i,1};
          iL = iLCOLL(1,:);
          if iL(1,3) ~= 0
              fprintf(fid, '\n********************************');
              fprintf(fid, '**********************************');
              fprintf(fid, '**********\n');
              fprintf(fid,'Total overload for outage of line');
              fprintf(fid, ': from bus');
              fprintf(fid, '%3.0f to bus %3.0f is %6.5f\n',iL);
```

```
                fprintf(fid, '*********************************');
                fprintf(fid, '*********************************');
                fprintf(fid, '*******\n');
                fprintf(fid,' Following lines are overloaded');
                fprintf(fid,' in this outage\n');
                fprintf(fid, '*******************************');
                fprintf(fid, '*******************************');
                fprintf(fid, '***********\n');
                fprintf(fid, '  From bus  To bus');
                fprintf(fid, '          Overload (pu)\n');
                fprintf(fid, '  *******  ******        *****');
                fprintf(fid, '******');
                for j = 2:size (iLCOLL,1);
                    fprintf(fid, '\n %6.0f %7.0f %18.5f\n',...
                        iLCOLL(j,:));
                end
            end
        end
    end
    LAngle = Angle{size (Angle,1),1};
    fprintf(fid,'\n****************************************\n');
    fprintf(fid,'\n***************Bus data*****************\n');
    fprintf(fid, '     No. bus          Voltage angle (Rad)\n');
    fprintf(fid, '***************       *******************\n');
    for i = 1:size (Busdata,1);
        fprintf(fid, '\n %10.0f %27.5f \n', i, LAngle(i,:));
    end
    fclose(fid);
    fid = fopen('results1.txt', 'wt');
    fprintf(fid, '--------------------------------------');
    fprintf(fid, '\n Added candidate lines are as follows:\n');
    fprintf(fid, '--------------------------------------\n');
    fprintf(fid, '    From bus     To bus\n');
    fprintf(fid, '    --------     ------');
    fprintf(fid, '\n %8.0f %11.0f', Adline');
    fprintf(fid, '\n\n******************************************');
    fprintf(fid, '*****************************');
    fprintf(fid, '\n                             Normal');
    fprintf(fid, ' condition\n');
    fprintf(fid, '********************************************');
    fprintf(fid, '************************\n');
    if (isempty(Noll) == 1)
        fprintf(fid, 'No overloaded in normal condition\n');
    else
        NNOLL = Noll{size (Noll,1),1};
        NL = size (Linedata,1);
        NS = length (find (Os~=0));
        fprintf(fid, '\Overload at normal');
        fprintf(fid, 'Total overload of normal condition is');
        fprintf(fid, '%3.5f pu\n',sum(NNOLL(:,4)));
        fprintf(fid, '*****************************************');
        fprintf(fid, '*******************************');
        fprintf(fid, '\n************************ Candidate');
        fprintf(fid, ' branches *************************\n');
```

```
        fprintf(fid, '***************************************');
        fprintf(fid, '**********************************\n');
        fprintf(fid, '\n  From bus       To bus        Line flow');
        fprintf(fid, ' (pu)       Overload (pu)\n');
        for i=1:NS
            fprintf(fid, '\n %6.0f %10.0f %20.5f %20.5f\n', ...
                NNOLL(i+NL,:));
        end
        fprintf(fid, '******************************************');
        fprintf(fid, '*********************************');
        fprintf(fid, '\n************************* Existing');
        fprintf(fid, ' branches **************************\n');
        fprintf(fid, '******************************************');
        fprintf(fid, '********************************\n');
        fprintf(fid, '\n  From bus       To bus        Line flow');
        fprintf(fid, ' (pu)       Overload (pu)\n');
        for i=1:NL
            fprintf(fid, '\n %6.0f %10.0f %20.5f %20.5f\n',...
                NNOLL(i,:));
        end
            fprintf(fid, '\n**********************************');
            fprintf(fid, '************************************');
            fprintf(fid, '*****\n');
    end
    nco=size (Coll,1);
    oc=Coll{nco,1};
    nc=size (oc,1);
    ocl=0;
    for i=1:nc
        occ=oc{i,1};
        ocl=ocl+occ(1,3);
    end
    if (ocl==0)
        fprintf(fid, '\n No overload in N-1 condition\n');
        fprintf(fid, '\n*************************************\n');
    else
        fprintf(fid, '\n****************************************');
        fprintf(fid, '**********************************\n');
        LCOLL=Coll{size (Coll,1),1};
        fprintf(fid, '                Overloaded lines in N-1');
        fprintf(fid, ' condition');
        for i=1: size (LCOLL,1)
            iLCOLL=LCOLL{i,1};
            iL=iLCOLL(1,:);
            if iL(1,3)~=0
                fprintf(fid, '\n*********************************');
                fprintf(fid, '***********************************');
                fprintf(fid, '*******\n');
                fprintf(fid,' Total overload for outage of line');
                fprintf(fid, ': from bus');
                fprintf(fid, ' %3.0f to bus %3.0f is%6.5f\n',iL);
                fprintf(fid, '***********************************');
                fprintf(fid, '***********************************');
                fprintf(fid, '*****\n');
```

```
                fprintf(fid,' Following lines are overloaded in');
                fprintf(fid, ' this outage\n');
                fprintf(fid, '**********************************');
                fprintf(fid, '**********************************');
                fprintf(fid, '*****\n');
                fprintf(fid, '  From bus  To bus');
                fprintf(fid, '        Overload (pu)\n');
                fprintf(fid, '  *******  ******        *******');
                fprintf(fid, '****');
                for j=2:size (iLCOLL,1);
                    fprintf(fid, '\n %6.0f %7.0f %18.5f\n',...
                        iLCOLL(j,:));
                end
            end
        end
    end
    LAngle=Angle{size (Angle,1),1};
    fprintf(fid,'\n****************************************\n');
    fprintf(fid,'\n**************Bus Data*****************\n');
    fprintf(fid, '     No. Bus          Voltage Angle (Rad)\n');
    fprintf(fid, '***************       ******************\n');
    for i=1:size (Busdata,1);
        fprintf(fid, '\n %10.0f %27.5f \n', i, LAngle(i,:));
    end
    fclose(fid);
    clc
    type results1.txt
    delete results1.txt
```

h) "ybus_calculation" M-file code

```
function [Ybus, linedata, busdata, nIs, nbus, bus_number]...
     = ybus_calculation(Busdata, Linedata, Solution, ...
     CandidateLinedata, LineType, Lg);
if isempty(Lg), Lg = 0; end
if isempty(Linedata)
    fprintf('Input argument "Linedata" containing the');
    fprintf(' information of network lines.');
    error('"Linedata" is undefined.');
end
if isempty(Busdata)
    fprintf('Input argument "Busdata" containing the');
    fprintf(' information of network buses.');
    error('"Busdata" is undefined.');
end
%if nargin<3 | isempty(Solution), linedata = Linedata; end ??

%% Problem outputs:

%% Ybus: admittance matrix
%% Bdata: data of network buses after considering load growth
%% Ldata: data of network lines after adding candidate lines
%% Nis: number of islands in the base network
%% Nbus: number of buses
```

```
%% Problem inputs:

%% Busdata: data of the network buses
%% Linedata: data of the network lines
%% Candid: candidate lines
%% Linetype: data of different line types
%% Lg: load growth rate

Bd = Busdata;
Ld = Linedata;
Sol = Solution;
Cl = CandidateLinedata;
Lt = LineType;
%% Finding suggested solutions %%
Iz = find (Solution~=0);
nIz = length(Iz);
nline = size (Linedata,1);

for i = 1:nIz
    can(1,1) = size (Linedata,1)+i; can(1,2) = Cl(Iz(i),2);
    can(1,3) = Cl(Iz(i),3);
    %candid(1,4)=(Lt((Cl(Iz(i),4)),2)*Cl(Iz(i),5))/...
    %     (Cl(Iz(i),6));
    can(1,4) = 0;
    can(1,5) = (Lt((Cl(Iz(i),4)),3)*Cl(Iz(i),5))/...
        (Cl(Iz(i),6));
    can(1,6) = (Lt((Cl(Iz(i),4)),4))*(Cl(Iz(i),6));
    can(1,7) = Cl(Iz(i),5);
    Ld(nline+i,:) = can(1,:);
end
linedata = Ld;
exl = size (Linedata,1);
%% Islanding detedtion and uodating busdata
busnumber = Bd(:,1);
nl = Ld(:,2);
nr = Ld(:,3);
nlr = union(nl,nr);
%Is = setdiff(nlr,busnumber);
Is = setxor(nlr,busnumber);
bus_number = setxor(busnumber,Is);
nbus = length(bus_number);
nIs = length (Is);
for i = 1:nbus
    busdata (i,:) = Bd(bus_number(i),:);
end
busdata(:,4) = busdata(:,4).*(1+Lg); busdata(:,5) = ...
    busdata(:,5).*(1+Lg);

j = sqrt(-1);
i = sqrt(-1);
X = Ld(:,5);
nbr = length(Ld(:,1));
Z = (j*X);
```

```
y = ones(nbr,1)./Z;          % Branch admittance
Ybus = zeros(nbus,nbus);     % Initialize Ybus to zero
%% Formation of the off diagonal elements
for k = 1:nbr;
       Ybus(nl(k),nr(k)) = Ybus(nl(k),nr(k))-y(k);
       Ybus(nr(k),nl(k)) = Ybus(nl(k),nr(k));
end
%% Formation of the diagonal elements
for n = 1:nbus
    for m = (n+1):nbus
        Ybus(n,n) = Ybus(n,n)-Ybus(n,m);
    end
    for m = 1:n-1
        Ybus(n,n) = Ybus(n,n)-Ybus(n,m);
    end
end
```

i) "dcpf" M-file code

```
function [angle_r,angle_d, PF, OL, SOL] = ...
    dcpf(busdata, linedata, Ybus)
if nargin<3 | isempty(Ybus)
    error('Input argument "Ybus" is undefined');
end
if nargin<2 | isempty(linedata)
    fprintf('Input argument "Linedata" containing the');
    fprintf(' information of lines.');
    error('"Linedata" is undefined.');
end
if isempty(busdata)
    fprintf('Input argument "busdata" containing the');
    fprintf(' information of buses.');
    error('"busdata" is undefined.');
end

%% Problem outputs:

%% angle_r: voltage angle based on radian
%% angle_d: voltage angle based on degree
%% PF: power flow data of lines
%% OL: overload information of lines
%% SOL: total overload of the network

%% Problem inputs:

%% busdata: required data of network buses
%% busdata: required data of network lines
%% Ybus: computed ybus of the netowrk

nbus = size (busdata,1);
nl = linedata(:,2);
nr = linedata(:,3);
Smax = linedata(:,6);
nbr = length(nl);
```

```
%% Computing net power of buses
Ps1 = (busdata(:,3)-busdata(:,4));
%% Finding non-slack buses in the busdata matrix
code = busdata(:,2);
[aa] = find(code~=3);
%% Forming Network suceptance matrix (B)
for n = 1:length(aa)
    for m = 1:length(aa)
            Ymn = Ybus(aa(n),aa(m));
            B(n,m) = -imag(Ymn);
    end
    Ps(n,1) = Ps1((aa(n)),1);
end
%% Computing voltage angle values of all buses
Binv = inv(B);
ang1 = Binv*Ps;
%% angle_r: volatge angle based on radian
angle_r = zeros(nbus,1);
for i=1: length(aa)
    aaa = aa(i);
    angle_r(aaa) = ang1(i);
end
%% angle_d: voltage angle based on degree
angle_d = angle_r*(180/pi);
%% Computing Power flow and overload of all lines
jay = sqrt(-1);
for i = 1:nbr
     PF(i,1) = nl(i); OL(i,1) = nl(i);
     PF(i,2) = nr(i); OL(i,2) = nr(i);
     PF(i,3) = (angle_r(nl(i))-angle_r(nr(i)))/...
          (linedata(i,5));
    if abs(PF(i,3))>Smax(i)
        OL(i,3) = abs(PF(i,3));
        OL(i,4) = abs(PF(i,3))-Smax(i);
    else
        OL(i,3) = PF(i,3);
        OL(i,4) = 0;
    end
end
%% Computing total overload of the network
SOL = sum(OL(:,4));
```

j) "Total_Cost" M-file code

```
function [TC]=Total_Cost(Isolnew, Solution, candid, LineType)
In=Isolnew;
TC=0;
for i=1:length (In)
    TC=TC+(LineType(candid(In(i),4),6))*candid(In(i),5)*...
        Solution(In(i));
end
```

k) "contingency" M-file code

```
function [COL, Cnis, OLD] = contingency(linedata, busdata)
```

```
if isempty(busdata)
    fprintf('Input argument "busdata" containing the');
    fprintf(' information of buses.');
    error('"busdata" is undefined.');
end
if isempty(linedata)
    fprintf('Input argument "linedata" containing the');
    fprintf(' information of lines.');
    error('"linedata" is undefined.');
end

%% Problem outputs

%% COL: total overload of each contingency
%% Cnis: total number of islands in each contingency
%% OLD: over load and power flow data of all lines
    % in each contingency

%% Problem inputs:

%% busdata: required data of network buses
%% linedata: required data of network lines

%% Computing overload and power flow data in each contingency
    % (each iteration) and summing all overloads (COL);
Cnis = 0;
COL = 0;
for i = 1:size (linedata, 1)
%% Updating linedata after outage of each line
    esl = setxor (linedata (:,1), i); % Exsiting lines
    ulinedata = linedata; ulinedata(i,4) = 10^10;
    ulinedata(i,5) = 10^10; ULD = ulinedata;
    ulinedata1 = linedata (esl,:); ULD1 = ulinedata1;
%% Computing number of islands in each contingency
    nl = ULD1(:,2); nr = ULD1(:,3);
%% Exsiting buses:
    nbs = intersect (busdata (:,1), union(nl,nr));
    Is = setxor(nbs,busdata (:,1));   % Islanded buses
    UBD = busdata;
    nbus = size(busdata,1);
    Cnis = Cnis+length(Is);   % Number of islands
%% Computing Ybus for updated bus data (UBD) and updated
    % line data (ULD) for each contingency
    [Ybus]= ybus_calculation(UBD, ULD, [], [],[], []);
%% Running dc power flow for UBD and ULD
    [angle_r,angle_d, PF, OL, SOL]= dcpf(UBD, ULD, Ybus);
%% Computing overload and power flow data of all lines
    % in each contingency (each iteration)
    COL=COL+SOL;
    OL(:,3)=[];
    idOL=find(OL(:,3)~=0);
    OLF=OL(idOL,:); IOL(1,1)=linedata(i,2);
    IOL(1,2)=linedata(i,3); IOL(1,3)=SOL;
    for j=2:size(OLF,1)+1
```

```
            IOL(j,:)=OLF(j-1,:);
        end
        OLD{i,1}=IOL;
        clear IOL
    end
```

L.5 DCLF.m

a) "DCLF" M-file code

```
clear
clc
%% Problem inputs:
Busdata = xlsread('DCLFDATA.xls', 'Busdata');
%% Busdata: Required bus data:
%% Busdata(:,1): bus number
%% Busdata(:,2): bus type 3=slack bus, 2=PV buses 1=PQ buses
%% Busdata(:,3): bus generation
%% Busdata(:,4): bus load

Linedata = xlsread('DCLFDATA.xls', 'Linedata');
%% Linedata: required branch data:
%% Linedata(:,1): branch ID
%% Linedata(:,2): branch source bus
%% Linedata(:,3): branch destination bus
%% Linedata(:,4): branch resistance
%% Linedata(:,5): branch reactance
%% Linedata(:,6): branch thermal loading
%% Linedata(:,7): branch circuit ID
%% Lg: load growth
Lg = xlsread('DCLFDATA.xls', 'Load growth');

%% Problem outputs:
% Normal condition
[Angle_r,Angle_d, Pf, Ol, Sol] = Dcpf(Busdata, Linedata, Lg);
%% Angle_r: voltage phase (radian)
%% Angle_d: voltage phase (degree)
%% Pf: flow of branches
%% Ol: over load amount of each branches
%% Sol: sum of all overloads

% N-1 condition
[Col, Old] = Contingency(Busdata, Linedata, Lg);
%% Col: total overload of each contingency
%% Old: over load and power flow data of all branchs
    % in each contingency

%% Printing the obtained results in both command window and
    % in result1.txt in the ANEP directory
print_DCLF
```

b) " Dcpf " M-file code

```
function [Angle_r,Angle_d, Pf, Ol, Sol] = ...
    Dcpf(Busdata, Linedata, Lg)
if nargin<3 | isempty(Lg), Lg = 0; end

%% Problem outputs:

%% Angle_r: voltage phase (radian)
%% Angle_d: voltage phase (degree)
%% Pf: flow of branches
%% Ol: over load amount of each branches
%% Sol: sum of all overloads

%% Problem inputs:

%% Busdata: required bus data:
%% Busdata(:,1): bus number
%% Busdata(:,2): bus type 3=slack bus, 2=PV buses 1=PQ buses
%% Busdata(:,3): bus generation
%% Busdata(:,4): bus load

%% Linedata: required branch data:
%% Linedata(:,1): branch ID
%% Linedata(:,2): branch source bus
%% Linedata(:,3): branch destination bus
%% Linedata(:,4): branch resistance
%% Linedata(:,5): branch reactance
%% Linedata(:,6): branch thermal loading
%% Linedata(:,7): branch circuit ID

%% Lg: load growth

%% Conversion block; to convert buses names
    % to consecutive numbers
Busname=Busdata(:,1);
nbus = length(Busname);
Busnumber = 1:nbus;
NL = Linedata(:,2);
NR = Linedata(:,3);
save namedata Busname Busnumber NL NR
for i = 1:length(NL)
    for j = 1:length(Busnumber);
        if NL(i) == Busname(j)
            nnl(i) = Busnumber(j);
        end
        if NR(i) == Busname(j)
            nnr(i) = Busnumber(j);
        end
    end
end
LD = Linedata; LD(:,2) = nnl; LD(:,3) = nnr';
BD = Busdata; BD(:,1) = Busnumber;
```

```
%% Ybus calculation
[Ybus, linedata, busdata] = Ybuscal(BD, LD, Lg);
%% Load flow calculation
nbus = size (busdata,1);
nl = linedata(:,2);
nr = linedata(:,3);
Smax = linedata(:,6);
nbr = length(nl);
Ps1 = (busdata(:,3)-busdata(:,4));
%% Finding non-slack buses in the busdata matrix
code = busdata(:,2);
[aa]=find(code~=3);
for n = 1:length(aa)
    for m = 1:length(aa)
            Ymn = Ybus(aa(n),aa(m));
            B(n,m) = -imag(Ymn);
    end
    Ps(n,1) = Ps1((aa(n)),1);
end
Binv = inv(B);
ang1 = Binv*Ps;
Angle_r = zeros(nbus,1);
%%%%%%%%%%%%%%%%%%%%%%%%%%%%%%%%end
%% Calculation Power flow and over load values
for i = 1 : length(aa)
    aaa = aa(i);
    Angle_r(aaa) = ang1(i);
end
Angle_d = Angle_r*(180/pi);
jay = sqrt(-1);
for i = 1:nbr
     Pf(i,1) = NL(i); Ol(i,1)=NL(i);
     Pf(i,2) = NR(i); Ol(i,2)=NR(i);
     Pf(i,3) = Linedata(i,7); Ol(i,3)=Linedata(i,7);
     Pf(i,4) = (Angle_r(nl(i))-Angle_r(nr(i)))/...
         (linedata(i,5));
    if abs(Pf(i,4))>Smax(i)
        Ol(i,4) = Pf(i,4);
        Ol(i,5) = abs(Pf(i,4))-Smax(i);
    else
        Ol(i,4) = Pf(i,4);
        Ol(i,5) = 0;
    end
end
Sol = sum(Ol(:,5));
```

c) "Contingency" M-file code

```
function [Col, Old] = Contingency(Busdata, Linedata, Lg)
if nargin<3 | isempty(Lg), Lg = 0; end

%% Problem inputs:

%% Busdata: required data of network buses:
```

```
%% Busdata(:,1): bus number
%% Busdata(:,2): bus type 3=slack bus, 2=PV buses 1=PQ buses
%% Busdata(:,3): bus generation
%% Busdata(:,4): bus load
%% Linedata: required data of network branches:
%% Linedata(:,1): branch ID
%% Linedata(:,2): branch source bus
%% Linedata(:,3): branch destination bus
%% Linedata(:,4): branch resistance
%% Linedata(:,5): branch reactance
%% Linedata(:,6): branch thermal loading
%% Linedata(:,7): branch circuit ID
%% Lg: load growth

%% Outputs

%% Col: total overload of each contingency
%% Cnis: total number of islands in each contingency
%% Old: over load and power flow data of all branchs
    % in each contingency

%% Computing overload and power flow data in each contingency
    % (each iteration) and summing all overloads (Col);
Col = 0;
for i = 1:size (Linedata, 1)
%%  Updating Linedata after outage of each branch
    esl = setxor (Linedata (:,1), i);     %Exsiting branchs
    ulinedata = Linedata; ulinedata(i,4) = 10^10;
    ulinedata(i,5) = 10^10; ULD = ulinedata;
    ulinedata1 = Linedata(esl,:); ULD1 = ulinedata1;
    UBD = Busdata; nbus = size(Busdata,1);
%%  Running dc power flow for updated bus data (UBD) and
    % updated line data (ULD)
    [angle_r,angle_d, PF, OL, SOL] = Dcpf(UBD, ULD, Lg);
%% Computing overload and power flow data of all branchs
    % in each contingency (each iteration)
    Col = Col+SOL;
    OL(:,4) = [];
    idOL = find(OL(:,4) ~= 0);
    OLF = OL(idOL,:);
    IOL(1,1) = Linedata(i,2); IOL(1,2) = Linedata(i,3);
    IOL(1,3) = Linedata(i,7); IOL(1,4) = SOL;
    for j = 2:size(OLF,1)+1
      IOL(j,:) = OLF(j-1,:);
    end
    Old{i,1} = IOL;
    clear IOL
end
```

d) "print_DCLF" M-file code

```
fid = fopen('results.txt', 'wt');
fprintf(fid, '**********************************************');
fprintf(fid, '**************************');
```

```
fprintf(fid,'\n                                    Normal');
fprintf(fid,' condition\n');
fprintf(fid, '**********************************************');
fprintf(fid,'**************************\n');
fprintf(fid,'\n***************Bus data****************\n');
fprintf(fid, '     No. bus            Voltage angle (Rad)\n');
fprintf(fid, '****************        ******************\n');
for i = 1:size (Busdata,1);
    fprintf(fid, ' %10.0f %27.5f \n', ...
        Busdata(i,1), Angle_r(i,1));
end
if Sol == 0
    fprintf(fid, '\n No overload in normal condition\n');
else
    NL = size (Linedata,1);
    fprintf(fid, '\n Overload at normal:');
    fprintf(fid,'\n Total overload of normal condition is');
    fprintf(fid,' %3.5f pu\n',Sol);
    fprintf(fid, '****************************************');
    fprintf(fid, '***************************************');
    fprintf(fid,'\n*************** Power flow and overload');
    fprintf(fid,' values of branches *****************\n');
    fprintf(fid, '******************************************');
    fprintf(fid, '*************************************\n');
    fprintf(fid, '\n From bus      To bus      Circuit ID ');
    fprintf(fid, '      Line flow (pu)       Overload (pu)\n');
    for i = 1 : NL
      fprintf(fid,'\n %1.0f %14.0f %15.0f %20.5f %20.5f\n'...
          , Ol(i,:));
    end
end
fprintf(fid, '\n\n****************************************');
fprintf(fid, '*********************************************');
fprintf(fid, '*******');
fprintf(fid,'\n                                N-1 condition\n');
fprintf(fid, '***********************************************');
fprintf(fid, '**********************************************');

if (Col == 0)
    fprintf(fid, '\n No overload in N-1 condition\n');
    fprintf(fid, '\n*************************************\n');
else
    fprintf(fid, '\n**************************************');
    fprintf(fid, '*****************************************');
    fprintf(fid, '**************\n');
    LCOL=Old{size (Col,1),1};
    fprintf(fid, '                    Overload values of ');
    fprintf(fid, 'branches in N-1 condition');
    for i = 1 : size (Old,1)
       iOLD = Old{i,1};
       iL = iOLD(1,:);
       fprintf(fid, '\n*********************************');
       fprintf(fid, '************************************');
       fprintf(fid, '************************\n');
```

```
            fprintf(fid,' Total overload for outage of line: ');
            fprintf(fid, '');
            fprintf(fid,'From bus %3.0f to bus %3.0f and',iL(1:2));
            fprintf(fid,' circuit ID %3.0f is %6.5f pu\n',iL(3:4));
            fprintf(fid, '*************************************');
            fprintf(fid, '*************************************');
            fprintf(fid, '*****************\n');
            fprintf(fid,' following lines are overloaded in');
            fprintf(fid,' this outage\n');
            fprintf(fid, '**************************************');
            fprintf(fid,'***************************************');
            fprintf(fid,'***************\n');
            fprintf(fid, '  From Bus To Bus  Circuit ID         ');
            fprintf(fid,'Overload (pu)\n');
            fprintf(fid, '  *******  ******  **********         ');
            fprintf(fid,'************');
            for j = 2 : size (iOLD,1);
                fprintf(fid, '\n %5.0f %10.0f %8.0f %18.5f\n',...
                    iOLD(j,:));
            end
        end
    end
    fclose(fid);
    %% Print in the command window
    fprintf('**************************************************');
    fprintf('*********************');
    fprintf('\n                             Normal condition\n');
    fprintf('**************************************************');
    fprintf('*********************\n');
    fprintf('\n***************Bus data******************\n');
    fprintf('     No. bus          Voltage angle (Rad)\n');
    fprintf('****************       ********************\n');
    for i = 1 : size (Busdata,1);
        fprintf(' %10.0f %27.5f \n', Busdata(i,1), Angle_r(i,1));
    end
    if Sol == 0
        fprintf('\n No overload in normal Condition\n');
    else
        NL = size (Linedata,1);
        fprintf('\n overload at Normal:');
        fprintf('\n Total overload of normal condition is ');
        fprintf('%3.5f pu\n',Sol);
        fprintf('*********************************************');
        fprintf('*********************************');
        fprintf('\n*************** Power flow and overload ');
        fprintf('values of branches ****************\n');
        fprintf('********************************************');
        fprintf('**********************************\n');
        fprintf('\n From bus      To bus      Circuit ID    ');
        fprintf('   Line flow (pu)      Overload (pu)\n');
        for i = 1 : NL
            fprintf('\n %6.0f %12.0f %14.0f %16.5f %19.5f\n',...
                Ol(i,:));
        end
```

```
end
fprintf('\n\n***********************************************');
fprintf('********************************************');
fprintf('\n                                 N-1 condition\n');
fprintf('***************************************************');
fprintf('***************************************\n');
if (Col == 0)
    fprintf('\n No overload in N-1 condition\n');
    fprintf('\n*************************************\n');
else
    fprintf('\n**********************************************');

fprintf('*********************************************\n');
    LCOL=Old{size (Col,1),1};
    fprintf('                Overload values of branches in');
    fprintf(' N-1 condition');
    for i = 1 : size (Old,1)
        iOLD = Old{i,1};
        iL = iOLD(1,:);
        fprintf('\n*****************************************');
        fprintf('******************************************');
        fprintf('*********\n');
        fprintf(' Total overload for outage of line: from ');
        fprintf('bus %3.0f to bus %3.0f and ',iL(1:2));
        fprintf('circuit ID %3.0f is %6.5f pu\n',iL(3:4));
        fprintf('******************************************');
        fprintf('******************************************');
        fprintf('*******\n');
        fprintf(' Following lines are overloaded ');
        fprintf('in this outage\n');
        fprintf('******************************************');
        fprintf('******************************************');
        fprintf('*******\n');
        fprintf('  From Bus To Bus  Circuit ID         ');
        fprintf('Overload (pu)\n');
        fprintf('  *******  ******  **********         ');
        fprintf('************');
        for j = 2 : size (iOLD,1);
            fprintf('\n %5.0f %10.0f %8.0f %18.5f\n',...
                iOLD(j,:));
        end
    end
end
```

e) "Ybuscal" M-file code

```
function [Ybus, linedata, busdata] = ...
    Ybuscal(busdata, linedata, Lg);
if nargin<3 | isempty(Lg), Lg = 0; end
%%
busdata(:,4) = busdata(:,4).*(1+Lg);
busdata(:,5) = busdata(:,5).*(1+Lg);
%% Computation of admittance of all branches
j = sqrt(-1);
```

```
i = sqrt(-1);
X = linedata(:,5);
nbr = length(linedata(:,1));
nbus = size (busdata,1);
nl = linedata(:,2); nr = linedata(:,3);
Z = (j*X);
y = ones(nbr,1)./Z;          % Branch admittance
Ybus = zeros(nbus,nbus);     % Initialize Ybus to zero
%% Formation of the off diagonal elements
for k = 1 : nbr;
       Ybus(nl(k),nr(k)) = Ybus(nl(k),nr(k))-y(k);
       Ybus(nr(k),nl(k)) = Ybus(nl(k),nr(k));
end
%% Formation of the diagonal elements
for n = 1 : nbus
    for m = (n+1) : nbus
        Ybus(n,n) = Ybus(n,n)-Ybus(n,m);
    end
    for m = 1 : n-1
        Ybus(n,n) = Ybus(n,n)-Ybus(n,m);
    end
end
```

L.6 ACLF.m

a) "ACLF" M-file code

```
clear
clc
%% Load Data
Linedata = xlsread('ACLFDATA.xls', 'Linedata');
Busdata = xlsread('ACLFDATA.xls', 'Busdata');
Setdata = xlsread('ACLFDATA.xls', 'Loadflowsetting');
Basemva = Setdata (1,1);            % Base MVA
MIter = Setdata (1,2);              % Maximum iteration
Acc = Setdata (1,3);                % Accuracy
%% Voltage acceptable deadband
Vmin = Setdata (1,4); Vmax = Setdata (1,5);

%% busdata(:,1): bus number
%% busdata(:,2): type of bus 1-Slack, 2-PV, 3-PQ
%% busdata(:,3): voltage of PV buses
%% busdata(i,5): active power Load in bus i
%% busdata(i,6): reactive power Load in bus i
%% busdata(i,7): active power generation in bus i
%% busdata(i,8): reactive power generation in bus i
%% busdata(i,9): Qmin; minimum reactive power limit of bus i
%% busdata(i,10): Qmax; maximum reactive power limit of bus i
%% busdata(i,11): injected reactive power to bus i

for i = 1 : size (Busdata,1)
```

```
        if Busdata(i,2) == 3
            Vini(i) = 1.0;
        else
            Vini(i) = Busdata(i,3);
        end
    end

    [Vb0, Fij0, nfij0,Vprof0, SID0] = Acpf(Busdata, ...
        Linedata, Basemva, MIter, Acc, 0, Vmin, Vmax, Vini);
    %% Calculating voltage stability index (Pstab)
        % in normal condition
    slstep = 0.005;                     % Small step length
    llstep = 0.05;                      % Large step length
    mstep = 1000;                       % Mmaximum step
    if  SID0 == 0
        fprintf('\n ****************************************');
        fprintf('*****************************************');
        fprintf('******************')
        fprintf('\nWARNING: The load flow solution did not ');
        fprintf('converged At Base Case ')
        fprintf('\n ****************************************');
        fprintf('*****************************************');
        fprintf('****************** \n')
    else
        i = 1;
        SID = 1;
        while i <= mstep & SID == 1
            LR = i*llstep;
            [Vb, Fij, nfij, Vprofc, SID] = Acpf(Busdata,...
                Linedata, Basemva, MIter, Acc, LR, Vmin,...
                Vmax, Vini);
            DelV{i,1} = Vb(:,2); DelV{i,2} = Vprofc;
            if SID ~= 1
                SID = 1;
                LR = (LR-llstep);
                j = 0;
                j = i+j;
                while SID == 1
                    LR = LR+slstep;
                    [Vb, Fij, nfij, Vprofc, SID] = Acpf(...
                        Busdata, Linedata, Basemva, MIter, ...
                        Acc, LR, Vmin, Vmax, Vini);
                    DelV{j,1} = Vb(:,2); DelV{j,2} = Vprofc;
                    j = j+1;
                end
                save DelV DelV
                Pstab = LR+1;
                break
            end
            i = i+1;
        end
        for j=1:size (DelV,1)
            A=DelV{j,1};
            for i=1:size (Busdata,1)
```

```
            C(i,1)=Busdata (i,1);
            k=j+1;
            C(i,k)=A (i,1);
        end
    end

%% Calculating voltage profile index (Vprof) voltage
    % stability index (Pstab) in N-1 condition
    linenumber = size(Linedata, 1);
    lineno = 1 : linenumber;
    for i = 1 : linenumber
        esl = setxor (lineno, i);   % Exsiting lines
        ulinedata = Linedata; ulinedata(i,3) = 10^10;
        ulinedata(i,4) = 10^10; ulinedata(i,5) = 0;
        ULD = ulinedata; ulinedata1 = Linedata (esl,:);
        ULD1 = ulinedata1; UBD = Busdata;
        nbus = size(Busdata,1);
        [Vbc0, Fijc0, nfijc0,Vprofc0, SID0] = Acpf(UBD,...
            ULD1, Basemva, MIter, Acc, 0, Vmin, Vmax,Vini);

        if  SID0 == 0
           TC{i,1} = Linedata(i,1);
           TC{i,2} = Linedata(i,2);
           TC{i,3} = -1;
           TC{i,4} = 0;
           TC{i,5} = C;
           continue
        else
           ii = 1;
           SID = 1;
           while ii <= mstep & SID == 1
                  LRc = ii*llstep;
                  [Vbc, Fijc, nfijc,Vprofc, SID] = Acpf(...
                      UBD, ULD1, Basemva, MIter, Acc, LRc,...
                      Vmin, Vmax,Vini);
                  DelVc{ii,1} = Vbc(:,2);
                  DelVc{ii,2} = Vbc(:,3);
                  if SID ~= 1
                     SID = 1;
                     LRc = (LRc-llstep);
                     jj = 0;
                     jj = ii+jj;
                     while SID == 1
                         LRc = LRc+slstep;
                         [Vbc, Fijc, nfijc,Vprofc, SID] =...
                             Acpf(UBD, ULD1, Basemva, ...
                             MIter, Acc, LRc, Vmin,...
                             Vmax,Vini);
                         DelVc{jj,1} = Vbc(:,2);
                         DelVc{jj,2} = Vbc(:,2);
                         jj = jj+1;
                     end
                     save DelVc DelVc
                     break
```

```
                    end
                    ii = ii+1;
                end
            end
            clc
            Pstabc = LRc+1;
            for j = 1 : size (DelVc,1)
                A = DelVc{j,1};
                %% Voltage profile in different iteration
                B = DelVc{j,2};
                for iii = 1:size(Busdata,1)
                    C(iii,1) = Busdata (iii,1);
                    k = j+1;
                    C(iii,k) = A(iii,1);
                end
            end

            TC{i,1} = Linedata(i,1);
            TC{i,2} = Linedata(i,2);
            TC{i,3} = Vprofc0;
            TC{i,4} = Pstabc;
            TC{i,5} = C;

        end
        print_rpp
    end
```

b) "Acpf" M-file code

```
function[Vb, Fij, nfij,Vprof, convergence] = Acpf(Busdata...
    , Linedata, baseMVA, MIter, Acc, LR, Vmin, Vmax,Vini)
%% Program for Newton-Raphson load flow analysis
%% Assumption, bus 1 is considered as slack bus

%% Calling ybusppg.m to get bus admittance matrix
%% Y = ybusppg();

%% Calling busdata30.m to get bus datas
%% busdata = busdata30();

%% Base MVA
%% baseMVA = 100;

%% Outputs
%% Vb: voltage of buses
%% Fij: line flow data
%% vprof: voltage profile
%% convergence: load flow convergence indication

%% Inputs
%% Basemva: Base MVA
%% MIter: maximum iteration of solving load flow
%% Acc: load flow solving telorance
%% Vmin<V< Vmax; voltage acceptable dead band for calculating
```

```
    % voltage profile index
%% LR: load growth
%% Linedata: network line data
%% Linedata: network bus data
%% Conversion block for converting  buses names to numbers
Busname = Busdata(:,1);
nbus = length(Busname);
Busnumber = 1 : nbus;
nl = Linedata(:,1);
nr = Linedata(:,2);
save namedata Busname Busnumber nl nr
for i = 1 : length(nl)
    for j = 1 : length(Busnumber);
        if nl(i) == Busname(j)
            nnl(i) = Busnumber(j);
        end
        if nr(i) == Busname(j)
            nnr(i) = Busnumber(j);
        end
    end
end
linedata = Linedata; linedata(:,1) = nnl;
linedata(:,2) = nnr'; busdata = Busdata;
busdata(:,1) = Busnumber;
%% Ybus calculation
[Y] = LFYBUS(linedata,busdata, baseMVA);
%% Data retrivial from busdata
bus = busdata(:,1);          % Bus number
type = busdata(:,2);         % Type of bus 1-Slack, 2-PV, 3-PQ
%% Type of bus 1-Slack, 0-PV, 2-PQ
%% type = busdata(:,2);
V = busdata(:,3);            % Specified voltage
del = busdata(:,4);          % Voltage angle
Pli = busdata(:,5);          % PLi
Qli = busdata(:,6);          % QLi
Pg = busdata(:,7);           % PGi
Qg = busdata(:,8);           % QGi
pv = find(type == 2);        % Index of PV buses
pq = find(type == 3);        % Index of PQ buses
Pl = Pli*(1+LR);             % Load growth consideration
Ql = Qli*(1+LR);
npv = length(pv);            % Number of PV buses
npq = length(pq);            % Number of PQ buses
Qmin = busdata(:,9);         % Minimum reactive power limit
Qmax = busdata(:,10);        % Maximum reactive power limit
nbus = max(bus);             % To get no. of buses
%% Computing net power of each bus
P = Pg - Pl;                 % Pi = PGi - PLi
Q = Qg - Ql;                 % Qi = QGi - QLi
%% P = Pl - Pg;              % Pi = PGi - PLi
%% Q = Ql - Qg;              % Qi = QGi - QLi
P = P/baseMVA;               % Converting to p.u.
Q = Q/baseMVA;
Qmin = Qmin/baseMVA;
```

```
Qmax = Qmax/baseMVA;
%% Tol = 10;                % Tolerence kept at high value
%% Iter = 1;                % Iteration starting
%% Pre-specified value of active and reactive power
Psp = P;
Qsp = Q;
G = real(Y);                % Conductance
B = imag(Y);                % Susceptance

%% Beginning of the load flow calculation
convergence = 1;
Tol = 10;                   % Tolerence kept at high value
Iter = 1;                   % Iteration starting
IIII = 1;
%% Iteration starting
while (Tol > Acc | IIII == 1) & Iter <= MIter
    P = zeros(nbus,1);
    Q = zeros(nbus,1);
%% Calculate P and Q
    for i = 1:nbus
        for k = 1:nbus
            P(i) = P(i)+V(i)*V(k)*(G(i,k)*cos(del(i)...
                -del(k)) + B(i,k)*sin(del(i)-del(k)));
            Q(i) = Q(i)+V(i)*V(k)*(G(i,k)*sin(del(i)...
                -del(k)) - B(i,k)*cos(del(i)-del(k)));
        end
    end

%% Checking Q-limit violations
%% Only checked up to 7th iterations
%% if Iter <= 7 && Iter > 4
    if Iter >= 5
        IIII = 0;
        for n = 2 : nbus
            if type(n) == 2
                QG = Q(n)+Ql(n)/baseMVA;
                % QG = Q(n);
                if QG < Qmin(n)
                    V(n) = V(n) + 0.001;
                    IIII = 1;
                elseif QG > Qmax(n)
                    V(n) = V(n) - 0.001;
                    IIII = 1;
                end
            end
        end
    end

%% Calculate change from specified value
    dPa = Psp-P;
    dQa = Qsp-Q;
    k = 1;
    dQ = zeros(npq,1);
    for i = 1:nbus
```

```
            if type(i) == 3
                dQ(k,1) = dQa(i);
                k = k+1;
            end
        end
        dP = dPa(2:nbus);
        M = [dP; dQ];           % Mismatch vector

    %% Jacobian
    %% J1: derivative of real power injections with angles
        J1 = zeros(nbus-1,nbus-1);
        for i = 1:(nbus-1)
            m = i+1;
            for k = 1:(nbus-1)
                n = k+1;
                if n == m
                    for n = 1:nbus
                        J1(i,k) = J1(i,k) + V(m)* V(n)*...
                        (-G(m,n)*sin(del(m)-del(n)) + B(m,n)...
                        *cos(del(m)-del(n)));
                    end
                    J1(i,k) = J1(i,k) - V(m)^2*B(m,m);
                else
                    J1(i,k) = V(m)* V(n)*(G(m,n)*sin(del(m)...
                        -del(n)) - B(m,n)*cos(del(m)-del(n)));
                end
            end
        end

    %% J2: derivative of real power injections with V
        J2 = zeros(nbus-1,npq);
        for i = 1:(nbus-1)
            m = i+1;
            for k = 1:npq
                n = pq(k);
                if n == m
                    for n = 1:nbus
                        J2(i,k) = J2(i,k) + V(n)*(G(m,n)*...
                            cos(del(m)-del(n)) + B(m,n)*...
                            sin(del(m)-del(n)));
                    end
                    J2(i,k) = J2(i,k) + V(m)*G(m,m);
                else
                    J2(i,k) = V(m)*(G(m,n)*cos(del(m)-del(n))...
                        + B(m,n)*sin(del(m)-del(n)));
                end
            end
        end

    %% J3: derivative of reactive power injections with angles
        J3 = zeros(npq,nbus-1);
        for i = 1:npq
            m = pq(i);
            for k = 1:(nbus-1)
```

```
                n = k+1;
                if n == m
                    for n = 1:nbus
                        J3(i,k) = J3(i,k) + V(m)* V(n)*(G(m,n)...
                            *cos(del(m)-del(n)) + B(m,n)*...
                            sin(del(m)-del(n)));
                    end
                    J3(i,k) = J3(i,k) - V(m)^2*G(m,m);
                else
                    J3(i,k) = V(m)* V(n)*(-G(m,n)*cos(del(m)...
                        -del(n)) - B(m,n)*sin(del(m)-del(n)));
                end
            end
        end

    %% J4: derivative of reactive power injections with V
        J4 = zeros(npq,npq);
        for i = 1:npq
            m = pq(i);
            for k = 1:npq
                n = pq(k);
                if n == m
                    for n = 1:nbus
                        J4(i,k) = J4(i,k) + V(n)*(G(m,n)*sin...
                            (del(m)-del(n)) - B(m,n)*cos...
                            (del(m)-del(n)));
                    end
                    J4(i,k) = J4(i,k) - V(m)*B(m,m);
                else
                    J4(i,k) = V(m)*(G(m,n)*sin(del(m)-del(n))...
                        - B(m,n)*cos(del(m)-del(n)));
                end
            end
        end
        J = [J1 J2; J3 J4];     % Jacobian
        X = inv(J)*M;           % Correction vector
        dTh = X(1:nbus-1);      % Change in voltage angle
        dV = X(nbus:end);       % Change in voltage magnitude

    %% Updating state vectors
        del(2:nbus) = dTh + del(2:nbus);    % Voltage angle
        k = 1;
        for i = 2:nbus
            if type(i) == 3
                V(i) = dV(k) + V(i);        % Voltage magnitude
                k = k+1;
            end
        end
        Tol = max(abs(M));                  % Tolerance.
        if Iter==MIter & Tol > Acc
            convergence=0;
            break
        else
        Iter = Iter + 1;
```

```
        end
    end
    Iter;                          % Number of iterations took
    Vs = V;                        % Bus voltage magnitudes in p.u.
    Del = 180/pi*del;              % Bus voltage angles in degree

    %% Outputs

    %% Line power flow data
    jay = sqrt (-1);
    Vmr = V.*cos(del); Vmi = V.*sin(del);
    Vm = Vmr + jay*(Vmi);
    Iij = zeros(nbus,nbus);    % Line current
    Sij = zeros(nbus,nbus);    % Line flow
    Si = zeros(nbus,1);        % Bus power injections
    busdata(:,3) = Vs;
    [Y] = LFYBUS(linedata,busdata, baseMVA);
    %% Line power flows
    ii = 0;
    Fij = zeros (nbus,4);
    for m = 1:nbus
        for n = m+1:nbus
            Iij(m,n) = -(Vm(m) - Vm(n))*Y(m,n);
            Iij(n,m) = -Iij(m,n);
            Sij(m,n) = Vm(m)*conj(Iij(m,n));
            Sij(n,m) = -Sij(m,n);
            if Sij(m,n) ~= 0
                ii = ii+1;
                %% Fij (ii,1) = m; Fij (ii,2) = n;
                %% Fij (ii,3) = real (Sij(m,n));
                %% Fij (ii,4) = imag (Sij(m,n));
                Fij (ii,1) = Busname(m);
                Fij (ii,2) = Busname(n);
                Fij (ii,3) = real (Sij(m,n));
                Fij (ii,4) = imag (Sij(m,n));
            end
        end
    end
    nfij=ii;
    %% Bus power injections..
    for i = 1 : nbus
        for k = 1 : nbus
            Si(i) = Si(i) + conj(Vm(i))* Vm(k)*Y(i,k);
        end
    end
    Pi = real(Si); Qi = -imag(Si);
    %% Bus data information
    for i = 1 : nbus
        Vb(i,1) = Busname(i); Vb(i,2) = V(i); Vb(i,3) = del(i);
        Vb(i,4) = Pi(i); Vb(i,5) = Qi(i);
    end
    %% Computing voltage profile
    Vprof = 0;
    for i = 1 : length(busdata(:,1))
```

```
        if Vs(i) >= Vmax
           Vprof = Vprof+((Vs(i)-Vini(i))^2);
        else
          if Vs(i) <= Vmin
              Vprof = Vprof+((Vs(i)-Vini(i))^2);
          end
        end
    end
```

c) "LFYBUS" M-file code

```
function[Ybus,nbr,nl,nr,nbus] = ...
    LFYBUS(linedata,busdata, baseMVA);
j = sqrt(-1);
i = sqrt(-1);
ai = sqrt(-1);
nl = linedata(:,1);
nr = linedata(:,2);
R = linedata(:,3);
X = linedata(:,4);
Bc = j*linedata(:,5);
a = linedata(:,6);
nbr = length(linedata(:,1));
nbus = max(max(nl), max(nr));
Z = R + j*X;
y = ones(nbr,1)./Z;        % Branch admittance
v = busdata(:,3);
Qinj = busdata(:,11)./baseMVA;
rrb = ai.*(Qinj./(v.^2));
for n = 1 : nbr
if a(n) <= 0  a(n) = 1; else end
Ybus = zeros(nbus,nbus);
%% Obtaining nondiagonal elements
for k = 1 : nbr;
       Ybus(nl(k),nr(k)) = Ybus(nl(k),nr(k))-y(k)/a(k);
       Ybus(nr(k),nl(k)) = Ybus(nl(k),nr(k));
    end
end
%% Formation of the diagonal elements
for  n = 1 : nbus
     for k = 1 : nbr
         if nl(k) == n
         Ybus(n,n) = Ybus(n,n) + y(k)/(a(k)^2) + Bc(k);
         elseif nr(k) == n
         Ybus(n,n) = Ybus(n,n) + y(k) + Bc(k);
         else, end
     end
     Ybus(n,n) = Ybus(n,n) + rrb(n);
end
clear Pgg
```

d) "print_rpp" M-file code

```
%clc
fid = fopen('results.txt', 'wt');
```

```
NL = size (Fij,1);
fprintf(fid, '\n\n\n************************** Normal ');
fprintf(fid, 'condition ***********************');
if Vprof0<0
    fprintf(fid,' The load flow does not converge ');
    fprintf(fid, 'in normal condition\n');
    fprintf(fid, '------------------------------------');
    fprintf(fid, '--------------\n\n');
else
    fprintf(fid,' \n Voltage profile index(Vprof) in normal');
    fprintf(fid, 'condition is %6.5f\n',Vprof0);
    fprintf(fid,' \n Voltage stability index (Pstab) in ');
    fprintf(fid, 'normal condition is %6.5f\n',Pstab);
    fprintf(fid, '-------------------------------------------');
    fprintf(fid, '---------------------------\n\n');
    fprintf(fid, '*************************** Load flow ');
    fprintf(fid, 'results **********************\n');
    fprintf(fid, '*******************************************');
    fprintf(fid, '***************************\n');
    fprintf(fid, '                    Bus data                    \n');
    fprintf(fid, '----------------------------------------\n');
    fprintf(fid, '    Bus number      Voltage      Phase\n');
    fprintf(fid, '    ----------      -------      -----');
    for i=1:size (Busdata,1);
        fprintf(fid, '\n %10.0f %13.3f %13.3f', Vb0(i,1),...
            Vb0(i,2), Vb0(i,3));
    end
    fprintf(fid, '\n-----------------------------------------');
    fprintf(fid, '\n\n********************** Power flow ');
    fprintf(fid, 'of branches**********************\n');
    fprintf(fid, '-------------------------------------------');
    fprintf(fid, '---------------------------');
    fprintf(fid, '\n  From bus      To bus       Active ');
    fprintf(fid, 'power (Pu)      Reactive power (Pu)');
    fprintf(fid, '\n  --------      ------       ---------');
    fprintf(fid, '-------      ----------------');
    for i=1:NL
        fprintf(fid, '\n %5.0f %14.0f %15.3f %22.3f',...
            Fij0(i,:));
    end
    fprintf(fid, '\n  ------------------------------------');
    fprintf(fid, '------------------------------');
    fprintf(fid, '\n\n\n************************************');
    fprintf(fid, '*************');
end
fprintf(fid, '\n                 N-1 condition             ');
fprintf(fid, '                          \n');
fprintf(fid, '*******************************************');
fprintf(fid, '***\n\n');
for i = 1 : NL
    if TC{i,3} < 0
        fprintf(fid,' !!!!\n');
        fprintf(fid,' For outage of the line from bus ');
        fprintf(fid,'%3.0f to bus %3.0f\n',TC{i,1},TC{i,2});
```

```
            fprintf(fid,' The load flow does not converge\n');
            fprintf(fid, '------------------------------------');
            fprintf(fid, '-----------\n\n');
        else
            fprintf(fid,' For outage of the line from bus ');
            fprintf(fid,'%3.0f to bus %3.0f \n',TC{i,1},TC{i,2});
            fprintf(fid,' Voltage profile index (Vprof) is ');
            fprintf(fid,'%6.5f and\n',TC{i,3});
            fprintf(fid,' Voltage stability index (Pstab) is');
            fprintf(fid,' %6.5f\n',TC{i,4});
            fprintf(fid, '------------------------------------');
            fprintf(fid, '-----------\n\n');
        end
    end
    %% Printing in the command window
    fprintf('\n************************* Normal condition ***');
    fprintf('*********************');
    if Vprof0<0
        fprintf(' The load flow does not converge in normal ');
        fprintf('condition\n');
        fprintf('-------------------------------------------');
        fprintf('-------\n\n');
    else
        fprintf(' \n Voltage profile index(Vprof) in normal ');
        fprintf('condition is %6.5f\n', Vprof0);
        fprintf(' \n Voltage stability index (Pstab) in normal');
        fprintf('condition is %6.5f\n',Pstab);
        fprintf('-------------------------------------------');
        fprintf('---------------------\n\n');
        fprintf('************************* Load flow ');
        fprintf('results *********************\n');
        fprintf('*******************************************');
        fprintf('**********************\n');
        fprintf('                   Bus data                 \n');
        fprintf('---------------------------------------\n');
        fprintf('    Bus number      Voltage      Phase\n');
        fprintf('    ----------      -------      -----');
        for i = 1 : size (Busdata,1);
            fprintf('\n %10.0f %13.3f %13.3f',...
                Vb0(i,1), Vb0(i,2), Vb0(i,3));
        end
        fprintf('\n---------------------------------------');
        fprintf('\n\n********************** Power flow of ');
        fprintf('branches*********************\n');
        fprintf('-------------------------------------------');
        fprintf('------------------------');
        fprintf('\n  From bus      To bus       Active power ');
        fprintf('(Pu)       Reactive power (Pu)');
        fprintf('\n  --------      ------       --------------');
        fprintf('--        ----------------');
        for i = 1 : NL
            fprintf('\n %5.0f %14.0f %15.3f %22.3f', Fij0(i,:));
        end
        fprintf('\n  ------------------------------------------');
```

```
        fprintf('------------------------');
        fprintf('\n\n\n**************************************');
        fprintf('********');
    end
    fprintf('\n               N-1 condition                   ');
    fprintf('                 \n');
    fprintf('********************************************');
    fprintf('**\n\n');
    for i = 1 : size(TC,1)
        if TC{i,3} < 0
            fprintf(' !!!!\n');
            fprintf(' For outage of the line from bus ');
            fprintf('%3.0f to bus %3.0f\n',TC{i,1},TC{i,2});
            fprintf(' The load flow does not converge\n');
            fprintf('--------------------------------------');
            fprintf('--------\n\n');
        else
            fprintf(' For outage of the line from bus ');
            fprintf('%3.0f to bus %3.0f \n',TC{i,1},TC{i,2});
            fprintf(' Voltage profile index (Vprof) is ');
            fprintf('%6.5f and\n',TC{i,3});
            fprintf(' Voltage stability index (Pstab) is ');
            fprintf('%6.5f\n',TC{i,4});
            fprintf('--------------------------------------');
            fprintf('--------\n\n');
        end
    end
```

Index